Silage Preparation, Processing and Efficient Utilization

Silage Preparation, Processing and Efficient Utilization

Guest Editor

Siran Wang

MDPI

Basel • Beijing • Wuhan • Barcelona • Belgrade • Novi Sad • Cluj • Manchester

Guest Editor
Siran Wang
Institute of Animal Science
Jiangsu Academy of
Agricultural Sciences
Nanjing
China

Editorial Office
MDPI AG
Grosspeteranlage 5
4052 Basel, Switzerland

This is a reprint of the Special Issue, published open access by the journal *Agriculture* (ISSN 2077-0472), freely accessible at: https://www.mdpi.com/journal/agriculture/special_issues/BSO9E2NJV4.

For citation purposes, cite each article independently as indicated on the article page online and as indicated below:

Lastname, A.A.; Lastname, B.B. Article Title. *Journal Name* **Year**, *Volume Number*, Page Range.

ISBN 978-3-7258-3411-2 (Hbk)
ISBN 978-3-7258-3412-9 (PDF)
https://doi.org/10.3390/books978-3-7258-3412-9

Contents

About the Editor . . . . **vii**

Siran Wang
Silage Preparation, Processing and Efficient Utilization
Reprinted from: *Agriculture* **2025**, *15*, 128, https://doi.org/10.3390/agriculture15020128 . . . . . **1**

Siran Wang, Chenglong Ding, Jipeng Tian, Yunhui Cheng, Nengxiang Xu, Wenjie Zhang, et al.
Fermentation Profile, Bacterial Community Structure, Co-Occurrence Networks, and Their Predicted Functionality and Pathogenic Risk in High-Moisture Italian Ryegrass Silage
Reprinted from: *Agriculture* **2024**, *14*, 1921, https://doi.org/10.3390/agriculture14111921 . . . . **6**

Feng Chen, Jiao Wang, Sujiang Zhang, Abdul Shakoor Chaudhry and Hassan Khanaki
Assessing Fermentation Quality, Aerobic Stability, In Vitro Digestibility, and Rumen Degradation Characteristics of Silages Mixed with Sweet Sorghum and Aerial Parts of Licorice
Reprinted from: *Agriculture* **2024**, *14*, 212, https://doi.org/10.3390/agriculture14020212 . . . . . **20**

Chen Zhang, Zimo Zhang, Mengfan Zhu, Yongliang Wang, Tangtang Zhou, Fachun Wan, et al.
Isolation of *Bacillus velezensis* from Silage and Its Effect on Aerobic Stability and In Vitro Methane Production of Whole-Plant Corn Silage
Reprinted from: *Agriculture* **2024**, *14*, 830, https://doi.org/10.3390/agriculture14060830 . . . . . **43**

Mikael Neumann, Ellen Baldissera, Livia Alessi Ienke, André Martins de Souza, Paulo Eduardo Piemontez de Oliveira and Valter Harry Bumbieris Junior
Nutritional Value Evaluation of Corn Silage from Different Mesoregions of Southern Brazil
Reprinted from: *Agriculture* **2024**, *14*, 1055, https://doi.org/10.3390/agriculture14071055 . . . . **57**

Na Zhao, Xia Hao, Maozhe Yin, Changqing Li, Chao Wang and Hongyan Han
Influence of Growth Stages and Additives on the Fermentation Quality and Microbial Profiles of Whole-Plant Millet Silage
Reprinted from: *Agriculture* **2024**, *14*, 1323, https://doi.org/10.3390/agriculture14081323 . . . . **71**

Mengwei Li, Faiz-ul Hassan, Muhammad Uzair Akhtar, Lijuan Peng, Fang Xie, Qian Deng, et al.
Fibrolytic Enzymes and Lactic Acid Bacteria Improve the Ensiling Characteristics of Ramie and Elephant Grass Mixed Silage
Reprinted from: *Agriculture* **2024**, *14*, 1746, https://doi.org/10.3390/agriculture14101746 . . . . **87**

Siriwan D. Martens, Wolfgang Wagner, Mariana Schneider, Klaus Hünting, Susanne Ohl and Christof Löffler
Screening Lactic Acid Bacteria Strains for Their Tolerance to Increased Osmotic Pressure and Their Suitability to Ensile High Dry Matter Forages
Reprinted from: *Agriculture* **2024**, *14*, 825, https://doi.org/10.3390/agriculture14060825 . . . . . **101**

Ronnie Coêlho de Andrade, Marco Antonio Previdelli Orrico Junior, Yara América da Silva, Marciana Retore, Tatiane Fernandes, Ana Carolina Amorim Orrico, et al.
Impact of Monensin Sodium and Essential Limonene Oil on the Fermentation and Chemical Composition of Total Mixed Ration Silages with Moisture Variations
Reprinted from: *Agriculture* **2024**, *14*, 1319, https://doi.org/10.3390/agriculture14081319 . . . . **112**

Søren Ugilt Larsen, Helle Hestbjerg, Uffe Jørgensen and Anne Grete Kongsted
Ensiling of Willow and Poplar Biomass Is Improved by Ensiling Additives
Reprinted from: *Agriculture* **2024**, *14*, 1853, https://doi.org/10.3390/agriculture14101853 . . . . **128**

Yifei Huang, Siling Zhang, Maoling Hu, Shuyuan Yang, Ailing Hui, Wencheng Zhang and Zeyu Wu
Effects of Biological Additives on the Fermentation Quality and Microbial Community of High-Moisture Oat Silage
Reprinted from: *Agriculture* **2024**, *14*, 2275, https://doi.org/10.3390/agriculture14122275 . . . . **150**

Mingli Zheng, Qin Li, Peichun Mao, Xiaoxia Tian, Yuxia Guo and Lin Meng
Identification and Correlation Analysis of Key Clostridia and LAB Species in Alfalfa Silages Prepared with Different Cultivars and Additives
Reprinted from: *Agriculture* **2024**, *14*, 1963, https://doi.org/10.3390/agriculture14111963 . . . . **166**

About the Editor

Siran Wang

Siran Wang obtained a PhD degree in Agronomy from the Nanjing Agricultural University in 2020. Now, he is an associate researcher and works in the Institute of Animal Science in Jiangsu Academy of Agricultural Sciences. He has chaired or participated in seven Natural Science Foundation of China projects. To date, Siran Wang has contributed to more than 100 scientific publications, containing original research articles, editorials, international conference papers, e-books, patents, and so on. His main research interests include forage processing and its efficient utilization, the screening and application of functional lactic acid bacteria, and the development and utilization of unconventional feed resources.

agriculture

Editorial

Silage Preparation, Processing and Efficient Utilization

Siran Wang [1,2]

1 Institute of Animal Science, Jiangsu Academy of Agricultural Sciences, Nanjing 210014, China; wangsiran@jaas.ac.cn
2 Key Laboratory for Crop and Animal Integrated Farming, Ministry of Agriculture and Rural Affairs, Nanjing 210014, China

Ensiling is a technique for the long-term preservation of feed which is both effective and widespread, and it is characterized by low costs and ease of operation [1]. At the initial stage of ensiling, water-soluble carbohydrates (WSCs) are broken down into carbon dioxide, water and energy under the respiration of aerobic bacteria. As oxygen is depleted, lactic acid bacteria (LAB) attached to forage multiply and convert WSCs into organic acids, thereby creating an anaerobic and acidic environment, which in turn inhibits the activities of undesirable bacteria, such as clostridia, and reduces the risk of forage spoilage [2]. Ensilage is a complicated process involving the actions of microbes and biochemical variations. It is one of the most significant methods of preserving herbages [3]. Animals that consume silage have been shown to enhance the bioavailability of animal protein and decrease methane emissions [4]. A plethora of silage resources are available on Earth. However, the utilization rates remain low, resulting in significant resource wastage and substantial environmental degradation. Conversely, the accelerated growth in animal production has resulted in a persistent annual deficit of animal roughage. Consequently, it is now imperative to investigate the current state of silage preparation, processing and efficient utilization. The Special Issue, entitled "Silage Preparation, Processing and Efficient Utilization", comprises 11 original research articles. Collectively, these contributions illustrate the promising future of silage application, with the objective of advancing the frontiers of animal feed.

The scarcity of quality roughage resources is the primary factor hindering the advancement of the sheep industry in numerous developing countries. In China, southern Xinjiang is distinguished by its extensive saline soil and desertification areas, characterized by minimal natural rainfall and infertile soil. Consequently, the availability of quality roughage resources in this region is notably limited. Sweet sorghum (SS) has been identified as a promising forage, capable of thriving in high salinity, semi-arid and arid environments. Licorice (LC) has been characterized as a drought-tolerant, saline-resistant perennial leguminous herbage. In this study, Chen et al. (contribution 1) evaluated the ensiling characteristics, fermentative products, in vitro digestibility, aerobic stability and ruminal degradation characteristics of silage mixtures with various proportions of SS and aerial sections of LC. The ensiling of SS and LC mixtures was found to enhance silage quality, particularly at a ratio of 50:50 on a dry matter basis.

The most effective ensilage practice is wilting in the field. This process has been shown to reduce ensilage effluent, decrease proteolysis and respiration, increase the levels of WSC as a fermentative substrate and favor LAB over clostridia and enterobacteria, thereby reducing the production of butyric acids under anaerobic conditions. It is imperative that the LAB strains not only survive, but also perform under enhanced osmotic pressure during ensiling, a process that arises from the optimal wilting of forage. Consequently, simple laboratory protocols are needed to isolate suitable LAB strains as bacterial additives for

Received: 24 December 2024
Accepted: 2 January 2025
Published: 8 January 2025

Citation: Wang, S. Silage Preparation, Processing and Efficient Utilization. *Agriculture* **2025**, *15*, 128. https://doi.org/10.3390/agriculture15020128

high-dry-matter (DM) forages. Martens et al. (contribution 2) simulated a high-osmolality environment without inducing salt stress to choose an appropriate indicator of LAB performance. Finally, the developed liquid growth medium was shown to approximate high-DM conditions, thereby enabling the selection of osmotolerant homofermentative LAB strains.

Following the opening of a silo, the silage within is susceptible to a range of challenges, including contamination, rot and aerobic deterioration. Biocontrol bacteria, defined as a type of biological antiseptic, have been shown to be highly natural and effective. Zhang et al. (contribution 3) screened a strain named D-2, which demonstrated anti-microbial activity against silage spoilage microorganisms. The study examined the effects of this strain on the fermentative profile, in vitro digestion and gas production and aerobic stability of ensilage. The D-2 strain was isolated from rotten whole-crop corn ensilage and identified as *Bacillus velezensis*. Overall, the D-2 strain has been shown to possess a high level of resistance to both rot-causing and pathogenic microbes, thereby facilitating its adaptation to the conditions prevalent in ensilage. Furthermore, it has been demonstrated to enhance aerobic stability and mitigate the loss of nutritive value in ensilage, suggesting potential methods in the prevention of ensilage rot and greenhouse gas production.

In Brazil, corn (*Zea mays* L.) is the most widely cultivated crop for ensilage production. This is due to its high productivity, high energy content, good nutritive value, ease of fermentation inside the silo and good palatability for most ruminants. However, environmental conditions vary significantly among cultivation regions and directly affect the qualitative and productive performance of this forage, resulting in alterations in the quality and yield of the ensilage. Consequently, Neumann et al. (contribution 4) undertook an evaluation of the chemical–bromatological composition of 498 samples of corn silage from mesoregions in Southern Brazil during the 2022/2023 summer harvest. The study revealed that the silages from the Southwest-PR and West-SC regions exhibited higher ratios of rapidly degrading substrates and superior nutritive value. In contrast, the silages from the Central South-PR region demonstrated greater energy levels, while those from the North-PR, Northwest-RS and West-PR regions exhibited higher levels of low-digestible fibers and lower nutritional value.

The utilization of total mixed rations (TMRs) in the production of animal feeds has become a prevalent practice, enabling the provision of nutritionally balanced diets and a reduction in feed selection by animals. Monensin (MON) and essential oils (ELO) have been shown to possess antimicrobial characteristics that could affect the fermentative profile. In their research, de Andrade et al. (contribution 5) assessed the effects of using ELO and MON on the quality of the TMR ensilage process and identified the primary changes in TMR fermentation at various moisture levels. The study concluded that the strategic utilization of ELO and MON might effectively enhance the TMR fermentation quality by promoting acid production and improving aerobic stability and could benefit ruminants. The study proved that ELO interacts with the ensilage moisture level and affects its impact on ensiling characteristics. Consequently, when the DM level of silages exceeds 40%, it is advised to administer a greater dose (ELO600, 600 mg of essential limonene oil per kg of DM), and for TMR silages with a DM level below 30%, a lower dose (ELO300, 300 mg of essential limonene oil per kg of DM) is recommended.

Millet (*Setaria italica*) is one of the most significant crops in the world, and it is notable for its low water needs compared to other cereal crops and its ability to thrive in poor soils. Millet is characterized by its high nutritional content, specifically its abundance of crude protein and fiber, which renders it a highly suitable feedstock. However, its high moisture level could pose significant challenges during storage, and the process of producing quality whole-crop millet ensilage is intricate due to the impacts of multiple factors. Zhao et al. (contribution 6) investigated the effects of varying growth stages (harvested at different stages) and the incorporation of *Lactiplantibacillus plantarum* and soluble carbohydrates on the fermentative profiles and bacterial communities of whole-plant millet. The results suggested that the dough stage possessed the greatest WSC and crude protein levels. The dough stage was found to have higher lactic acid and crude protein contents than other stages, and the lowest pH was recorded during this stage. The study concluded with the recommendation that millet should be harvested at the dough stage and that the addition of a mixture of sugar and *Lactiplantibacillus plantarum* should be made to enhance ensilage quality and aerobic stability.

It is imperative for dairy producers to understand the influences of adding enzyme and bacteria on mixed ensilage to make informed decisions. Ramie (*Boehmeria nivea* L. Gaud.) is a non-conventional fiber plant that is applied as animal feed due to its important nutritive value. It is cultivated in numerous regions worldwide. Li et al. (contribution 7) assessed the chemical variations in ensilage prepared from various elephant grass and ramie proportions in response to enzyme and bacteria additives. The best-quality ensilage was found when the proportion of elephant grass to ramie was 70:30 and the doses of compound enzyme and bacteria and enzyme were 20 mg/kg and 200 mg/kg, respectively. The results indicated that fermentative products are influenced by forage ratios/types, while enzymes and bacteria interact not only with each other but also with forage types to influence the compositions of mixed ensilage prepared from elephant grass and ramie.

The ensiling of biomass from poplar and willow harvested for feeds in the growth season has been demonstrated as a viable method of preservation; however, the extent of research conducted on this subject is limited. Larsen et al. (contribution 8) focused on storing green tree biomass collected from poplar and willow in the growth season by ensilage, i.e., via decreasing the pH to a low value of about 4.0. The laboratory-scale ensilage experiments indicate that the pH is usually not decreased sufficiently in poplar and willow biomass during ensilage without the application of a feed additive. However, a low pH value could be observed in willow biomass by adding a dose of 2–5 kg per tonne of fresh weight (FW) of formic acid (78%) or by applying molasses and/or LAB, which could ensure an adequate and rapid pH decrease. In the case of poplar biomass, LAB and molasses appear to be less effective, while formic acid seems to be a more effective silage additive. This research provides comprehensive guidelines for ensuring a low pH during the ensilage of poplar and willow biomass.

Ensiling high-moisture herbage poses a considerable challenge in obtaining good fermentation, because the ensilage process is susceptible to butyric acid fermentation and nutrition loss. The most effective method of lowering the moisture content of herbage is field wilting. However, harvesting herbage is not without its challenges, particularly in rainy and wet regions. Consequently, farmers may be obliged to harvest herbage and produce ensilage when its moisture level is high. Italian ryegrass (*Lolium multiflorum* Lam.) is an extensively planted herbage with good palatability and high amounts of yield for livestock, and it flourishes during the spring and winter months. As a high-energy grass, it is well suited for the feeds of ruminants, and it could be utilized as fresh grass, hay or ensilage. Nevertheless, research on high-moisture Italian ryegrass ensilage is scarce. Thus, Wang et al. (contribution 9) assessed the fermentative profiles, bacterial communities, co-occurrence networks and their functional shifts and pathogenic risks in high-moisture Italian ryegrass (IR) ensilage. The results indicated that the fermentation process decreased the complexity of the bacterial networks in IR ensilage. The metabolism of amino acid and carbohydrate was limited during the initial stage of fermentation. In addition, 1-phosphofructokinase and pyruvate kinase were important in accelerating the lactic acid production. A higher proportion in the "potentially pathogenic" category was observed in the bacterial communities of IR ensilage than in fresh IR. It was concluded that the high-moisture IR ensilage performed well in terms of fermentation quality, whereas risks for bacterial contamination and pathogens existed after fermentation.

Alfalfa (*Medicago sativa* L.) is the most extensively planted legume globally, with an estimated production of about 450 million tons across 30 million hectares. In recent years, alfalfa production in China has been increasing annually, with 546,700 hectares being cultivated in 2020. Nevertheless, clostridial fermentation remains the predominant process responsible for the deterioration of direct-cut alfalfa silage. In response to this challenge, LAB inoculants are often applied to promote lactic acid fermentation, because other technical ways show some inevitable defects. In this study, Zheng et al. (contribution 10) screened target-based LAB strains and identified and correlated the key *Clostridia* and LAB species in alfalfa silage. The study concluded that *Enterococcus faecalis*, *Lentilactobacillus buchneri* and *Lactiplantibacillus pentosus* could be utilized as potential LAB inoculants for the targeted inhibition of clostridial fermentation. This study has enhanced our comprehension of the clostridial fermentation mechanism of ensilage and may be instrumental in the isolation of target-based LAB additives to produce quality alfalfa ensilage.

Oat (*Avena sativa* L.) is regarded as a major food crop and feed resource on a global scale. It is abundant in vitamins, dietary fiber, carbohydrates, quality protein and minerals, thus rendering it an optimal selection for livestock feeds. Nevertheless, the seasonal nature of oat production can lead to fluctuations in the consistent year-round supply of herbage, and insufficient storage methods result in enormous losses of fresh herbage. To address this challenge, Huang et al. (contribution 11) investigated the impact of biological additives, namely Bacillus subtilis (BS), xylanase (XT), Streptococcus bovis (SB) and their combinations, on the fermentative products and microbial community of high-moisture oat ensilage. In comparison with the control (no additives), the SB and SBBS groups promoted the lactic acid production and decreased the levels of propionic acid, acetic acid, ammonia nitrogen and butyric acid in ensilage. In contrast, the XT, SBXT and BSXT treatments resulted in a decrease in acid and neutral detergent fiber contents, accompanied by an enhancement in WSC level. The conclusion drawn was that adding SBBS and SB helps to improve the silage quality of forage oat, but BSXT, SBXT and XT performed superiorly in degrading lignocellulose of herbage.

This Special Issue comprises a series of original studies in the field of silage, including research on the exploration of silage additives and novel raw materials, the utilization of co-ensiling technology, the investigation of microbial community structure and pathogenic risks and associated subjects. The findings of these studies contribute to the increase in ensilage raw materials and improvements in silage fermentation quality. Future research should focus on studying microbial interactions, metabolites and the identification and construction of core microorganisms during ensiling and aerobic exposure.

Funding: This study was financially supported by National Natural Science Foundation of China (32301500), National Key R&D Program of China (2024YFD1301200), and Natural Science Foundation of Jiangsu Province (BK20220747).

Acknowledgments: We appreciate the support of the Special Issue "Silage Preparation, Processing and Efficient Utilization" in *Agriculture* and thank editors, authors whose valuable work was published under this issue and thus contributed to the success of the edition.

Conflicts of Interest: The authors declare no conflicts of interest.

List of Contributions:

1. Chen, F.; Wang, J.; Zhang, S.; Chaudhry, A.S.; Khanaki, H. Assessing Fermentation Quality, Aerobic Stability, In Vitro Digestibility, and Rumen Degradation Characteristics of Silages Mixed with Sweet Sorghum and Aerial Parts of Licorice. *Agriculture* **2024**, *14*, 212.

2. Martens, S.D.; Wagner, W.; Schneider, M.; Hünting, K.; Ohl, S.; Löffler, C. Screening Lactic Acid Bacteria Strains for Their Tolerance to Increased Osmotic Pressure and Their Suitability to Ensile High Dry Matter Forages. *Agriculture* **2024**, *14*, 825.
3. Zhang, C.; Zhang, Z.; Zhu, M.; Wang, Y.; Zhou, T.; Wan, F.; Zhang, Y.; Chen, L. Isolation of *Bacillus velezensis* from Silage and Its Effect on Aerobic Stability and In Vitro Methane Production of Whole-Plant Corn Silage. *Agriculture* **2024**, *14*, 830.
4. Neumann, M.; Baldissera, E.; Alessi Ienke, L.; Martins de Souza, A.; Piemontez de Oliveira, P.E.; Harry Bumbieris Junior, V. Nutritional Value Evaluation of Corn Silage from Different Mesoregions of Southern Brazil. *Agriculture* **2024**, *14*, 1055.
5. de Andrade, R.C.; Orrico Junior, M.A.P.; da Silva, Y.A.; Retore, M.; Fernandes, T.; Orrico, A.C.A.; Vargas Junior, F.M.d.; Amaral, I.P.d.O. Impact of Monensin Sodium and Essential Limonene Oil on the Fermentation and Chemical Composition of Total Mixed Ration Silages with Moisture Variations. *Agriculture* **2024**, *14*, 1319.
6. Zhao, N.; Hao, X.; Yin, M.; Li, C.;Wang, C.; Han, H. Influence of Growth Stages and Additives on the Fermentation Quality and Microbial Profiles of Whole-Plant Millet Silage. *Agriculture* **2024**, *14*, 1323.
7. Li, M.; Hassan, F.-u.; Akhtar, M.U.; Peng, L.; Xie, F.; Deng, Q.; Zhong, H.; Wei, K.; Yang, C. Fibrolytic Enzymes and Lactic Acid Bacteria Improve the Ensiling Characteristics of Ramie and Elephant Grass Mixed Silage. *Agriculture* **2024**, *14*, 1746.
8. Larsen, S.U.; Hestbjerg, H.; Jørgensen, U.; Kongsted, A.G. Ensiling of Willow and Poplar Biomass Is Improved by Ensiling Additives. *Agriculture* **2024**, *14*, 1853.
9. Wang, S.; Ding, C.; Tian, J.; Cheng, Y.; Xu, N.; Zhang, W.; Wang, X.; Nazar, M.; Liu, B. Fermentation Profile, Bacterial Community Structure, Co-Occurrence Networks, and Their Predicted Functionality and Pathogenic Risk in High-Moisture Italian Ryegrass Silage. *Agriculture* **2024**, *14*, 1921.
10. Zheng, M.; Li, Q.; Mao, P.; Tian, X.; Guo, Y.; Meng, L. Identification and Correlation Analysis of Key Clostridia and LAB Species in Alfalfa Silages Prepared with Different Cultivars and Additives. *Agriculture* **2024**, *14*, 1963.
11. Huang, Y.; Zhang, S.; Hu, M.; Yang, S.; Hui, A.; Zhang, W.; Wu, Z. Effects of Biological Additives on the Fermentation Quality and Microbial Community of High-Moisture Oat Silage. *Agriculture* **2024**, *14*, 2275.

References

1. Wang, S.; Ding, C.; Tian, J.; Cheng, Y.; Xu, N.; Zhang, W.; Wang, X.; Nazar, M.; Liu, B. An evaluation of storage length on ensiling characteristics, bacterial community compositions, co-occurrence networks, and their functional shifts and pathogenic risk in high-moisture oat silage. *Chem. Biol. Technol. Agric.* **2024**, *11*, 173. [CrossRef]
2. Zieliński, M.; Rusanowska, P.; Zielińska, M.; Dudek, M.; Nowicka, A.; Purwin, C.; Fijałkowska, M.; Dębowski, M. Influence of preparation of *Sida hermaphrodita* silages on its conversion to methane. *Renew. Energy* **2021**, *163*, 437–444. [CrossRef]
3. McDonald, P.; Henderson, A.R.; Heron, S. *The Biochemistry of Silage*; Chalcombe Publications: Abersytwyth, UK, 1991.
4. Bayat, A.R.; Tapio, I.; Vilkki, J.; Shingfield, K.J.; Leskinen, H. Plant oil supplements reduce methane emissions and improve milk fatty acid composition in dairy cows fed grass silage-based diets without affecting milk yield. *J. Dairy Sci.* **2018**, *101*, 1136–1151. [CrossRef] [PubMed]

agriculture

MDPI

Article

Fermentation Profile, Bacterial Community Structure, Co-Occurrence Networks, and Their Predicted Functionality and Pathogenic Risk in High-Moisture Italian Ryegrass Silage

Siran Wang [1,2], Chenglong Ding [1,2], Jipeng Tian [1,2], Yunhui Cheng [1,2], Nengxiang Xu [1,2], Wenjie Zhang [1,2], Xin Wang [1,2], Mudasir Nazar [1,2] and Beiyi Liu [1,2,*]

1 Institute of Animal Science, Jiangsu Academy of Agricultural Sciences, Nanjing 210014, China; wangsiran@jaas.ac.cn (S.W.); dingcl@jaas.ac.cn (C.D.); tianjipeng2231@jaas.ac.cn (J.T.); chengyh@jaas.ac.cn (Y.C.); 20000001@jaas.ac.cn (N.X.); 20160033@jaas.ac.cn (W.Z.); 18855068256@163.com (X.W.); mudasirnazar50@yahoo.com (M.N.)

2 Key Laboratory for Crop and Animal Integrated Farming, Ministry of Agriculture and Rural Affairs, Nanjing 210014, China

* Correspondence: byliu@jaas.ac.cn

Abstract: This study aimed to assess the fermentation characteristics, bacterial community structure, co-occurrence networks, and their predicted functionality and pathogenic risk in high-moisture Italian ryegrass (IR; *Lolium multiflorum* Lam.) silage. The IR harvested at heading stage (208 g dry matter (DM)/kg fresh weight) was spontaneously ensiled in plastic silos (10 L scale). Triplicated silos were opened after 1, 3, 7, 15, 30, and 60 days of fermentation, respectively. The bacterial community structure on days 3 and 60 were investigated using high-throughput sequencing technology, and 16S rRNA-gene predicted functionality and phenotypes were determined by PICRUSt2 and BugBase tools, respectively. After 60 days, the IR silage exhibited good ensiling characteristics indicated by large amounts of acetic acid (~58.7 g/kg DM) and lactic acid (~91.5 g/kg DM), relatively low pH (~4.20), acceptable levels of ammonia nitrogen (~87.0 g/kg total nitrogen), and trace amounts of butyric acid (~1.59 g/kg DM). *Psychrobacter* was prevalent in fresh IR, and *Lactobacillus* became the most predominant genus after 3 and 60 days. The ensilage process reduced the complexity of the bacterial community networks in IR silage. The bacterial functional pathways in fresh and ensilaged IR are primarily characterized by the metabolism of carbohydrate and amino acid. The pyruvate kinase and 1-phosphofructokinase were critical in promoting lactic acid fermentation. A greater ($p < 0.01$) abundance of the "potentially pathogenic" label was noticed in the bacterial communities of ensiled IR than fresh IR. Altogether, the findings indicated that the high-moisture IR silage exhibited good ensiling characteristics, but the potential for microbial contamination and pathogens still remained after ensiling.

Keywords: Italian ryegrass; ensiling; silage; fermentation quality; bacterial community

Citation: Wang, S.; Ding, C.; Tian, J.; Cheng, Y.; Xu, N.; Zhang, W.; Wang, X.; Nazar, M.; Liu, B. Fermentation Profile, Bacterial Community Structure, Co-Occurrence Networks, and Their Predicted Functionality and Pathogenic Risk in High-Moisture Italian Ryegrass Silage. *Agriculture* **2024**, *14*, 1921. https://doi.org/10.3390/agriculture14111921

Academic Editor: Marco Mariotti

Received: 26 September 2024
Revised: 24 October 2024
Accepted: 25 October 2024
Published: 29 October 2024

1. Introduction

Ensiling is regarded as an effective method for preserving green grass, as it conserves the nutritive composition of the grass and enhances its palatability for livestock. During ensilage, fermentable substrates are metabolized by lactic acid bacteria into lactic acid, and the pH is reduced, thus suppressing the undesirable microbial activity. Hence, the silage quality could be affected by several factors, including moisture level, epiphytic microorganisms, and the chemical compositions of the raw material, the storage temperature, and the air infiltration during ensilage [1].

Generally, ensiling high-moisture forage presents a significant challenge in achieving optimal fermentation, as the process is prone to nutrient loss and butyric acid fermentation, which can result in substantial economic losses and a low nutritive value of silage [2].

The most economical method of decreasing the moisture level of forage is field wilting, which is largely contingent on natural environmental conditions, including rainfall, vapor pressure deficit, solar radiation, and wind speed [3]. It is not always sunny when herbage is harvested, particularly in wet and rainy areas. This results in the phenomenon whereby farmers are occasionally compelled to harvest grass and make silage when its moisture content is high. Moreover, the nutritive value and chemical compositions of silage are significantly influenced by those post-harvesting conditions [4]. Hence, it is necessary to evaluate the fermentative profile of high-moisture grasses.

In the last decade, high-throughput sequencing technology has been extensively utilized to describe the microbial community of silage. Zhao et al. [5] studied the influences of growth stages and additives on the bacterial community dynamics of ensiled millet. They found that *Klebsiella* was evidently decreased in the additive-treated groups compared to control, and *Lentilactobacillus* predominated the treated groups. Furthermore, Wu et al. [6] investigated the influences of *Lacticaseibacillus casei* and molasses on the microbial compositions of *Caragana korshinskii*. They observed that the molasses and *L. casei* additions could promote the amino acid and carbohydrate metabolism. Nevertheless, as far as we know, most studies have merely described the alterations in bacterial community compositions and fermentation products in Italian ryegrass silage without undertaking an evaluation of the functional shifts and pathogenic risk of the bacterial community.

Italian ryegrass (*Lolium multiflorum* Lam.) is a widely cultivated grass with a great yield and good palatability for livestock, and it grows well during the winter and spring seasons [7]. As a high-energy herbage, it is well suited for the feeding of dairy cows, and it can be consumed as fresh grass, hay, or silage. However, research on high-moisture Italian ryegrass silage is limited. Hence, the objective of our study was to assess the fermentative products, bacterial community compositions, and co-occurrence networks and their predicted functionality and pathogenic risk during the ensilage of high-moisture Italian ryegrass.

2. Materials and Methods

2.1. Preparing Raw Material

Italian ryegrass (IR; *Lolium multiflorum* Lam.) was cultivated in experimental fields (31°52′ N, 119°21′ E, mean temperature 16.2 °C, mean elevation 25.1 m, and mean precipitation 1105 mm) in the Lishui District of Nanjing (Jiangsu, China). The experimental field (area 150 m^2; size 10 × 15 m) was equally divided into three replicated blocks. The sowing density, sowing depth, and row spacing of IR (Variety: Tetragold) were 30 kg/hm^2, 2–3 cm, and 30 cm, respectively. There was no additional fertilization during the growth of IR. Fresh IR was harvested at the early heading stage by a small forage harvester on 10 May 2022. The first-cutting height and stubble height were about 80 cm and 5 cm, respectively. The microbial populations and chemical components of fresh IR are presented in Table 1. The relatively lower crude protein content (97.6 g/kg DM) of fresh IR could be due to the moderate soil condition and no additional fertilization during planting. The harvested IR was chopped into a length of approximately 20 mm using a paper cutter and thoroughly mixed for silage production, without wilting or any additives. Then, about 7.50 kg of fresh IR was manually tightly packed into a cylindrical plastic silo (10 L) without a headspace, which was then hermetically sealed with plastic tapes and a screw top. The silos were stored at ambient temperature (26–34 °C). In total, 18 silos (3 replicates × 6 storage time) were prepared. Samples were completely poured out of silos, mixed on the ground, and then analyzed after 1, 3, 7, 15, 30, and 60 d for the bacterial community structure and chemical compositions, with three replicates.

Table 1. Chemical compositions and microbial populations of fresh Italian ryegrass (mean ± SD).

Items	Italian Ryegrass
Chemical compositions	
pH	6.08 ± 0.01
Dry matter (g/kg FW)	208 ± 2.86
Water-soluble carbohydrates (g/kg DM)	100 ± 1.25
Buffering capacity (mEq/kg DM)	93.4 ± 1.57
Neutral detergent fiber (g/kg DM)	554 ± 3.28
Acid detergent fiber (g/kg DM)	301 ± 2.32
Acid detergent lignin (g/kg DM)	38.4 ± 1.32
Crude protein (g/kg DM)	97.6 ± 1.48
Microbial populations	
Lactic acid bacteria (log_{10} cfu/g FW)	4.18 ± 0.06
Aerobic bacteria (log_{10} cfu/g FW)	7.59 ± 0.01
Yeasts (log_{10} cfu/g FW)	6.48 ± 0.06
Enterobacteriaceae (log_{10} cfu/g FW)	8.35 ± 0.12

Note: DM: dry matter; FW: fresh weight; mEq: milligram equivalent; cfu: colony-forming units. Data are mean values and standard deviations for triplicate samples.

2.2. Microbial and Chemical Compositions of Fresh and Ensiled Italian Ryegrass

At sampling, the raw material or ensilage samples were thoroughly mixed and collected. Subsequently, a preliminary subsample (30.0 g) was blended with 270 mL of sterile water for 60 s with a high-speed juice extractor (HR1855/40, Royal Dutch Philips Electronics Ltd., Amsterdam, The Netherlands), after which it was filtered using sterile gauze. The filtrate was subjected to an immediate determination of pH by means of an electronic pH meter (S400-Basic, Mettler-Toledo Ltd., Toledo, OH, USA). The buffering capacity of the raw material was detected in accordance with the methodology described by Playne and McDonald [8]. Subsequently, the filtrate was preserved at −20 °C to facilitate the measurement of organic acid (acetic acid (AA), γ-butyric acid, butyric acid, propionic acid, and lactic acid (LA)), ammonia nitrogen (NH_3-N), and ethanol levels. The NH_3-N levels were analyzed based on the reports described by Broderick and Kang [9]. The levels of ethanol and organic acid were measured using HPLC according to the report of Wang et al. [10].

To determine the dry matter (DM) contents, a second subsample (200 g) of fresh or ensiled IR was subjected to freeze drying (−20 °C~−40 °C) for 72 h using a freeze dryer (SZFD-300A, Shanghai Shunzhi Instrument Manufacturing Co., Ltd., Shanghai, China). Thereafter, the sample was milled to pass through a 1 mm sieve for subsequent tests. The milled samples were prepared for the determination of water-soluble carbohydrates (WSC), total nitrogen (TN), fiber, disaccharides, and monosaccharides contents. The WSC and TN contents were analyzed based on the descriptions of Murphy [11] and Krishnamoorthy et al. [12], respectively. The crude protein (CP) level was obtained by multiplying total nitrogen value by 6.25. The contents of acid detergent lignin (ADL), neutral detergent fiber (NDF), and acid detergent fiber (ADF) of raw material were analyzed using the methods described by Van Soest et al. [13]. Subsamples of ground lyophilized material were determined for disaccharides and monosaccharides (fructose, sucrose, and glucose) contents based on the descriptions of Desta et al. [14].

A third subsample (10 g) of raw material or ensilage mass was homogenized with 90 mL of sterile saline water for microbial enumeration. Following serial dilution, the population of yeast, lactic acid bacteria (LAB), *Enterobacteriaceae*, and aerobic bacteria was quantified through a culture-based method as reported by Wang et al. [10]. The microbial population was reported as colony-forming units (cfu) and converted to a logarithmic scale based on fresh weight (FW).

2.3. Compositions, Diversity, and Co-Occurrence Networks in the Bacterial Community

The abovementioned mother liquid, employed for counting microorganisms, was utilized to extract bacterial community genomic DNA from raw material (IRFM) and ensiled samples following 3 (IR_3) and 60 (IR_60) days of fermentation. This was achieved through the E.Z.N.A.® DNA Kit (Omega Bio-tek, Norcross, GA, USA), following the provided instruction. The V_3–V_4 section of the bacterial 16S rRNA gene was amplified using the primers 806-R and 338-F with the platform of Illumina MiSeq PE300. The UPARSE pipeline was employed to assign operational taxonomic units (OTUs) to the 16S rRNA on USEARCH (ver. 7.1). The OTU table was then subjected to manual filtering, with sequences derived from chloroplasts and chondriosomes removed from all samples. The taxonomy of each OTU representative sequence was analyzed by RDP Classifier (ver. 2.2) against the 16S rRNA gene database (Silva ver. 138).

The α-diversity of the bacterial communities was analyzed through QIIME (ver. 1.9.1) software, which calculated the following indices: Ace, Chao, Coverage, Shannon, Simpson, and Sobs. The β-diversity in bacterial community was studied through principal coordinate analysis (PCoA) by R software (ver. 4.4.0). A Venn plot of bacterial communities was constructed graphically using R software (ver. 4.4.0) at the OTU level. The compositions and relative abundances of bacterial community were determined at the genus and phylum levels, respectively.

NetworkX (ver. 2.8.0) software was employed to analyze multiple abundance correlations and build co-occurrence patterns among bacterial species. A co-occurrence was deemed robust if the p-value was lower than 0.05, while the coefficient of Spearman's correlation (ρ) was higher than or equal to 0.5. The co-occurrence networks were visualized by Gephi (ver. 0.10.1), wherein the edges denote correlations between nodes, and nodes represent bacterial species. The topological features of the co-occurrence networks were studied, as they contain the positive and negative correlation numbers as well as the numbers of edges and nodes.

2.4. Predicted Functionality and Phenotypes in the Bacterial Communities

The PICRUSt2 (Phylogenetic Investigation of Communities by Reconstruction of Unobserved States) tool was employed to predict the KEGG (Kyoto Encyclopedia of Genes and Genomes) functionalities of bacterial communities. It can be used for predicting functionality of 16S amplicon sequencing results [15].

The BugBase tool was employed to predict the phenotype of bacterial community. It is an algorithm that predicts organism-level coverage of functional pathways and biologically interpretable phenotypes [16]. BugBase first normalized the OTUs by the predicted 16S copy number and then used the provided pre-calculated file to predict bacterial phenotypes. The BugBase algorithm is based on the analysis of databases such as KEGG, PATRIC (Pathosystems Resource Integration Center), and Integrated Microbial Genomes (IMG), which are used for phenotype prediction and identification of the bacterial contributors.

2.5. Statistical Analysis

This experiment was designed using a completely randomized methodology comprising six ensiling times and three replicates. All statistical analyses were conducted using the statistical software package SPSS (version 19.0, SPSS Inc., Chicago, IL, USA). Data on fermentative parameters, microbial numbers, chemical components, and predicted enzymes were subjected to one-way analysis of variance (ANOVA). One-way ANOVA and ANOSIM (analysis of similarities) with 999 permutations were employed for the analysis of α-diversity indices and β-diversity, respectively. A Kruskal–Wallis H test was employed to compare the relative abundances of predicted phenotypes in the bacterial communities. A Student's t-test was employed to compare the "potentially pathogenic" category between fresh and ensiled IR. The statistical significance of the observed differences was examined through Tukey's multiple comparison test, and a value of $p < 0.05$ was considered as statistically significant.

3. Results and Discussion

3.1. Fermentation Characteristics and Microbial Numbers

The rapid increase in LA content and the concomitant decline in pH observed on day 3 (Table 2) suggest that LA was produced in abundance during the early period of ensilage. It is possible that the herbages were chopped into shorter sections to facilitate the rapid release of herbage juice, thereby promoting the reproduction of LAB during the initial stage of ensiling. The decline ($p < 0.05$) in LA on day 60 indicates that LA may have been converted into acetic acid with the prolongation of ensilage time, which agrees with the reports of Jahanzad et al. [17].

Table 2. Changes in ensiling characteristics and microbial populations during the ensilage of Italian ryegrass.

Items	Ensilage Days (d)						SEM	*p*-Value
	1	3	7	15	30	60		
Chemical compositions								
pH	5.93a	4.56b	4.16c	4.20c	4.18c	4.20c	0.155	<0.001
Lactic acid (g/kg DM)	15.6e	84.5d	95.2c	104b	120a	91.5c	8.001	<0.001
Acetic acid (g/kg DM)	6.33f	26.0e	38.1d	47.5c	52.6b	58.7a	4.306	<0.001
LA/AA	2.46b	3.26a	2.50b	2.18b	2.28b	1.56c	0.124	<0.001
Propionic acid (g/kg DM)	0.673c	0.827c	1.53b	2.55a	2.39a	2.50a	0.194	<0.001
Isobutyric acid (g/kg DM)	0.880b	0.947b	0.970b	1.15b	1.06b	1.89a	0.901	<0.001
Butyric acid (g/kg DM)	0.343b	1.29a	1.61a	1.52a	1.67a	1.59a	0.120	<0.001
Ethanol (g/kg DM)	7.65b	14.7a	15.7a	16.7a	15.4a	14.6a	0.777	<0.001
VFAs (g/kg DM)	8.23f	29.1e	42.2d	52.8c	57.7b	64.6a	4.636	<0.001
Dry matter (g/kg FW)	201a	196b	192c	190c	187d	186d	1.305	<0.001
NH_3-N (g/kg TN)	25.8e	44.6d	53.0c	77.3b	84.3ab	87.0a	5.495	<0.001
WSC (g/kg DM)	93.6a	41.9b	21.4c	17.3d	12.7e	7.38f	7.138	<0.001
Microbial populations								
Lactic acid bacteria ($\log_{10}$ cfu/g FW)	6.41d	7.66c	9.52b	10.2a	10.4a	7.35c	0.366	<0.001
Enterobacteriaceae ($\log_{10}$ cfu/g FW)	8.67a	8.99a	6.48b	4.34c	3.38d	2.47e	0.612	<0.001
Yeasts ($\log_{10}$ cfu/g FW)	6.38a	5.83b	4.48c	4.35c	4.09d	4.04d	0.259	<0.001

Note: Means within the same column with different letters differ significantly from each other ($p < 0.05$). SEM: standard error of the mean; DM: dry matter; FW: fresh weight; VFAs: volatile fatty acids; LA/AA: ratio of lactic acid to acetic acid; NH_3-N: ammonia nitrogen; TN: total nitrogen; WSC: water-soluble carbohydrate; cfu: colony-forming units; d: day.

The primary organisms responsible for producing acetic acid during ensiling were identified as enterobacteria, heterofermentative LAB, and *Propionibacterium* [1]. The considerable production of acetic acid in IR silage indicated the activity of certain heterofermentative LAB. Nishino et al. [18] also observed great acetic acid concentrations in grass silage with a high moisture content. An LA-to-AA ratio (LA/AA) of less than 3.0 would be indicative of a more dominant hetero-lactic fermentation [19]. The LA/AA ratio during ensiling was found to be less than 3.0 after 7 days, indicating that a stronger hetero-lactic fermentation had occurred in IR silage. The fermentation process yielded minor quantities of butyric acid, propionic acid, and isobutyric acid. This was ascribed to a rapid reduction in pH resulting from the accelerated production of LA, which effectively inhibited the activity of clostridia and other deleterious microbes.

The DM levels of IR ensilage were observed to decline during the fermentation process, indicating microbial breakdown of nutrients into CO_2 and H_2O. The NH_3-N level of high-quality silage should be lower than 100 g/kg TN [1]. Herein, the NH_3-N levels satisfied the requisite criteria of less than 100 g/kg TN, which might be linked with the rapid decline in pH observed at the initial stage. Zahiroddini et al. [20] also found the inhibitory influence of a rapid decline in ensilage pH on the proteolytic activity of forage enzymes and aerobic microbes. The considerable reduction ($p < 0.05$) in residual WSC levels on day 3 indicated that the fermentable substrates in IR were rapidly converted into LA during the initial period of ensilage.

Lactic acid bacteria are capable of utilizing a range of substrates and rapidly producing substantial quantities of LA. The decline ($p < 0.05$) in the population of LAB from day 30 to day 60 may be attributed to the insufficient WSC contents and the acidic environment that is characteristic of silage. The decline in *Enterobacteriaceae* numbers during ensilage was primarily related to the rapid reduction in pH at the initial period of ensilage. The yeast numbers remained at a relatively high level (>4.00 $\log_{10}$ cfu/g FW) at the conclusion of fermentation, indicating that some yeasts in IR silage can survive under acidic conditions, even at pH 3.5 [21]. Therefore, it is imperative to devote greater attention to enhancing the aerobic stability of high-moisture IR silage.

The glucose content declined rapidly during the initial period of ensiling (Figure 1). These findings align with those of Shao et al. [7], who concluded that the glucose content showed the greatest decline at the initial stage of ensilage. After 3 days, a more pronounced decline was noticed in glucose content (71.2%) compared to the sucrose (10.0%) and fructose (11.2%) contents. This directly indicates that glucose was the primary fermentable substrate and was more susceptible to degrade compared to other substrates during ensilage, which agrees with the reports of Shao et al. [22]. It is noteworthy that the fructose contents showed a gradual decline over the three-day ensiling period. This phenomenon has been attributed to the hydrolysis of fructans and sucrose, which is a consequence of the action of plant enzymes during the initial stages of ensiling [7].

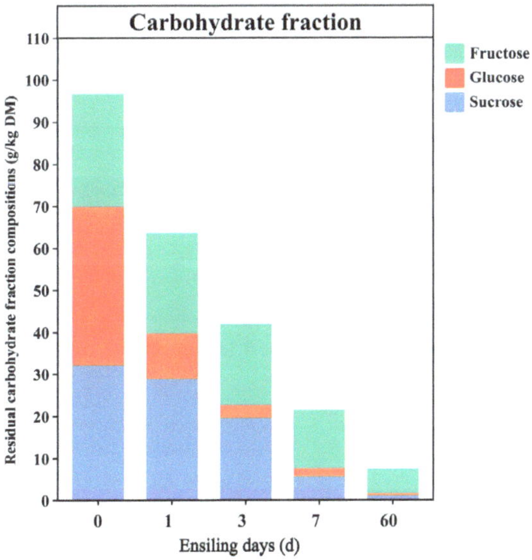

Figure 1. Changes in mono-and di-saccharides in fresh and ensiled Italian ryegrass. DM: dry matter; d: days.

3.2. Bacterial Community Diversity and Compositions

The coverage index of the bacterial communities in all samples was greater than 99.97% (Figure 2A), suggesting that the majority of bacterial communities were successfully obtained. In comparison to IR_3 and IR_60, the elevated ($p < 0.05$) indices of Chao, Sobs, and Ace in IRFM indicated that the diversity of bacterial community in IRFM was higher than that observed in IR_3 and IR_60 silage. Numerous studies have demonstrated that a reduction in pH can lead to a decline in microbial community diversity in a range of silage types [23,24]. The rarefaction curves for the bacterial communities reached saturation (Figure 2B), suggesting that the sequencing data were sufficient for reliable comparative analyses. A separation between IRFM and IR silage marks (Figure 2C) suggested the existence of different bacterial community compositions between fresh and ensiled IR. This was attributed to the presence of a more acidic microenvironment in IR silage than in fresh IR.

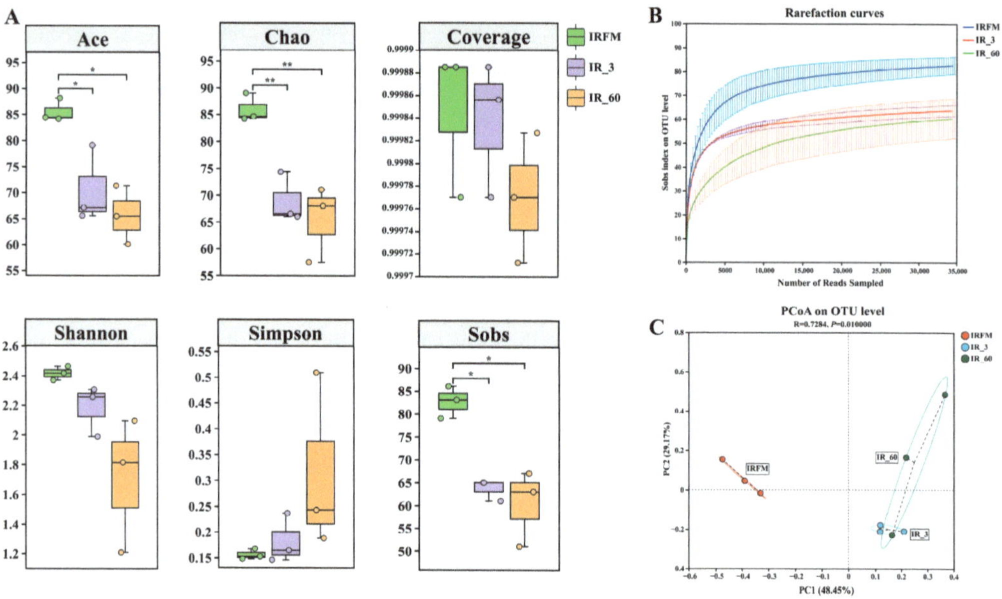

Figure 2. Bacterial community diversity of fresh and ensiled Italian ryegrass. (**A**) Alpha diversity (Ace, Chao1, Shannon, Coverage, Simpson, and Sobs indices) of bacterial community. (**B**) Rarefaction curves based on the Sobs index. (**C**) Beta diversity of bacterial community, calculated by principal coordinate analysis (PCoA) based on the Bray−Curtis distance metric. IRFM: fresh Italian ryegrass; IR_3: spontaneously ensiled Italian ryegrass after 3 days; IR_60: spontaneously ensiled Italian ryegrass after 60 days. * $0.01 < p \leq 0.05$; ** $0.001 < p \leq 0.01$.

The observation of a greater number of unique OTUs in the bacterial community epiphytic on IRFM (Figure 3A) was consistent with the higher α-diversity in IRFM compared to IR silage. In the bacterial community of IRFM (Figure 3B), the dominant phylum Proteobacteria played an important role in degrading organic matter and nitrogen and carbon cycling during anaerobic fermentation [25]. After ensiling, the increase in Firmicutes and decrease in Proteobacteria could be ascribed to the anaerobic and acidic microenvironment in silage, which limited aerobic microorganisms and promoted LAB strains. Keshri et al. [26] reported that anaerobic or low-pH conditions during ensilage favor the reproduction of Firmicutes species.

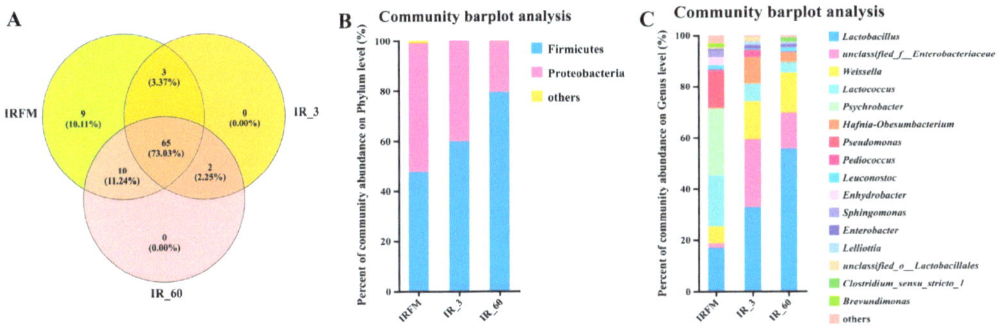

Figure 3. Bacterial community compositions of fresh and ensiled Italian ryegrass. (**A**) The Venn diagram of the core OTUs in fresh and ensiled Italian ryegrass. (**B**) The bar plots of bacterial community abundances at the phylum level. (**C**) The bar plots of bacterial community abundances at the genus level. IRFM: fresh Italian ryegrass; IR_3: spontaneously ensiled Italian ryegrass after 3 days; IR_60: spontaneously ensiled Italian ryegrass after 60 days.

The most prevalent genus within IRFM group was identified as *Psychrobacter* (Figure 3C), which may be associated with its capacity to thrive in humid and cold environments. Wang et al. [27] reported that the majority of *Psychrobacter* strains exhibited a preference for low temperatures, with an optimal growth temperature of approximately 20 °C. After ensilage, *Lactobacillus* rapidly became the most predominant after 3 days, maintaining this position until the conclusion of fermentation. The prevalence of *Lactobacillus* is associated with elevated LA concentrations in IR silage.

During the ensiling process, there was a notable decline in the relative abundance of *Enterobacteriaceae*, which declined from 26.5% on day 3 to 13.9% on day 60. *Enterobacteriaceae* are non-spore-forming, facultative anaerobic bacteria that are capable of fermenting LA to AA and other products, which can lead to a loss of nutritional value. The principal products of their fermentation process are acetic acid and carbon dioxide [28]. It can therefore be surmised that the elevated acetic acid concentrations were likely associated with the heightened abundance of *Enterobacteriaceae* in IR silage. While the production of acetic acid may enhance aerobic stability, prolonged fermentation could result in considerable energy and DM losses. The resulting silage is also less palatable to livestock.

It is noteworthy that the proportion of *Weissella* exhibited an increase from 14.8% on day 3 to 15.6% on day 60. This differs with the findings of Graf et al. [29], who found that *Weissella* was an initial colonizer and was then replaced by acid-resistant *Lactobacilli* with the decrease in pH and the ensilage progress. This may be ascribed to the enhanced competitiveness and vitality of *Weissella* epiphytic on fresh grass [30].

During the fermentation process, the relative abundance of *Hafnia-Obesumbacterium* was observed to decline from 10.2% on day 3 to 3.86% on day 60. As reported by Koivula et al. [31] and Hu et al. [32], *Hafnia-Obesumbacterium* has potential to cause contamination by enterobacteria and to degrade organic components. Therefore, it can be hypothesized that a higher concentration of ethanol, NH_3-N, and acetic acid during the ensiling process may be associated with an increased relative proportion of *Hafnia-Obesumbacterium*.

The relative proportion of *Lactococcus* exhibited a decline from 6.93% on day 3 to 4.08% on day 60. *Lactococcus* species are typically predominant in silage and are usually employed as inoculants to facilitate the LA fermentation at the early phase [33]. In this study, the observed decline in the relative abundance of *Lactococcus* during ensiling is likely attributable to the fact that cocci-LAB are unable to thrive in an acidic environment.

3.3. Co-Occurrence Networks in the Bacterial Community

Bacterial co-occurrence networks (Figure 4A–C) of fresh and fermented IR were conducted with the objective of achieving a comprehensive understanding of the impact of storage length on the interaction and correlation of the resulting microbiome. The ensiling process was observed to reduce the complexity of the bacterial networks in ensiled IR. The correlation structures of these networks were found to be more straightforward in IR silage than in IRFM, as indicated by lower edge and node numbers in IR silage than IRFM (Figure 4D). The greater complexity of the bacterial networks in IRFM in comparison to IR silage was in accordance with the higher bacterial community diversity in fresh IR.

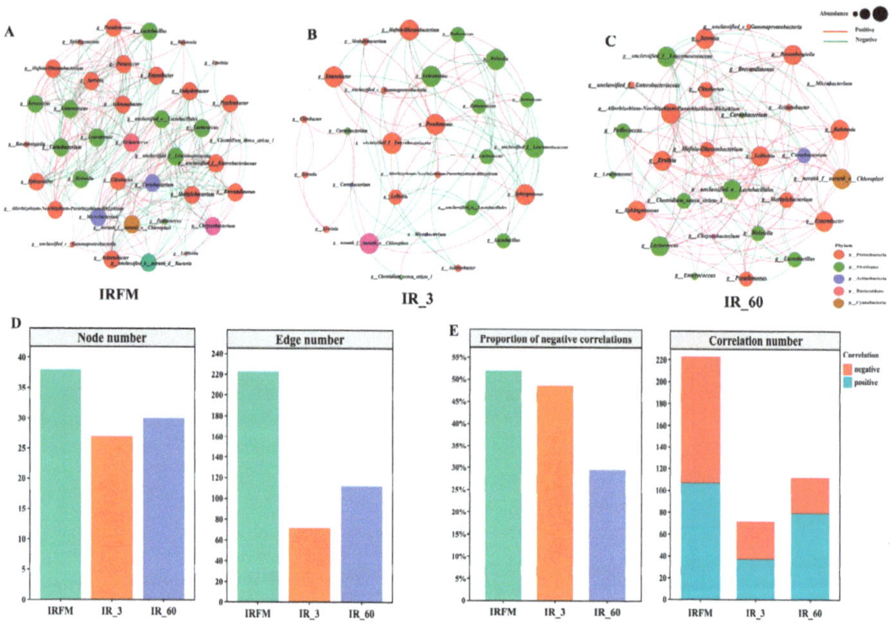

Figure 4. Bacterial co-occurrence networks of fresh and ensiled Italian ryegrass. Bacterial co-occurrence networks (Spearman correlation, the most abundant 300 species, p value < 0.05, and correlation > 0.5) of IRFM (**A**), IR_3 (**B**), and IR_60 (**C**). The node represents bacterial species, node color represents bacterial phylum, and node size represents the bacterial abundance. Edges are colored according to negative (green) and positive (red) correlations. (**D**) Bar plots of node and edge numbers, respectively. (**E**) Bar plots of negative correlation proportion and correlation number. IRFM: fresh Italian ryegrass; IR_3: spontaneously ensiled Italian ryegrass after 3 days; IR_60: spontaneously ensiled Italian ryegrass after 60 days.

The existence of negative correlations in the microbial co-occurrence networks indicates the potential for competition for resources. Conversely, positive correlations indicate the presence of collaborative or mutualistic associations among microbial taxa [34]. Herein, the proportions of negative correlation were found to be greater in the bacterial co-occurrence network of IRFM and IR_3 in comparison to IR_60 (Figure 4E). This suggests that the competition among bacterial species intensified at the early stage and diminished at the late stage of fermentation. A relatively greater positive correlation number was observed in the bacterial co-occurrence network of IR_60, indicating a greater degree of cooperative interaction among bacterial species in IR silage after 60 days of ensilage.

It is noteworthy that the alterations in ensilage over time exhibited different trends when comparing the transition from fresh IR to day 3 with that from day 3 to day 60. The simplest bacterial community composition was observed on day 3, with an enlargement of the composition on day 60, as evidenced by the edge and node numbers in Figure 4D and the correlation number in Figure 4E. It was hypothesized that the different direction was caused by the proliferation of *Clostridium_sensu_stricto_1*. *Clostridium_sensu_stricto_1* was observed on day 60 with a relative abundance of 1.64%. It is regarded as an undesirable microorganism that can survive in silage because it can convert sugar and protein into butyric acids, which reduces the fermentation quality and increases the risk of failed fermentation [35]. *Clostridium* also has the potential to survive in high-moisture silages, including mulberry, stylo, and alfalfa [36]. During ensilage, the production of acetic acid and NH_3-N was related to the metabolization of *Clostridium*.

3.4. Predicted Functionality in the Bacterial Community

In recent years, extensive functional prediction through the 16S rRNA gene dataset has been analyzed via the PICRUSt tool. PICRUSt2 firstly appeared in 2020 and represents an improvement based on PICRUSt1. PICRUSt2 is more accurate than PICRUSt1 and other competing methods overall because PICRUSt2 includes a larger and updated database of reference genomes and gene families and shows interoperability with any OTU-picking or denoising algorithm [37]. Therefore, the superior PICRUSt2 tool was employed to examine the influence of storage time on the functionality of the bacterial community in fresh and fermented IR.

The bacterial functional pathways at level 2 in fresh and fermented IR are primarily characterized by the metabolism of carbohydrate and amino acid (Figure 5). This is to be expected, given that ensilage relies on the capacity of LAB strains to convert the WSC in forage to organic acid (mainly lactic acid) under anaerobic conditions [1]. Furthermore, amino acids play a critical role in primary metabolism and bacterial protein synthesis [38], which explains the enhanced metabolism of carbohydrate and amino acid during ensiling. The ensiling process effectively restricts the amino acid and carbohydrate metabolism of most harmful bacteria on day 3, thus leading to a reduction in the majority of bacterial carbohydrate and amino acid metabolism in comparison to IRFM.

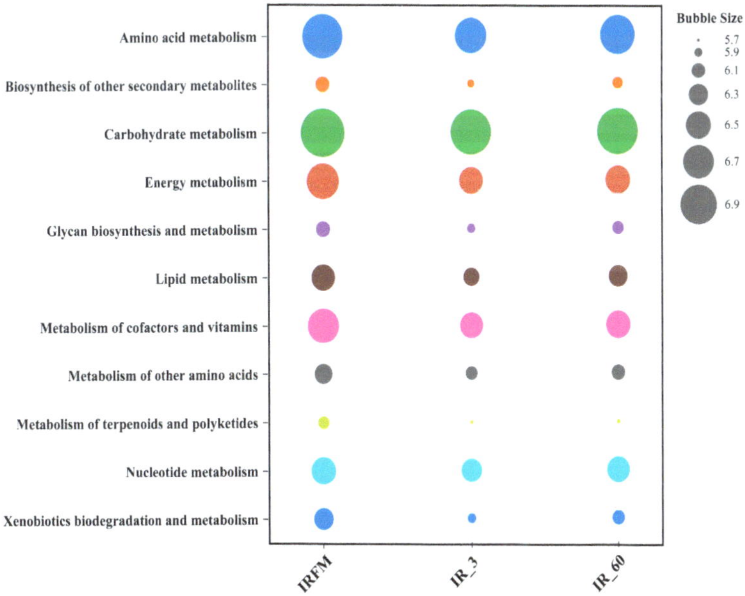

Figure 5. The 16S ribosomal RNA gene estimated Kyoto encyclopedia of genes and genomes functional profiles at the second pathway level. IRFM: fresh Italian ryegrass; IR_3: spontaneously ensiled Italian ryegrass after 3 days; IR_60: spontaneously ensiled Italian ryegrass after 60 days.

Herein, the energy metabolism was restricted during ensilage. This contrasts with the results reported by Xu et al. [39], who indicated that energy metabolism was enhanced in quality ensilage. Thus, further study is required to elucidate the effects of energy metabolism in ensilage. The metabolism of cofactors and vitamins was observed to decline in IR silage. It has been demonstrated that the application of LAB additives can directly stimulate the synthesis of vitamins during ensilage [30,40]. It was thus proposed that the metabolism of cofactors and vitamins in IR silage may be promoted using specific LAB additives.

3.5. Predicted Enzymes in the Bacterial Community

Homofermentative LAB played a pivotal role in the stimulation of LA fermentation. During the metabolic process of homofermentative LAB, glucose is converted into LA through the Embden–Meyerhof pathway (EMP). It is acknowledged that pyruvate kinase, 1-phosphofructokinase, and hexokinase are the key enzymes in EMP. The notable ($p < 0.05$) elevation in pyruvate kinase and 1-phosphofructokinase levels following ensiling (Figure 6A–C) suggests that LA fermentation in IR silage is more closely associated with these enzymes than hexokinase. The increase ($p < 0.05$) in D-/L-lactate dehydrogenase (Figure 6D,E) is in line with the substantial production of LA during ensilage.

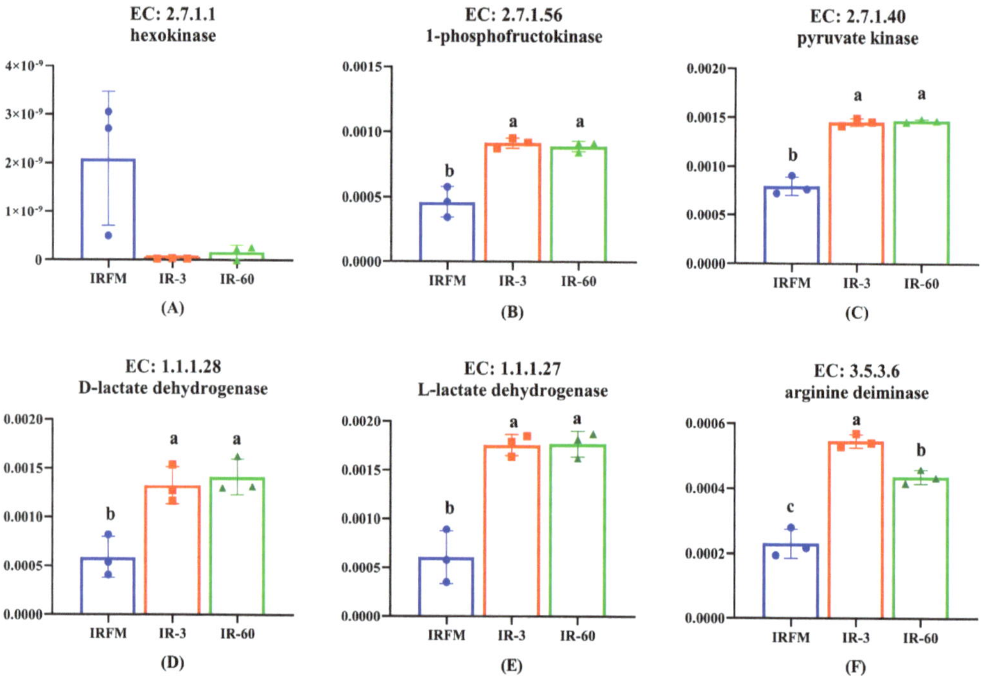

Figure 6. Changes of key enzymes involved in bacterial community metabolism during the ensiling of Italian ryegrass. (**A**) hexokinase; (**B**) 1-phosphofructokinase; (**C**) pyruvate kinase; (**D**) D-lactate dehydrogenase; (**E**) L-lactate dehydrogenase; (**F**) arginine deiminase. EC: reference metabolic pathway highlighting numbers; IRFM: fresh Italian ryegrass; IR_3: spontaneously ensiled Italian ryegrass after 3 days; IR_60: spontaneously ensiled Italian ryegrass after 60 days. Means with different letters differ significantly from each other ($p < 0.05$).

A reduction in pH may prompt certain systems in LAB to buffer LA acidity, including arginine deimination (ADI pathway). Therefore, the observed increase ($p < 0.05$) in arginine deiminase activity on day 3 (Figure 6F) might be ascribed to the elevated acidity levels in ensilage. Following a 60-day period, the notable ($p < 0.05$) decline in LAB populations may be associated with a substantial reduction in arginine deiminase levels on day 60. This is because the acidic environment in silage may not be effectively buffered and tolerated by LAB due to the decrease in arginine deiminase.

3.6. Predicted Phenotypes in the Bacterial Community

BugBase was employed as a means of predicting bacterial phenotypes, and it has already been applied in various fields [16]. It has been demonstrated to be an effective method for identifying microbiological discoveries that are readily interpretable and that are not achievable with current tools. In the context of the 16S rRNA gene sequencing analysis, BugBase offers a valuable avenue for gaining new insights into the pathogenic risks posed by silage microbes, thus facilitating the assessment of the safety and hygiene of animal feeds.

Herein, a reduction in the proportion of Gram-negative bacteria was found in IR silage compared to IRFM (Figure 7A), which is beneficial for fermentation quality, as Gram-negative bacteria, such as those containing lipopolysaccharide layers, have been reported to result in serious systemic infection [41]. Furthermore, a greater ($p < 0.01$) proportion of bacterial community was identified as "potentially pathogenic" in IR silage compared to fresh IR (Figure 7B), which is likely linked with the presence of some acid-resistant pathogens. The IR silage on days 3 and 60 was observed to increase the risks of pathogenic contamination. Those negative impacts cannot be mitigated by ensiling. It can be concluded that the addition of certain effective chemicals or LAB inoculants to high-moisture IR silage is crucial for accelerating acidification and reducing the risk of pathogenic contamination.

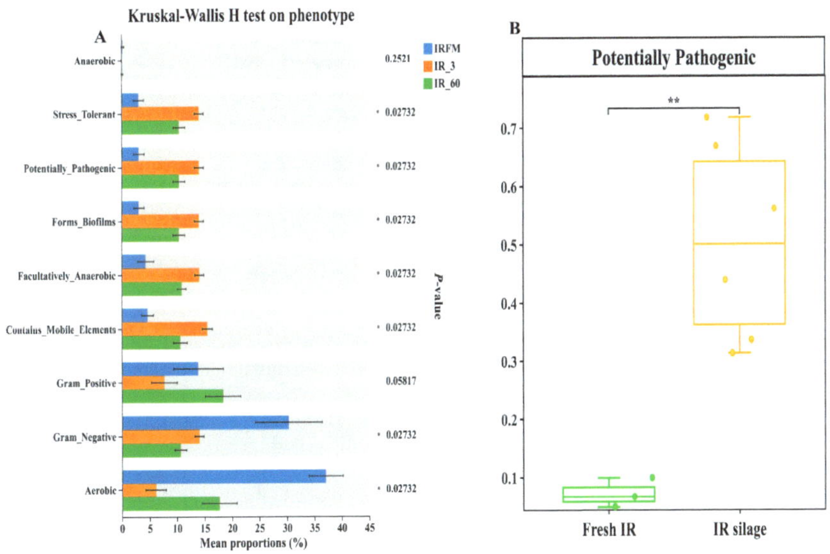

Figure 7. Bacterial phenotypes annotation of fresh and ensiled Italian ryegrass. (**A**) Phenotypes that reflect bacterial characteristics (Aerobic, Facultatively Anaerobic, Anaerobic, Gram_Negative, Gram_Positive, and Contains_Mobile_Elements) and bacterial resistance (Forms_Biofilms, Potentially_Pathogenic, and Stress_Tolerant). (**B**) Comparison of the Potentially_Pathogenic category between fresh and ensiled Italian ryegrass. The statistical analysis was conducted using a two-sided *t*-test. * $0.01 < p \leq 0.05$; ** $0.001 < p \leq 0.01$.

4. Conclusions

The high-moisture Italian ryegrass silage exhibited good fermentation quality after 60 days. *Lactobacillus* was dominant at the initial and final stages of fermentation. The ensiling process reduced the complexity of the bacterial community networks in ensiled IR. The metabolism of carbohydrate and amino acid was restricted at the early stage of ensilage. The pyruvate kinase and 1-phosphofructokinase were critical in promoting the lactic acid fermentation. A greater abundance in the "potentially pathogenic" category was found in the bacterial community of IR silage compared to fresh IR. Overall, the findings indicate that the high-moisture IR silage exhibited good ensiling characteristics, but the potential for microbial contamination and pathogens remained after ensiling.

Author Contributions: Methodology, X.W.; formal analysis, J.T. and M.N.; investigation, S.W., Y.C., N.X. and W.Z.; resources, C.D.; data curation, S.W.; writing—original draft, S.W.; visualization, S.W.; project administration, B.L.; funding acquisition, B.L. All authors have read and agreed to the published version of the manuscript.

Funding: This study was financially supported by National Natural Science Foundation of China (32301500), the earmarked fund for Jiangsu Agricultural Industry Technology System (JATS [2023]398), and the open project of Key Laboratory for Crop and Animal Integrated Farming, Ministry of Agriculture and Rural Affairs (202303).

Institutional Review Board Statement: Not applicable.

Data Availability Statement: Data are contained within the article.

Conflicts of Interest: The authors have no conflicts of interest to declare.

References

1. McDonald, P.; Henderson, A.R.; Heron, S. *The Biochemistry of Silage*; Chalcombe Publications: Abersytwyth, UK, 1991.
2. Li, Y.; Nishino, N. Effects of ensiling fermentation and aerobic deterioration on the bacterial community in Italian ryegrass, guinea grass, and whole-crop maize silages stored at high moisture content. *Asian Australas. J. Anim. Sci.* **2013**, *26*, 1304–1312. [CrossRef] [PubMed]
3. Tian, J.; Yin, X.; Zhang, J. Effects of wilting during a cloudy day and storage temperature on the fermentation quality and microbial community of Napier grass silage. *J. Sci. Food Agric.* **2022**, *102*, 4384–4391. [CrossRef] [PubMed]
4. Higgs, R.J.; Chase, L.E.; Ross, D.A.; Van Amburgh, M.E. Updating the Cornell Net Carbohydrate and Protein System feed library and analyzing model sensitivity to feed inputs. *J. Dairy Sci.* **2015**, *98*, 6340–6360. [CrossRef] [PubMed]
5. Zhao, N.; Hao, X.; Yin, M.; Li, C.; Wang, C.; Han, H. Influence of growth stages and additives on the fermentation quality and microbial profiles of whole-plant millet silage. *Agriculture* **2024**, *14*, 1323. [CrossRef]
6. Wu, B.; Ren, T.; Dan, N.; Yu, P.; Wang, H.; Wei, M.; Hu, Z.; Siqin, T.; Wu, T.; Niu, H. The responses of fermentation characteristics and microbial community structure and co-occurrence network to molasses and *Lacticaseibacillus casei* additions during anaerobic fermentation of *Caragana korshinskii*. *LWT-Food Sci. Technol.* **2024**, *206*, 116607. [CrossRef]
7. Shao, T.; Zhang, Z.X.; Shimojo, M.; Wang, T.; Masuda, Y. Comparison of fermentation characteristics of Italian ryegrass (*Lolium multiflorum* Lam.) and guineagrass (*Panicum maximum* Jacq.) during the early stage of ensiling. *Asian Australas. J. Anim. Sci.* **2005**, *18*, 1727–1734. [CrossRef]
8. Playne, M.J.; McDonald, P. The buffering constituents of herbage and of silage. *J. Sci. Food Agric.* **1966**, *17*, 264–268. [CrossRef]
9. Broderick, G.A.; Kang, J.H. Automated simultaneous determination of ammonia and total amino acids in ruminal fluid and in vitro media. *J. Dairy Sci.* **1980**, *63*, 64–75. [CrossRef]
10. Wang, S.; Wang, Y.; Zhao, J.; Dong, Z.; Li, J.; Nazar, M.; Kaka, N.A.; Shao, T. Influences of growth stage and ensiling time on fermentation profile, bacterial community compositions and their predicted functionality during ensiling of Italian ryegrass. *Anim. Feed Sci. Technol.* **2023**, *298*, 115606. [CrossRef]
11. Murphy, R.P. A method for the extraction of plant samples and the determination of total soluble carbohydrates. *J. Sci. Food Agric.* **1958**, *9*, 714–717. [CrossRef]
12. Krishnamoorthy, U.; Muscato, T.V.; Sniffen, C.J.; Van Soest, P.J. Nitrogen fractions in selected feedstuffs. *J. Dairy Sci.* **1982**, *65*, 217–225. [CrossRef]
13. Van Soest, P.J.; Robertson, J.B.; Lewis, B.A. Methods for dietary fiber, neutral detergent fiber, and nonstarch polysaccharides in relation to animal nutrition. *J. Dairy Sci.* **1991**, *74*, 3583–3597. [CrossRef]
14. Desta, S.T.; Yuan, X.J.; Li, J.F.; Shao, T. Ensiling characteristics, structural and nonstructural carbohydrate composition and enzymatic digestibility of Napier grass ensiled with additives. *Bioresour. Technol.* **2016**, *221*, 447–454.
15. Douglas, G.M.; Maffei, V.J.; Zaneveld, J.R.; Yurgel, S.N.; Brown, J.R.; Taylor, C.M.; Huttenhower, C.; Langille, M.G. PICRUSt2 for prediction of metagenome functions. *Nat. Biotechnol.* **2020**, *38*, 685–688. [CrossRef] [PubMed]
16. Ward, T.; Larson, J.; Meulemans, J.; Hillmann, B.; Lynch, J.; Sidiropoulos, D.; Spear, J.R.; Caporaso, G.; Blekhman, R.; Knight, R.; et al. BugBase predicts organism-level microbiome phenotypes. *BioRxiv* **2017**, 133462. [CrossRef]
17. Jahanzad, E.; Sadeghpour, A.; Hashemi, M.; Afshar, R.; Hosseini, M.B.; Barker, A. Silage fermentation profile, chemical composition and economic evaluation of millet and soya bean grown in monocultures and as intercrops. *Grass Forage Sci.* **2016**, *71*, 584–594. [CrossRef]

18. Nishino, N.; Li, Y.; Wang, C.; Parvin, S. Effects of wilting and molasses addition on fermentation and bacterial community in guinea grass silage. *Lett. Appl. Microbiol.* **2012**, *54*, 175–181. [CrossRef]
19. Ferrero, F.; Piano, S.; Tabacco, E.; Borreani, G. Effects of conservation period and *Lactobacillus hilgardii* inoculum on the fermentation profile and aerobic stability of whole corn and sorghum silages. *J. Sci. Food Agric.* **2019**, *99*, 2530–2540. [CrossRef]
20. Zahiroddini, H.; Baah, J.; Absalom, W.; McAllister, T.A. Effect of an inoculant and hydrolytic enzymes on fermentation and nutritive value of whole crop barley silage. *Anim. Feed Sci. Technol.* **2004**, *117*, 317–330.
21. Wang, Y.; He, L.; Xing, Y.; Zheng, Y.; Zhou, W.; Pian, R.; Yang, F.; Chen, X.; Zhang, Q. Dynamics of bacterial community and fermentation quality during ensiling of wilted and unwilted *Moringa oleifera* leaf silage with or without lactic acid bacterial inoculants. *MSphere* **2019**, *4*, e00341-19.
22. Shao, T.; Ohba, N.; Shimojo, M.; Masuda, Y. Changes in mono-and disaccharides compositions of guineagrass (*Panicum maximum* Jacq.) silage during early stages of ensiling. *J. Fac. Agric. Kyushu Univ.* **2003**, *47*, 333–339. [CrossRef] [PubMed]
23. Méndez-García, C.; Peláez, A.I.; Mesa, V.; Sánchez, J.; Golyshina, O.V.; Ferrer, M. Microbial diversity and metabolic networks in acid mine drainage habitats. *Front. Microbiol.* **2015**, *6*, 475.
24. Eikmeyer, F.G.; Köfinger, P.; Poschenel, A.; Jünemann, S.; Zakrzewski, M.; Heinl, S.; Mayrhuber, E.; Grabherr, R.; Pühler, A.; Schwab, H.; et al. Metagenome analyses reveal the influence of the inoculant *Lactobacillus buchneri* CD034 on the microbial community involved in grass ensiling. *J. Biotechnol.* **2013**, *167*, 334–343. [PubMed]
25. Ma, S.; Fang, C.; Sun, X.; Han, L.; He, X.; Huang, G. Bacterial community succession during pig manure and wheat straw aerobic composting covered with a semi-permeable membrane under slight positive pressure. *Bioresour. Technol.* **2018**, *259*, 221–227. [CrossRef]
26. Keshri, J.; Chen, Y.; Pinto, R.; Kroupitski, Y.; Weinberg, Z.G.; Sela, S. Microbiome dynamics during ensiling of corn with and without *Lactobacillus plantarum* inoculant. *Appl. Microbiol. Biotechnol.* **2018**, *102*, 4025–4037. [CrossRef]
27. Wang, S.; Zhao, J.; Dong, Z.; Li, J.; Shao, T. Sequencing and microbiota transplantation to determine the role of microbiota on the fermentation type of oat silage. *Bioresour. Technol.* **2020**, *309*, 123371. [CrossRef]
28. Ni, K.; Wang, F.; Zhu, B.; Yang, J.; Zhou, G.; Pan, Y.I.; Tao, Y.; Zhong, J. Effects of lactic acid bacteria and molasses additives on the microbial community and fermentation quality of soybean silage. *Bioresour. Technol.* **2017**, *238*, 706–715. [CrossRef]
29. Graf, K.; Ulrich, A.; Idler, C.; Klocke, M. Bacterial community dynamics during ensiling of perennial ryegrass at two compaction levels monitored by terminal restriction fragment length polymorphism. *J. Appl. Microbiol.* **2016**, *120*, 1479–1491. [CrossRef] [PubMed]
30. Bai, J.; Ding, Z.; Ke, W.; Xu, D.; Wang, M.; Huang, W.; Zhang, Y.; Liu, F.; Guo, X. Different lactic acid bacteria and their combinations regulated the fermentation process of ensiled alfalfa: Ensiling characteristics, dynamics of bacterial community and their functional shifts. *Microb. Biotechnol.* **2021**, *14*, 1171–1182. [CrossRef]
31. Koivula, T.T.; Juvonen, R.; Haikara, A.; Suihko, M.-L. Characterization of the brewery spoilage bacterium *Obesumbacterium* proteus by automated ribotyping and development of PCR methods for its biotype 1. *J. Appl. Microbiol.* **2006**, *100*, 398–406. [CrossRef]
32. Hu, Y.; Shen, Y.; Wang, J. Pretreatment of antibiotic fermentation residues by combined ultrasound and alkali for enhancing biohydrogen production. *J. Clean. Prod.* **2020**, *268*, 122190. [CrossRef]
33. Muck, R.E. Recent advance in silage microbiology. *Agric. Food Sci.* **2013**, *22*, 3–15.
34. Banerjee, S.; Kirkby, C.A.; Schmutter, D.; Bissett, A.; Kirkegaard, J.A.; Richardson, A.E. Network analysis reveals functional redundancy and keystone taxa amongst bacterial and fungal communities during organic matter decomposition in an arable soil. *Soil Biol. Biochem.* **2016**, *97*, 188–198. [CrossRef]
35. Wu, P.W.; Li, L.H.; Jiang, J.F.; Sun, Y.M.; Yuan, Z.H.; Feng, X.D.; Guo, Y.F. Effects of fermentative and non-fermentative additives on silage quality and anaerobic digestion performance of Pennisetum purpureum. *Bioresour. Technol.* **2020**, *297*, 122425. [PubMed]
36. Driehuis, F.; Wilkinson, J.M.; Jiang, Y.; Ogunade, I.; Adesogan, A.T. Silage review: Animal and human health risks from silage. *J. Dairy Sci.* **2018**, *101*, 4093–4110. [CrossRef]
37. Wang, S.; Wang, Y.; Liu, H.; Li, X.; Zhao, J.; Dong, Z.; Li, J.; Kaka, N.A.; Nazar, M.; Shao, T. Using PICRUSt2 to explore the functional potential of bacterial community in alfalfa silage harvested at different growth stages. *Chem. Biol. Technol. Agric.* **2022**, *9*, 98. [CrossRef]
38. Jing, Y.; Mu, C.; Wang, H.; Shen, J.; Zoetendal, E.G.; Zhu, W. Amino acid utilization allows intestinal dominance of *Lactobacillus amylovorus*. *ISME J.* **2022**, *16*, 2491–2502. [CrossRef]
39. Xu, D.; Wang, N.; Rinne, M.; Ke, W.; Weinberg, Z.G.; Da, M.; Bai, J.; Zhang, Y.; Li, F.; Guo, X. The bacterial community and metabolome dynamics and their interactions modulate fermentation process of whole crop corn silage prepared with or without inoculants. *Microb. Biotechnol.* **2021**, *14*, 561–576. [CrossRef]
40. Zhang, X.; Ke, C.; Bai, J.; Li, H.; Xu, M.; Ding, T.; Guo, S. The effect of *Pediococcus acidilactici* J17 with high-antioxidant activity on antioxidant, α-tocopherol, β-carotene, fatty acids, and fermentation profiles of alfalfa silage ensiled at two different dry matter contents. *Anim. Feed Sci. Technol.* **2020**, *268*, 114614. [CrossRef]
41. Wang, Z.; Li, Q.; Li, J.; Shang, L.; Li, J.; Chou, S.; Lyu, Y.; Shan, A. pH-responsive antimicrobial peptide with selective killing activity for bacterial abscess therapy. *J. Med. Chem.* **2022**, *65*, 5355–5373. [CrossRef]

agriculture

MDPI

Article

Assessing Fermentation Quality, Aerobic Stability, In Vitro Digestibility, and Rumen Degradation Characteristics of Silages Mixed with Sweet Sorghum and Aerial Parts of Licorice

Feng Chen [1,2], Jiao Wang [1], Sujiang Zhang [1,*], Abdul Shakoor Chaudhry [3] and Hassan Khanaki [4]

1 Key Laboratory of Tarim Animal Husbandry Science and Technology, College of Animal Science and Technology, Tarim University, Alar 843300, China; chenfengdky@163.com (F.C.); wang20220910@126.com (J.W.)
2 College of Animal Science and Technology, Northwest Agriculture and Forestry University, Yangling 712100, China
3 School of Natural and Environmental Sciences, Newcastle University, Newcastle upon Tyne NE1 7RU, UK; abdul.chaudhry@newcastle.ac.uk
4 Faculty of Veterinary and Agricultural Sciences, The University of Melbourne, Dookie College, VIC 3647, Australia; h.khanaki@unimelb.edu.au
* Correspondence: zsjdky@126.com; Tel.: +86-152-9230-0910

Citation: Chen, F.; Wang, J.; Zhang, S.; Chaudhry, A.S.; Khanaki, H. Assessing Fermentation Quality, Aerobic Stability, In Vitro Digestibility, and Rumen Degradation Characteristics of Silages Mixed with Sweet Sorghum and Aerial Parts of Licorice. *Agriculture* **2024**, *14*, 212. https://doi.org/10.3390/agriculture14020212

Academic Editor: Siran Wang

Received: 6 December 2023
Revised: 25 January 2024
Accepted: 26 January 2024
Published: 28 January 2024

Abstract: (1) Aim: This study aimed to evaluate the fermentation quality, chemical composition, aerobic stability, in vitro digestibility, and rumen degradation characteristics of silage mixtures with different ratios of sweet sorghum (SS) and aerial parts of licorice (LC). (2) Methods: Five mixtures were produced on a dry matter (DM) basis: (i) 0%SS + 100%LC (0%SS); (ii) 25%SS + 75%LC (25%SS); (iii) 50%SS + 50%LC (50%SS); (iv) 75%SS + 25%LC (75%SS); and (v) 100%SS + 0%LC (100%SS). First, the chemical composition of the silages was measured before and after fermentation. Next, the aerobic stability, dynamic microbial colonization and dynamic volatile fatty acids of the mixed silage after fermentation were determined for 0, 5, 10, 15, 20, and 25 days. Finally, the parameters related to gas production and the characteristics of the gas production were determined. At the same time, the rate of degradation of the chemical composition of the mixed silage in the rumen was studied. (3) Results: (a) As the proportion of SS increased, pH, ammonia, butyric acid, acetate, and aerobic stability showed a decreasing trend, but lactic acid content gradually increased. (b) The content of the fermentation and gas production parameters were significantly higher in 100%SS and 50%SS than others ($p < 0.05$). (c) The rate of degradation of DE, ME, Neg, DM, CP, ADF, NDF, and ADL of 50%SS in the rumen of sheep was significantly higher than others ($p < 0.05$). (4) Conclusions: In conclusion, ensiling SS and LC mixtures can improve silage quality, especially if the SS and LC are ensiled together at a ratio of 50:50.

Keywords: sweet sorghum; licorice; mixed silage; fermentation; aerobic stability; rumen; degradation characteristics

1. Introduction

The shortage of high-quality roughage resources is the main factor restricting the development of the sheep industry in many developing countries. Southern Xinjiang is the largest saline soil area and desertification area in China, which has relatively little natural rainfall and poor soil. Due to natural limitations, high-quality roughage resources in southern Xinjiang are extremely scarce. There is 0.067×10^8 hm^2 of marginal land in Xinjiang, among which the total area of saline land is 1.336×10^7 hm^2, accounting for 36.8% of saline–alkali land area in China [1]. The saline soil ecosystem is very fragile and the secondary soil salinization is serious in Xinjiang. The main solution is to popularize the cultivation of saline-tolerant forage crops.

Sweet sorghum (SS) is a promising forage that grows in arid, semi-arid, and high-salinity areas [2]. It is also known to be stress-resistant, drought-tolerant, and highly water-use efficient [3]. These excellent characteristics are very suitable for promoting planting in the southern Xinjiang region. SS is a resilient C_4, which has the characteristics of high fermentability, nutrient digestibility, and palatability [4,5]. However, the crude protein content in SS silage is insufficient to maximize growth efficiency in most ruminant production systems [6–8].

Licorice (LC) is a saline-resistant, drought-tolerant perennial herb of the legume family [9]. Licorice root and its extract are important Chinese herbal medicines with high medicinal value. It is mainly found in China, Mongolia, Central Asia, and Russia [10]. China is one of the countries rich in licorice. It is mainly found in the arid and semi-arid regions of north-eastern, northern, and north-western China. Xinjiang is the most common region [11]. Cui et al. (2023) [12] reported that licorice could prevent soil pollution and desertification, reduce soil erosion, and protect the ecological environment. The wild licorice resources of Xinjiang rank first in China, with southern Xinjiang accounting for over 70%. Licorice, with abundant protein, glycyrrhizin, flavonoids, and polysaccharides, has been a promising source of green fodder for animal feeding [13]. Abarghuei et al. (2021) [14] reported that the aerial parts of Glycyrrhiza are high-quality feeds for cattle and sheep. During the process of preparing hay, licorice stems and leaves fall off severely, resulting in severe nutrient loss. However, making a single licorice silage is not easy to achieve [15]. Therefore, mixing sweet sorghum with licorice stems and leaves to make silage can overcome their respective shortcomings. Liu et al. (2023) [16] reported that mixed silage with different proportions of alfalfa and maize can improve the nutritional value of silage. Ni et al. (2018) [17] found that forage soybean ensiled with corn or sorghum could be an alternative approach to improve forage soybean silage quality. Recently, some studies found that mixed silages with different forages could be a feasible way to improve the silage quality and aerobic stability of the fermentation system compared with the sole fermentation of various forages [18,19]. In conclusion, a mixed silage of legumes and gramineous plants has a better success rate than single-legume silage, and the mixed silage has improved fermentation quality, nutritional value, digestion, and metabolism for ruminant feeding. And the quality of mixed silage is closely related to the mixing ratio.

Recent studies have mainly reported the use of SS and alfalfa mixed silage in the total mixing ratio [20], the bioaugmentation effect of rumen fluid on SS silage [21], the nutritional value and fermentation characteristics of silage of different SS varieties [22], or different ratios of SS silage and corn silage in lactating dairy cows [23]. This study hypothesized that by using appropriate proportions, mixed silage with SS and LC might provide comprehensive nutrition and could also improve the fermentation quality of mixed silages. Therefore, this study examined the fermentation quality, aerobic stability, in vitro digestibility, and rumen degradation characteristics of silages mixed with sweet sorghum and aerial parts of licorice. We expected that the results obtained in this research could provide useful information for the practical use of mixed silages to feed ruminants.

2. Materials and Methods

2.1. Silage Mixture Preparation

The Cowley variety of SS with a 22.5% Brix value and wild LC were collected at the Agricultural Research Station of Tarim University, Xinjiang, China (longitude 81°31′ E, latitude 40°56′ N). The whole plant of SS at the maturity stage and aerial parts of licorice at the setting stage (50% setting rate) were harvested and chopped into 2–3 cm particle sizes by a multi-functional chopper (9DF53, Yanbei Animal Husbandry Machinery Group Co. Ltd., Beijing, China). A silage wrapping machine (D5552, Qufu Tianliang Trading Co. Ltd., Shandong, China) was used for making five types of SS-LC silage mixtures with different SS-to-LC ratios: (i) 0%SS + 100%LC (0%SS); (ii) 25%SS + 75%LC (25%SS); (iii) 50%SS + 50%LC (50%SS); (iv) 75%SS + 25%LC (75%SS); (v) 100%SS + 0%LC (100%SS). After careful blending, 5.5 kg of each mixture was placed in a 10 L lab hopper (polyethylene flask fitted with an air-tight topper and sealed with screwcaps and plastic bands). Each processing group had 6 replicates. This anaerobic fermentation process was conducted at room temperature (20–25 °C) for 150 days.

2.2. Chemical and Microbial Analyses of Mixed Silages with Sweet Sorghum and Licorice at Different Proportions of Sweet Sorghum and Licorice

At the time of sampling, the pre-ensiled samples (fresh ingredients) were divided into two sub-samples, and each ensiled sample was divided into three sub-samples. The first sub-sample was oven-dried at 60 °C for 48 h to determine the dry matter (DM) content. The obtained powder samples were stored for later analysis of crude ash (Ash), acid detergent fiber (ADF), ether extract (EE), neutral detergent fiber (NDF), total nitrogen (TN), and water-soluble carbohydrates (WSCs) [24]. TN × 6.25 was used to calculate crude protein (CP). After reaction with an anthrone reagent [25], WSC content was determined by colorimetry. Van Soest et al.'s (1991) procedures were used to determine ADF and NDF contents [26].

A 20 g sample of silage mixture from each treatment group was weighed separately, thoroughly mixed with 180 mL of distilled water, and left for 24 h at 4 °C. Two layers of gauze and filter paper were then used to filter the extract samples. The filtrate was used to determine ammonia nitrogen (NH_3-N), organic acid, and pH content. The pH was measured with a HANNAHI2221 pH-measuring instrument. The supernatant was centrifuged at 10,000× *g* for 10 min and collected for NH_3-N analysis [27].

After opening the lid for sampling and mixing, a sample of about 20 g was mixed with 180 mL distilled water in a 250 mL conical bottle, which was sealed before placing it into a refrigerator at 4 °C for 24 h. The suspension was then filtered first with four layers of gauze followed by a quantitative filter paper, and the filtered liquid was stored for later use. The extracts were centrifuged at 1500 r/min and mixed with 25% metaphoric acid solution at 5:1. The acidified supernatants were then loaded on a Thermo Scientific UltiMate 3000 high-performance liquid chromatograph (UltiMate XB-C18 column; Column temperature: 35 °C; Mobile phase: 0.1 mol/L potassium dihydrogen phosphate (KH_2PO_4); Flow rate: 1 mL/min) to determine the contents of lactic acid (LA), acetic acid (AA), propionic acid (PA), and butyric acid (BA) [28].

Representative and replicated silage samples were extracted on 0, 5, 10, 15, 20, and 25 days of aerobic exposure after successful ensiling. Each extract was diluted to 10^{-1}–10^{-8} g/mL by multiple dilution methods. About 1 mL of each dilution was taken from different dilution ratios and coated on different media including potato glucose agar (PDA) medium, high-salt Chach culture medium, MRS Medium, and ordinary agar medium. After incubating at 37 °C, 28 °C, 28 °C, and 28 °C for 48 h, yeast, mold, lactic acid bacteria (LAB), and aerobic bacteria (AB) were counted, respectively. The identification of the colonies was then carried out using special media, colony characterization, and microbiological microscopy. There were 3 replicates for each dilution ratio. The average colony count multiplied by the dilution of the appropriate dilution ratio was calculated as the number of microorganisms for the exposure time (CFU/g FM), and the results were expressed as log(CFU/g FM) [29,30].

The V-score method was used to evaluate the silage quality [31]. The V-score was calculated on a 100-point scale as follows: <60 (poor), 60–80 (fair), and 80–100 (good). The values of lactic acid/total acid, acetic acid/total acid, butyric acid/total acid, and NH_3-N/total nitrogen were comprehensively evaluated. The total score of organic acid was 100 points, the ratio of NH_3-N to total nitrogen was 50 points, and the comprehensive evaluation score was =(organic acid score)/2 + (NH_3-N ratio score of total nitrogen).

2.3. Aerobic Stability Analysis

A total of 180 silos (5 treatments × 6 exposure days × 6 replicates per treatment) were used for the aerobic stability test after 150 days of ensiling. Briefly, the mixed silages were removed from each silo, completely mixed, and loosely placed into a larger 10-L open-top polyethylene bottle. To prevent contamination by impurities such as fruit flies, the bottles were stored at room temperature (20–25 °C) and wrapped in a double layer of gauze. Multi-channel temperature recorders (MDL-1048A high-precision temperature recorder, Shanghai Tianhe Automation Instrument Co., Ltd., Shanghai, China.) were used to measure temperature changes by placing their sensors in a central position. Six probes were placed as blanks in the environment. Temperatures were recorded every hour, with aerobic stability defined as the time required for the sample to rise 2 °C above room temperature [32]. After 0, 5, 10, 15, 20, and 25 days of aerobic exposure, the dynamics of microbe counts, NH_3-N, organic acids, and WSCs in the samples were analyzed. The determination method was the same as above.

2.4. In Vitro Incubation

The research undertaken complies with the current animal welfare laws in China. The experiment was carried out at the Animal Research Station of Tarim University, Xinjiang, China. All procedures used in this study were performed according to the Guidelines for the Care and Use of Animals for Research in China (GB 14925-2001).

Rumen fluid was collected from the rumens of three rumen-cannulated male *Duolang* sheep prior to feeding in the morning. The rumen fluid was immediately filtered using four layers of medical gauze, transferred to the laboratory, and stored in a water bath at 39 °C. The ingredient and nutrient compositions of the base diet (air-dried foundation) for sheep are shown in Table 1. Rumen fluid was immediately filtered, moved to the laboratory, and stored at 39 °C in a water bath. Prior to use, the rumen fluid was mixed with a 1:2 (rumen fluid:artificial rumen fluid) buffer solution, as described by Menke and Steingass (1988) [33]. The whole process of preparing the buffered rumen fluid was carried out under continuous flushing with CO_2.

Table 1. Ingredient and nutrient composition of the basal diet (dry matter foundation).

Items	Content
composition of raw material (% DM [c])	
Alfalfa	25.00
Maize straw	25.00
Corn	29.37
Soybean meal	2.84
Expanded soybean	2.84
Wheat bran	4.27
Cottonpulp	1.89
Spouting corn husks	2.37
Distillers Dried Grains with Solubles	1.42
Premix [a]	5.00
Total	100.00
Nutrient composition [b] (% FM [c])	
Dry matter	92.56
Crude protein	13.33
Neutral detergent fiber	38.82
Acid detergent fiber	22.32
Ca	0.65
P	0.32

[a] Premix is provided for each kilogram of diet: vitamin A 15,000 IU; vitamin D 5000 IU; vitamin E 50 mg; iron 90 mg; copper 12.5 mg; manganese 50 mg; zinc 100 mg; selenium 0.3 mg; iodine 0.8 mg; cobalt 0.5 mg. [b] Nutritional components were measured values. [c] FM, fresh matter; DM, dry matter.

In vitro fermentation was performed in serum flasks according to the method by Contreras-Govea et al. (2011) [34] with some modifications. Gas production, modeling of gas production, and assaying of relevant parameters of gas production refer to Ørskov et al. (1979) [35]. The equations described by Menke et al. (1979) [36] were used to calculate the metabolizable energy (ME) and digestible organic matter (DOM).

2.5. The Degradation Characteristics of a Mixture Consisting of Sweet Sorghum and Licorice Stem and Leaf Silage in the Rumen of Sheep Were Evaluated Using the Nylon Bag Method

The experiment was conducted using six Duolang sheep with permanent fistulae. For each group of test feed, six parallel samples were established and inserted into the rumen of each of the six sheep. Each parallel group consisted of a 5.00 g portion of test feed placed in a nylon bag of known weight. One replicate was used for each parallel group. The nylon bags, made from 300-mesh nylon sieve silk with double stitching, measured 50 mm × 100 mm. They were introduced into the rumen through the sheep's rumen fistula, and after 4, 8, 12, 24, 48, and 72 h of digestion, six parallel samples were simultaneously retrieved. The samples were rinsed with tap water and dried at 65 °C, and then undamaged degradation residues from each nylon bag were transferred to sample vials for testing. The Duolang sheep were fed and managed according to conventional practices, and the composition and nutrient content of the basal diet can be found in Table 1.

The rumen degradation rate and degradation parameters of a component of the feed to be tested at a given time point were determined with reference to Mirzaci-Aghsaghali et al. (2008) [37] and Sehu et al. (2010) [38];

$$A = (B - C)/B \times 100\% \quad (1)$$

Here, A is the rumen degradation rate (%) of an ingredient of the feed to be tested at a given point in time; B is the mass of an ingredient of the feed to be tested (%); and C is the mass of an ingredient in the residue (%).

Degradation rate curve:

$$P = a + b\left(1 - e^{-ct}\right) \quad (2)$$

Here, P is the rumen degradation rate (%) of the feed to be tested at a given time point; a is the fast-degradation fraction (%); b is the slow-degradation fraction (%); c is the degradation rate constant for b (%/h); and t is the incubation time point of the sample in the rumen (h) [35].

Effective rate of degradation:

$$ED = a + (b \times c)/(c + k) \quad (3)$$

Here, k is the rumen chyme efflux rate, taken as k = 0.031%/h.

Digestible energy (DE), metabolizable energy (ME), and net energy for gain (NEg) metrics in mixed-silage diets were estimated using prediction equations for the energy value of sheep (kJ/kg) [36].

Predictive modeling of DE:

$$DE = 19.509 - 0.170 \times NDF - 0.006 \times OM - 0.097 \times CP\left(R^2 = 0.973, p < 0.001\right) \quad (4)$$

Predictive modeling of ME:

$$ME = 0.046 + 0.820 \times DE\left(R^2 = 0.972, p < 0.001\right) \quad (5)$$

Predictive modeling of NEg:

$$NEg = ME \times 0.4571 \quad (6)$$

2.6. Statistical Analysis

Analysis of variance (ANOVA) was performed using the general linear model procedure of SPSS 26.0. A one-way ANOVA was performed on aerobic stability, chemical composition, fermentation quality, and in vitro incubation data. A two-way ANOVA was performed on chemical composition and microbial counts after aerobic exposure. The statistical difference between the mean values was determined using Tukey's multiple comparisons test and was considered to be significant at $p < 0.05$.

3. Results

3.1. Chemical and Microbial Compositions of Pre-Ensiled and Ensiled Materials of the Mixed Silage of Sweet Sorghum and Aerial Parts of Licorice

The chemical composition of the mixed silage ingredients is shown in Table 2. Among all roughages, SS had a lower DM, CP, NDF, ADF, Ash, and ADL content and a higher WSC content. The content of EE was similar between SS and LC. After fermenting mixed silages for 150 days, the DM, CP, NDF, ADF, Ash, and ADL content decreased with the increasing level of SS silages (Table 3, $p < 0.05$). However, the WSC content increased in all mixed silages, which was between 3.60 and 8.48 (% DM). The 100%SS and 75%SS had significantly higher WSC content than that of 0%SS ($p < 0.05$). The NDF and ADF contents of all mixed silages were between 40.96 and 44.88 (% DM) and between 20.85 and 26.81 (% DM), respectively. The 0%SS and 25%SS groups were significantly higher than those in the other three groups ($p < 0.05$), but there was no significant difference between the 0%SS and 25%SS groups ($p > 0.05$). There were no significant differences in EE contents among all the silage mixtures ($p > 0.05$). The LAB population in all silage mixtures was greater than 10^5 cfu/g FM, even if the content of SS in the SS-LC mixed silages was increased ($p < 0.05$). No significant difference was observed in the aerobic bacteria and yeast population among the five silage mixtures ($p > 0.05$), but lactic acid bacteria showed a significant upward trend.

Table 2. Chemical composition of ingredients used for the mixed silage of sweet sorghum and aerial parts of licorice.

Items	Treatments	
	Sweet Sorghum	Licorice
Dry matter (% FM)	33.78	37.15
Crude protein (% DM)	7.14	13.75
Neutral detergent fiber (% DM)	39.18	47.59
Acid detergent fiber (% DM)	19.24	31.64
Crude ash (% DM)	7.01	13.98
Acid detergent lignin (% DM)	5.51	10.54
Water soluble carbohydrate (% DM)	18.63	7.86
Ether extract (% DM)	2.13	2.14

Table 3. Effect of chemical compositions of the mixed silage of sweet sorghum and aerial parts of licorice after 150 days of ensiling.

Items	Treatments					Mean	SEM [a]	*p*-Value
	0%SS	25%SS	50%SS	75%SS	100%SS			
Chemical composition								
Dry matter, DM (% FM)	39.76 ^{A}	37.91 ^{A}	35.90 ^{B}	35.11 ^{B}	34.24 ^{B}	36.59	0.54	<0.001
Crude protein, CP (% DM)	14.17 ^{A}	12.68 ^{A}	9.35 ^{B}	7.70 ^{B}	6.49 ^{C}	10.08	0.78	<0.001
Neutral detergent fiber, NDF (% DM)	44.88 ^{A}	43.88 ^{A}	41.20 ^{B}	42.82 ^{B}	40.96 ^{C}	42.75	0.40	<0.001
Acid detergent fiber, ADF (% DM)	26.81 ^{A}	27.37 ^{A}	23.52 ^{B}	22.93 ^{B}	20.85 ^{C}	24.03	0.66	<0.001
Crude ash, Ash (% DM)	12.74 ^{A}	12.18 ^{B}	10.64 ^{C}	9.43 ^{D}	6.72 ^{E}	10.34	0.58	<0.001
Acid detergent lignin, ADL (% DM)	10.33 ^{A}	9.17 ^{A}	8.87 ^{B}	7.74 ^{CD}	5.01 ^{D}	8.22	0.48	<0.001
Water soluble carbohydrate, WSC (% DM)	3.60 ^{C}	4.95 ^{B}	5.72 ^{B}	7.34 ^{A}	8.48 ^{A}	6.02	0.46	<0.001
Ether extract, EE (% DM)	2.12	2.11	2.13	2.12	2.12	2.12	0.01	0.996
Microbial composition (Log10 cfu/g FM [b])								
Lactic acid bacteria	6.64 ^{C}	7.12 ^{B}	7.52 ^{B}	8.21 ^{A}	8.52 ^{A}	7.60	0.19	<0.001
Aerobic bacteria	5.42	5.39	5.46	5.33	5.43	5.41	0.02	0.350
Yeasts	3.08	3.15	3.11	3.05	3.01	3.08	0.02	0.099

$^{A–E}$ The same letter indicates non-significant differences ($p > 0.05$) and different letters indicate significant differences ($p < 0.05$). Capital letters are significant differences between treatment groups. [a] SEM, standard error of means. [b] cfu, colony-forming units.

3.2. Fermentation Characteristics of the Mixed Silages of Sweet Sorghum and Aerial Parts of Licorice

Table 4 shows that all mixed silages and their interactions had a significant effect on AA, LA, NH_3-N, and pH contents ($p < 0.05$). With the increasing level of SS, the LA content continuously increased, and the pH and NH_3-N gradually decreased after ensiling for 150 days. Although the content of SS in silage increased the LA content, it decreased the pH value and the NH_3-N, and AA content. The 50%SS, 75%SS, and 100%SS silages were better preserved than 0%SS and 25%SS silages on the basis of the V-score (Table 4). There was a fluctuating upward trend in the ratio of lactic acid to acetic acid (LA/AA). The SS-LC mixed silages, particularly 0%SS and 25%SS, had significantly ($p < 0.05$) lower LA/AA values than that of the 100%SS mixed silage. With the increasing level of the SS ratio, the content of BA in all mixed silages showed a downward trend, and the content of BA of 0%SS and 25%SS was significantly ($p < 0.05$) higher than those of the 50%SS, 75%SS, and 100%SS. However, there were no significant ($p > 0.05$) differences in PA content among all silage mixtures. With the increasing level of the SS ratio, the V-score value continuously increased after ensiling for 150 days. The V-score values of 50%SS, 75%SS, and 100%SS were 82.00, 85.50, and 88.00, respectively, which were all higher than 80 points. Therefore, the 50%SS, 75%SS, and 100%SS silages had high fermentation quality.

Table 4. Effect of fermentation characteristics of the mixed silage of sweet sorghum and aerial parts of licorice.

Items	Treatments					Mean	SEM [a]	*p*-Value
	0%SS	25%SS	50%SS	75%SS	100%SS			
pH	5.01 ^{A}	4.34 ^{B}	4.15 ^{B}	4.05 ^{B}	4.01 ^{C}	4.31	0.10	<0.001
Ammonia nitrogen (% TN [b])	4.85 ^{A}	4.61 ^{AB}	3.97 ^{B}	3.86 ^{B}	3.54 ^{B}	4.17	0.13	<0.001
Lactic acid (% FM)	1.42 ^{C}	1.89 ^{BC}	2.56 ^{B}	3.14 ^{A}	3.79 ^{A}	2.56	0.23	<0.001
Acetic acid (% FM)	1.16 ^{A}	0.96 ^{B}	0.85 ^{B}	0.67 ^{C}	0.56 ^{D}	0.84	0.06	0.003
Propanoic acid (% FM)	0.08	0.08	0.08	0.09	0.08	0.08	0.01	0.782
Butyric acid (% FM)	0.18 ^{A}	0.10 ^{A}	0.06 ^{B}	0.05 ^{B}	0.03 ^{C}	0.08	0.06	<0.001
Lactic acid/total acid	34.23 ^{E}	40.75 ^{D}	45.07 ^{C}	52.36 ^{B}	54.89 ^{A}	45.46	2.02	<0.001
Acetic acid/total acid	28.72 ^{A}	25.16 ^{B}	22.36 ^{C}	20.34 ^{D}	19.05 ^{E}	23.13	0.93	<0.001
Butyric acid/total acid	16.56 ^{A}	15.27 ^{B}	10.35 ^{C}	9.45 ^{D}	6.74 ^{E}	10.58	1.06	<0.001
V-score	75.00	78.50	82.00	85.50	88.00	-	-	-
Grade	fair	fair	good	good	good	-	-	-

$^{A–E}$ The same letter indicates non-significant differences ($p > 0.05$) and different letters indicate significant differences ($p < 0.05$). [a] SEM, standard error of means. [b] TN, total nitrogen.

3.3. The Chemical Compositions of the Mixed Silage during Aerobic Exposure of the Mixed Silage of Sweet Sorghum and Aerial Parts of Licorice

As shown in Table 5, among all the treatment groups, with the increase in the SS ratio, the content of DM, pH, and AA showed a decreasing trend, which was significant in all treatment groups ($p < 0.05$). However, the content of WSC and LA showed an increasing trend, which was significant in all treatment groups ($p < 0.05$). With the increase in aerobic exposure time, the content of DM and pH showed an upward trend. However, there was a downward trend in AA, LA, and WSC contents. The relevant data of each aerobic exposure showed different degrees of significance ($p < 0.05$).

Table 5. Effect of aerobic exposure days on fermentative characteristics of the mixed silage of sweet sorghum and aerial parts of licorice.

Items	Treatments	Aerobic Exposure Days						Mean	SEM [a]	*p*-Value [b]		
		0	5	10	15	20	25			M	N	M × N
Dry matter (% FM)	0%SS	39.76 ^{Ca}	42.09 ^{Ba}	44.76 ^{Ba}	46.68 ^{Ba}	50.53 ^{Aa}	53.62 ^{Aa}	46.24	1.24	<0.001	<0.001	<0.001
	25%SS	37.91 ^{Cab}	40.37 ^{Ba}	42.23 ^{Bb}	43.85 ^{Bb}	45.01 ^{Ab}	52.80 ^{Aa}	43.70				
	50%SS	35.90 ^{Cbc}	38.55 ^{Bb}	40.81 ^{Bc}	42.50 ^{Bbc}	43.31 ^{Ac}	48.24 ^{Ab}	41.55				
	75%SS	35.11 ^{Cc}	38.33 ^{Bb}	40.18 ^{Bc}	41.77 ^{Bc}	43.26 ^{Ac}	45.95 ^{Ac}	40.77				
	100%SS	34.24 ^{Cc}	36.53 ^{Bc}	37.74 ^{Bd}	38.93 ^{Bd}	39.35 ^{Bd}	43.02 ^{Ad}	38.30				
Water soluble carbohydrate (% FM)	0%SS	3.60 ^{Ac}	2.38 ^{Ad}	2.01 ^{Bd}	1.79 $^{B}{}_{c}$	1.68 ^{Ce}	1.53 ^{Cd}	2.17	0.34	0.028	<0.001	<0.001
	25%SS	4.95 ^{Ab}	4.26 ^{Ac}	3.89 ^{Ac}	3.15 ^{Bb}	2.77 ^{Bd}	2.25 ^{Cc}	3.55				
	50%SS	5.72 ^{Ab}	5.54 ^{Ac}	5.12 ^{Bb}	4.47 ^{Bb}	3.99 ^{Cc}	3.78 ^{Cb}	4.77				
	75%SS	7.34 ^{Aa}	6.79 ^{Ab}	6.24 ^{Ba}	5.62 ^{Ba}	4.68 ^{Cb}	4.16 ^{Cab}	5.81				
	100%SS	8.48 ^{Aa}	7.84 ^{Aa}	7.16 ^{Aa}	6.29 ^{Ba}	5.62 ^{Ba}	4.61 ^{Ca}	6.67				
pH	0%SS	5.01 ^{Ca}	6.56 ^{Ba}	6.89 ^{Ba}	7.17 ^{Aa}	7.59 ^{Aa}	7.83 ^{Aa}	6.84	0.04	<0.001	<0.001	<0.001
	25%SS	4.34 ^{Cb}	5.02 ^{Bb}	5.75 ^{Bb}	6.98 ^{Ab}	7.34 ^{Ab}	7.65 ^{Ab}	6.18				
	50%SS	4.15 ^{Cb}	4.68 ^{Bb}	5.50 ^{Bb}	6.72 ^{Abc}	7.56 ^{Abc}	7.75 ^{Ab}	6.06				
	75%SS	4.05 ^{Cb}	4.53 ^{Cbc}	5.48 ^{Bbc}	6.54 ^{Ab}	7.48 ^{Ab}	7.75 ^{Ab}	5.97				
	100%SS	4.01 ^{Cc}	4.34 ^{Cc}	5.36 ^{Bc}	6.51 ^{Ac}	7.40 ^{Ac}	7.74 ^{Ab}	5.89				
Lactic acid (% FM)	0%SS	1.42 ^{Ac}	1.12 ^{Ac}	0.96 ^{Ad}	0.83 ^{Bc}	0.80 ^{Bc}	0.76 ^{Bc}	0.98	0.21	<0.001	<0.001	<0.001
	25%SS	1.89 ^{Abc}	1.58 ^{Ab}	1.32 ^{Ac}	1.10 ^{Bc}	1.05 ^{Bc}	0.99 ^{Bc}	1.32				
	50%SS	2.56 ^{Ab}	2.28 ^{Ab}	2.01 ^{Bb}	1.85 ^{Bb}	1.62 ^{Cb}	1.34 ^{Cb}	1.94				
	75%SS	3.14 ^{Aa}	2.81 ^{Aa}	2.53 ^{Ba}	2.16 ^{Ba}	1.96 ^{Cb}	1.65 ^{Cb}	2.38				
	100%SS	3.79 ^{Aa}	3.24 ^{Aa}	2.92 ^{Ba}	2.74 ^{Ba}	2.43 ^{Ba}	2.10 ^{Ca}	2.87				
Acetic acid (% FM)	0%SS	1.16 ^{Aa}	0.89 ^{Aa}	0.75 ^{Aa}	0.63 ^{Ba}	0.54 ^{Ba}	0.41 ^{Ba}	0.73	0.05	<0.001	<0.001	<0.001
	25%SS	0.96 ^{Ab}	0.74 ^{Ab}	0.62 ^{Ab}	0.54 ^{Bb}	0.42 ^{Bb}	0.38 ^{Ba}	0.61				
	50%SS	0.85 ^{Ab}	0.62 ^{Ab}	0.54 ^{Ab}	0.43 ^{Bc}	0.37 ^{Bb}	0.32 ^{Bb}	0.52				
	75%SS	0.67 ^{Ac}	0.52 ^{Ac}	0.41 ^{Ac}	0.35 ^{Ad}	0.30 ^{Bc}	0.28 ^{Bc}	0.42				
	100%SS	0.56 ^{Ad}	0.47 ^{Ad}	0.43 ^{Ac}	0.36 ^{Ad}	0.30 ^{Bc}	0.27 ^{Bc}	0.40				

The same letter indicates non-significant differences ($p > 0.05$) and different letters indicate significant differences ($p < 0.05$). Uppercase letters show significance between different times. Lowercase letters show significance between different treatment groups. [a] SEM, standard error of means. [b] M, aerobic exposure days; N, treatments; M × N, the interaction between aerobic exposure days and treatments.

3.4. The Aerobic Stability of the Mixed Silage during Aerobic Exposure of the Mixed Silage of Sweet Sorghum and Aerial Parts of Licorice

Figure 1 shows the time taken for the temperature of mixed silage to rise by 2 °C above the ambient temperature after aerobic exposure. The 100%SS silage was the first to exceed 2 °C, followed by that of the 75%SS, 50%SS, 25%SS, and 0%SS. Figure 1 shows that the aerobic stability of 0%SS, 25%SS, 50%SS, 75%SS, and 100%SS silages decreased by 384 h, 360 h, 336 h, 288 h, and 240 h, respectively. The aerobic stability of 0%SS was significantly higher than that of the other four groups ($p < 0.05$), which were 25%SS, 50%SS, 75%SS, and 100%SS. There was no significant difference between the 25%SS and 50%SS silages ($p > 0.05$).

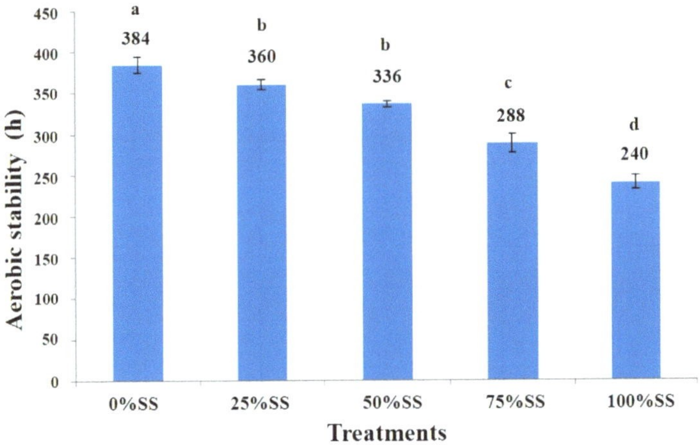

Figure 1. Aerobic stability of mixed silages during aerobic exposure ($n = 5$). The small black bars indicate the standard error of the mean. Lowercase letters show significance between different treatment groups.

3.5. The Microbial Changes of the Mixed Silage during Aerobic Exposure of the Mixed Silage of Sweet Sorghum and Aerial Parts of Licorice

As shown in Table 6, among all treatment groups, with the increase in the ratio of SS, the quantities of LAB, yeast, and AB showed an increasing trend, while the quantities of mold showed a decreasing trend. With the increase in aerobic exposure time, the number of LAB showed a decreasing trend because LAB is an anaerobic microorganism. The quantity of yeast, mold, and AB showed an increasing trend. Because they are aerobic microorganisms, the amount of air entering increases with an extension of aerobic exposure time.

3.6. Gas Production during In Vitro Fermentation of the Mixed Silages of Sweet Sorghum and Aerial Parts of Licorice

As shown in Figure 2, with an increase in in vitro fermentation time, the GP content of each treatment group increased gradually. The GP content of each treatment increased rapidly at 0–24 h. From 24 h to 48 h, the content of GP of each treatment group increased slowly until it stopped increasing at a later stage. Finally, the content of GP of each treatment group entered a plateau and was then relatively stable. The first plateau stage was seen in 0%SS followed by 25%SS, 75%SS, 50%SS, and 100%SS. At 72 h, the content of GP of 100%SS and 50%SS was significantly higher than that of 0%SS, 25%SS, and 75%SS. There was a clear difference between the 100% and 0% groups. However, the plateau of gas production for 100%SS occurred at 60 h and at 36 h for 0%SS.

Table 6. Effect of aerobic exposure days on microbial composition of the mixed silage of sweet sorghum and aerial parts of licorice.

Items [b]	Treatments	Aerobic Exposure Days						Mean	SEM [a]	p-Value		
		0	5	10	15	20	25			M	N	M × N
Lactic acid bacteria (Log_{10} cfu/g FM)	0%SS	6.64 ^{Ac}	6.13 ^{Ad}	5.86 ^{Bc}	5.35 ^{Bb}	4.15 ^{Cc}	3.65 ^{C}	5.30	0.12	<0.001	<0.001	<0.001
	25%SS	7.12 ^{Ab}	6.76 ^{Ac}	5.97 ^{Bb}	5.21 ^{Bb}	4.49 ^{Cb}	3.75 ^{C}	5.55				
	50%SS	7.52 ^{Ab}	7.02 ^{Ac}	6.16 ^{Bb}	5.64 ^{Ba}	4.78 ^{Ca}	3.85 ^{C}	5.83				
	75%SS	8.21 ^{Aa}	7.31 ^{Ab}	6.34 ^{Ba}	5.42 ^{Ba}	4.61 ^{Ca}	3.83 ^{C}	5.95				
	100%SS	8.52 ^{Aa}	7.64 ^{Aa}	6.69 ^{Ba}	5.05 ^{Bc}	4.45 ^{Cb}	3.81 ^{C}	6.03				
Yeast (Log_{10} cfu/g FM)	0%SS	3.08 ^{C}	3.20 ^{Cc}	3.86 ^{Bc}	4.27 ^{Bc}	4.88 ^{Ac}	5.05 ^{Ac}	4.06	0.16	<0.001	<0.001	<0.001
	25%SS	3.15 ^{C}	3.32 ^{Cb}	3.94 ^{Bc}	4.35 ^{Bc}	5.10 ^{Ac}	5.84 ^{Ab}	4.28				
	50%SS	3.11 ^{C}	3.46 ^{Cb}	4.01 ^{Bb}	4.62 ^{Bb}	5.89 ^{Ab}	6.31 ^{Ab}	4.57				
	75%SS	3.05 ^{C}	3.58 ^{Cb}	4.24 ^{Bb}	5.60 ^{Bb}	6.84 ^{Aa}	7.24 ^{Aa}	5.09				
	100%SS	3.01 ^{C}	3.99 ^{Ca}	4.74 ^{Ba}	5.86 ^{Ba}	7.14 ^{Aa}	7.68 ^{Aa}	5.40				
Mold (Log_{10} cfu/g FM)	0%SS	0	1.54 ^{C}	3.53 ^{Ba}	4.89 ^{Ba}	5.24 ^{Aa}	6.01 ^{Aa}	3.54	0.08	<0.001	<0.001	<0.001
	25%SS	0	1.23 ^{C}	3.24 ^{Ba}	3.95 ^{Ba}	4.87 ^{Ab}	5.81 ^{Aa}	3.18				
	50%SS	0	0	3.11 ^{Bb}	3.64 ^{Bb}	4.48 ^{Ab}	5.65 ^{Ab}	2.81				
	75%SS	0	0	3.02 ^{Bb}	3.24 ^{Bc}	4.15 ^{Ac}	5.24 ^{Ac}	2.61				
	100%SS	0	0	2.94 ^{Bc}	3.11 ^{Bc}	3.99 ^{Bc}	5.08 ^{Ac}	2.52				
Aerobic bacteria (Log_{10} cfu/g FM)	0%SS	5.42 ^{B}	5.68 ^{Bc}	5.96 ^{Ac}	6.13 ^{Ac}	6.24 ^{Ac}	6.34 ^{Ac}	5.96	0.10	<0.001	<0.001	<0.001
	25%SS	5.39 ^{B}	5.95 ^{Bb}	6.17 ^{Ab}	6.34 ^{Ab}	6.51 ^{Ab}	6.64 ^{Ab}	6.17				
	50%SS	5.46 ^{B}	6.00 ^{Bb}	6.34 ^{Ab}	6.57 ^{Ab}	6.68 ^{Ab}	6.95 ^{Ab}	6.33				
	75%SS	5.33 ^{C}	6.24 ^{Ba}	6.62 ^{Ba}	6.89 ^{Aa}	7.02 ^{Aa}	7.22 ^{Aa}	6.55				
	100%SS	5.43 ^{C}	6.59 ^{Ba}	7.01 ^{Aa}	7.16 ^{Aa}	7.30 ^{Aa}	7.42 ^{Aa}	6.82				

The same letter indicates non-significant differences ($p > 0.05$) and different letters indicate significant differences ($p < 0.05$). Uppercase letters show significance between different times. Lowercase letters show significance between different treatment groups. [a] SEM, standard error of means. [b] M, aerobic exposure days; N, treatments; M × N, the interaction between aerobic exposure days and treatments.

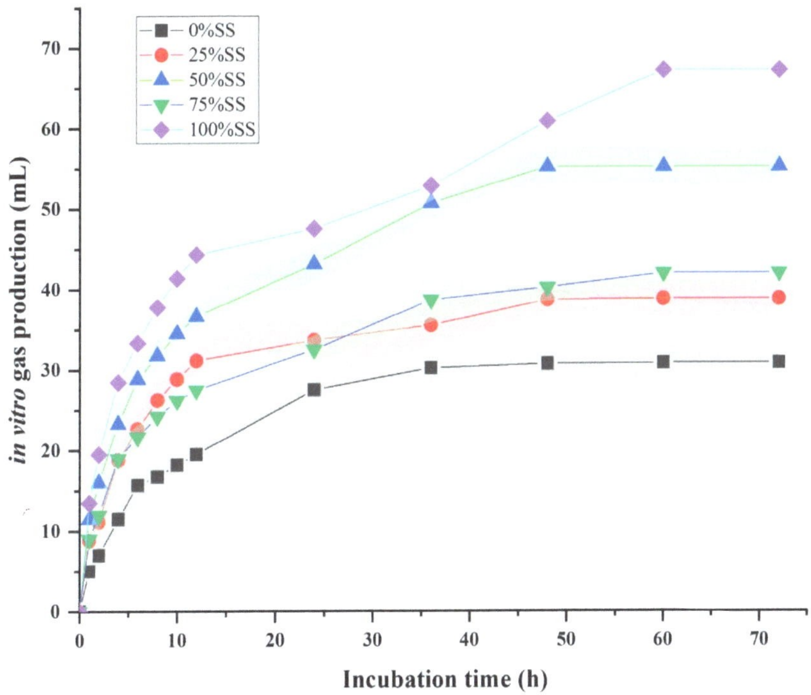

Figure 2. Gas production profiles (mL/g DM) from in vitro fermentation of the mixed silages for 72 h (n = 13). The width of the ribbon indicates the standard error.

3.7. In Vitro Parameters of the 150-Day Mixed Silages of the Mixed Silage of Sweet Sorghum and Aerial Parts of Licorice

As shown in Tables 7–10, there were significant differences in all index data among each treatment group. The content of potential GP, GP rate constant, DOM, ME, and IVDMD of 50%SS and 100%SS were significantly higher than those of the 0%SS, 25%SS, and 75%SS treatments ($p < 0.05$), and there was no significant difference between 50%SS and 100%SS ($p > 0.05$). Among all treatment groups, with the increasing level of the SS ratio, the pH showed a decreasing trend, ranging from 6.41 to 6.85. The content of NH_3-N in 50%SS and 100%SS was significantly lower than that in 0%SS, 25%SS, and 75%SS ($p < 0.05$). The content of NH_3-N in 50%SS and 100%SS was not significantly different ($p > 0.05$) and was 28.51 mg/dL and 25.32 mg/dL, respectively.

Table 7. Fermentation and gas production parameters from in vitro fermentation of TMR silage.

Items [a]	Treatments					Mean	SEM	p-Value
	0%SS	25%SS	50%SS	75%SS	100%SS			
potential GP (mL)	33.78 [C]	53.23 [B]	62.42 [A]	55.82 [B]	66.96 [A]	54.44	3.05	<0.001
GP rate constant (c, mL/h)	0.06 [C]	0.07 [B]	0.09 [A]	0.07 [B]	0.10 [A]	0.08	0.02	<0.001
pH	6.85 [A]	6.65 [B]	6.55 [B]	6.42 [C]	6.41 [C]	6.58	0.05	<0.001
NH_3-N (mg/dL)	46.48 [A]	36.68 [B]	28.51 [C]	33.17 [B]	25.32 [C]	34.03	1.96	<0.001
DOM (%)	57.33 [C]	62.06 [B]	69.33 [A]	61.16 [B]	72.66 [A]	64.51	1.50	<0.001
ME (MJ/kg DM)	6.50 [C]	7.47 [B]	8.95 [A]	7.28 [B]	9.63 [A]	7.79	0.31	<0.001
IVDMD (%)	38.44 [C]	55.18 [B]	64.98 [A]	56.98 [B]	68.56 [A]	56.83	2.79	<0.001

[A–C] The same letter indicates non-significant differences ($p > 0.05$) and different letters indicate significant differences ($p < 0.05$). Capital letters are significant differences between treatment groups. [a] potential GP, potential gas production; GP rate constant, gas production rate constant; NH_3-N, ammonia nitrogen; DOM, digestible organic matter; ME, metabolizable energy; IVDMD, in vitro dry matter digestibility; SEM, standard error of means.

Table 8. The content of IVDMD of mixed silage with different ratios of SS-LC in vitro fermentation.

Time/h	Treatments					Mean	SEM [a]	p-Value
	0%SS	25%SS	50%SS	75%SS	100%SS			
4	13.26 C	14.55 B	15.76 A	15.21 A	16.65 A	15.08	0.79	<0.001
8	15.54 C	20.06 B	23.42 A	18.22 B	25.35 A	20.51	0.47	<0.001
12	20.38 C	25.17 B	28.19 A	26.43 B	30.11 A	26.05	0.54	<0.001
24	28.46 C	38.63 B	47.63 A	42.24 B	50.65 A	41.52	0.82	<0.001
36	32.28 C	45.52 B	53.45 A	48.61 B	56.29 A	47.23	0.74	<0.001
48	35.16 C	48.33 B	56.27 A	50.53 B	61.42 A	50.34	0.47	<0.001
72	38.44 C	55.18 B	64.98 A	56.98 B	68.56 A	56.83	2.79	<0.001

A–C The same letter indicates non-significant differences ($p > 0.05$) and different letters indicate significant differences ($p < 0.05$). Capital letters are significant differences between treatment groups. [a] SEM, standard error of means.

Table 9. The content of pH of mixed silage with different ratios of SS-LC in vitro fermentation.

Time/h	Treatments					Mean	SEM [a]	p-Value
	0%SS	25%SS	50%SS	75%SS	100%SS			
4	7.45	7.39	7.34	7.28	7.45	7.34	0.03	<0.001
8	7.33	7.29	7.27	7.21	7.33	7.23	0.02	<0.001
12	7.18	7.10	7.04	6.92	7.18	7.01	0.02	<0.001
24	7.02	6.82	6.79	6.63	7.02	6.77	0.02	<0.001
36	6.92	6.87	6.70	6.56	6.92	6.71	0.02	<0.001
48	6.87	6.73	6.62	6.48	6.87	6.63	0.02	<0.001
72	6.85 A	6.65 B	6.55 B	6.42 C	6.41 C	6.58	0.05	<0.001

A–C The same letter indicates non-significant differences ($p > 0.05$) and different letters indicate significant differences ($p < 0.05$). Capital letters are significant differences between treatment groups. [a] SEM, standard error of means.

Table 10. The content of NH_3-N of mixed silage with different ratios of SS-LC in vitro fermentation.

Time/h	Treatments					Mean	SEM [a]	p-Value
	0%SS	25%SS	50%SS	75%SS	100%SS			
4	41.42 A	32.08 B	24.24 C	28.63 B	22.64 C	29.80	0.64	<0.001
8	43.38 A	34.37 B	26.16 C	30.11 B	24.52 C	31.71	0.56	<0.001
12	47.26 A	36.61 B	28.50 C	33.26 B	26.63 C	34.45	0.65	<0.001
24	50.16 A	39.18 B	31.61 C	35.37 B	28.48 C	36.96	0.63	<0.001
36	48.54 A	38.34 B	30.24 C	34.16 B	27.30 C	35.72	0.44	<0.001
48	47.34 A	37.52 B	29.11 C	33.37 B	26.45 C	34.76	0.56	<0.001
72	46.48 A	36.68 B	28.51 C	33.17 B	25.32 C	34.03	1.96	<0.001

A–C The same letter indicates non-significant differences ($p > 0.05$) and different letters indicate significant differences ($p < 0.05$). Capital letters are significant differences between treatment groups. [a] SEM, standard error of means.

3.8. Degradation Rates of DM, CP, ADF, NDF, and ADL in the Rumen of Sheep of Mixed Silage of Sweet Sorghum and Licorice with Aerial Parts

The results of Formulas (1)–(6) under subheading 2.5 of Materials and Methods are shown in Tables 11–16. The content of DE, ME, and Neg of 50%SS, 75%SS, and 100%SS were significantly higher than in 25%SS and 0%SS ($p < 0.05$). On the one hand, the rate of degradation of DM showed an increasing trend as the proportion of sweet sorghum increased and was significantly higher in 100%SS than in 0%SS ($p < 0.05$); the CP, NDF, ADF, and ADL contents were significantly higher in 25%SS and 50%SS than in 0%SS, 75%SS, and 100%SS ($p < 0.05$), and the difference was not significant in 25%SS and 50%SS ($p > 0.05$); the ADL content of 50%SS, 75%SS, and 100%SS was significantly higher than that of 0%SS and 25%SS ($p < 0.05$), and the difference between 50%SS, 75%SS, and 100%SS was not significant

($p > 0.05$). On the other hand, with the increase in rumen degradation time, the rates of degradation of DM, ADF, NDF, ADL, and CP showed an increasing trend in each treatment group. In conclusion, 25%SS and 50%SS showed the best degradation rate.

Table 11. Energy value of mixed silage with different ratios of SS-LC (dry matter basis).

Items/(kJ/kg) [a]	Treatments					Mean	SEM [a]	p-Value
	0%SS	25%SS	50%SS	75%SS	100%SS			
DE [a]	10.36 B	10.67 B	11.45 A	11.33 A	11.75 A	11.11	0.48	<0.001
ME [a]	8.54 B	8.79 B	9.43 A	9.34 A	9.68 A	9.15	0.54	<0.001
Neg [a]	3.90 B	4.02 B	4.31 A	4.27 A	4.42 A	4.18	0.25	<0.001

A,B The same letter indicates non-significant differences ($p > 0.05$) and different letters indicate significant differences ($p < 0.05$). Capital letters are significant differences between treatment groups. [a] DE, digestible energy; ME, metabolizable energy; NEg, net energy for gain; SEM, standard error of means.

Table 12. Degradation rate of the content of DM of mixed silage with different ratios of SS-LC in rumen of sheep.

Items	Treatments					Mean	SEM	p-Value
	0%SS	25%SS	50%SS	75%SS	100%SS			
DM								
4 (h)	19.22 B	23.42 B	27.58 A	29.11 A	30.48 A	25.96	0.13	<0.001
8 (h)	23.81 C	27.02 B	29.34 B	30.01 A	32.31 A	28.49	0.13	<0.001
12 (h)	29.93 C	33.31 B	35.26 B	37.24 A	40.75 A	35.29	0.14	<0.001
24 (h)	39.32 C	44.05 B	47.89 B	50.37 A	52.35 A	46.79	0.15	<0.001
48 (h)	42.48 C	48.22 B	50.75 B	54.71 A	56.51 A	50.53	0.15	<0.001
72 (h)	45.43 C	53.35 B	55.27 B	58.33 A	60.25 A	54.52	0.27	<0.001
Degradability parameters of DM [a]								
a (mL)	10.30 C	15.53 B	18.98 A	19.55 A	20.62 A	16.99	0.34	<0.001
b (mL)	34.71 C	37.63 B	36.44 B	39.54 A	39.54 A	37.57	0.53	<0.001
a + b (mL)	45.01 C	53.16 B	55.42 B	59.09 A	60.16 A	54.57	0.63	<0.001
c (mL/h)	0.07	0.05	0.05	0.05	0.06	0.06	0.01	<0.001
ED (%)	34.36 C	38.76 B	41.47 B	43.96 A	46.69 A	41.05	0.41	<0.001

A–C The same letter indicates non-significant differences ($p > 0.05$) and different letters indicate significant differences ($p < 0.05$). Capital letters are significant differences between treatment groups. [a] a, the rapidly degrading part; b, the slowly descending part; a + b, potentially degraded part; c, degradation rate constant; ED, effective degradation rate.

Table 13. Degradation rate of the content of CP of mixed silage with different ratios of SS-LC in rumen of sheep.

Items	Treatments					Mean	SEM	p-Value
	0%SS	25%SS	50%SS	75%SS	100%SS			
CP								
4 (h)	30.62 C	40.75 A	37.53 A	34.23 B	28.25 C	34.27	0.91	<0.001
8 (h)	33.02 C	46.64 A	43.12 A	38.27 B	30.38 D	38.28	0.69	<0.001
12 (h)	36.34 C	50.52 A	46.35 AB	42.16 B	32.58 C	41.59	0.86	<0.001
24 (h)	42.68 B	53.21 A	51.64 A	45.35 B	36.35 C	45.84	0.66	<0.001
48 (h)	50.46 C	56.53 A	54.28 A	52.16 B	43.08 D	51.30	0.52	<0.001
72 (h)	54.25 B	62.21 A	60.23 A	58.34 B	48.26 C	56.65	0.48	<0.001
Degradability parameters of CP								
a (mL)	26.96 C	38.15 A	34.09 A	32.94 B	26.73 C	31.77	0.36	<0.001
b (mL)	31.10 A	22.74 C	25.00 B	31.00 A	32.85 A	28.53	0.42	<0.001

Table 13. *Cont.*

Items	Treatments					Mean	SEM	*p*-Value
	0%SS	25%SS	50%SS	75%SS	100%SS			
a + b (mL)	58.06 [C]	60.89 [B]	59.09 [B]	63.94 [A]	59.58 [B]	60.31	0.69	<0.001
c (mL/h)	0.03 [B]	0.05 [A]	0.05 [A]	0.02 [B]	0.02 [B]	0.03	0.01	<0.001
ED (%)	42.26 [B]	52.19 [A]	49.52 [A]	45.10 [B]	39.6 [C]	45.73	0.43	<0.001

[A–D] The same letter indicates non-significant differences ($p > 0.05$) and different letters indicate significant differences ($p < 0.05$). Capital letters are significant differences between treatment groups. a, the rapidly degrading part; b, the slowly descending part; a + b, potentially degraded part; c, degradation rate constant; ED, effective degradation rate.

Table 14. Degradation rate of the content of NDF of mixed silage with different ratios of SS-LC in rumen of sheep.

Items	Treatments					Mean	SEM	*p*-Value
	0%SS	25%SS	50%SS	75%SS	100%SS			
NDF								
4 (h)	22.16 [B]	27.04 [A]	25.35 [A]	20.45 [B]	18.15 [C]	22.63	0.60	<0.001
8 (h)	27.24 [B]	32.16 [A]	30.24 [A]	25.36 [B]	23.23 [C]	27.64	1.39	<0.001
12 (h)	35.04 [B]	41.23 [A]	39.34 [A]	33.65 [B]	32.42 [C]	36.33	0.64	<0.001
24 (h)	47.85 [B]	51.65 [A]	50.22 [A]	45.55 [B]	42.04 [C]	47.46	1.49	<0.001
48 (h)	60.24 [B]	65.14 [A]	63.16 [A]	58.21 [B]	56.21 [C]	60.59	1.38	<0.001
72 (h)	64.34 [B]	70.24 [A]	67.34 [A]	62.02 [B]	60.37 [C]	64.86	0.84	<0.001
Degradability parameters of NDF								
a (mL)	12.89 [B]	18.80 [A]	16.48 [A]	11.34 [B]	9.99 [C]	13.90	0.72	<0.001
b (mL)	54.1	54.45	53.42	53.32	53.51	53.76	0.83	<0.001
a + b (mL)	66.99 [B]	73.25 [A]	69.90 [A]	64.66 [B]	63.50 [B]	67.66	0.98	<0.001
c (mL/h)	0.04	0.04	0.04	0.04	0.04	0.04	0.01	<0.001
ED (%)	43.37 [B]	49.48 [A]	46.58 [A]	41.38 [B]	40.14 [C]	44.19	1.172	<0.001

[A–C] The same letter indicates non-significant differences ($p > 0.05$) and different letters indicate significant differences ($p < 0.05$). Capital letters are significant differences between treatment groups. a, the rapidly degrading part; b, the slowly descending part; a + b, potentially degraded part; c, degradation rate constant; ED, effective degradation rate.

Table 15. Degradation rate of the content of ADF of mixed silage with different ratios of SS-LC in rumen of sheep.

Items	Treatments					Mean	SEM	*p*-Value
	0%SS	25%SS	50%SS	75%SS	100%SS			
ADF								
4 (h)	21.32 [B]	25.25 [A]	23.26 [A]	20.34 [B]	20.36 [B]	22.10	0.43	<0.001
8 (h)	27.13 [B]	30.34 [A]	28.24 [A]	26.15 [B]	25.42 [C]	27.45	0.42	<0.001
12 (h)	34.35 [B]	37.22 [A]	36.68 [A]	33.22 [BC]	32.51 [C]	34.79	0.48	<0.001
24 (h)	44.02 [B]	48.12 [A]	46.12 [A]	43.34 [B]	41.25 [C]	44.57	0.35	<0.001
48 (h)	53.16 [B]	57.34 [A]	55.22 [A]	52.46 [B]	50.55 [C]	53.74	0.52	<0.001
72 (h)	63.31 [B]	68.25 [A]	66.12 [A]	61.24 [BC]	59.28 [C]	63.64	0.42	<0.001
Degradability parameters of ADF								
a (mL)	16.20 [B]	19.99 [A]	17.99 [A]	14.42 [C]	15.31 [B]	16.78	0.47	<0.001
b (mL)	50.60 [B]	53.31 [A]	52.04 [A]	49.18 [B]	47.35 [C]	50.49	0.37	<0.001
a + b (mL)	66.80 [B]	73.30 [A]	70.04 [A]	63.60 [B]	62.66 [C]	67.28	0.45	<0.001
c (mL/h)	0.03	0.03	0.03	0.04	0.03	0.03	0.01	<0.001
ED (%)	41.89 [B]	45.76 [A]	43.58 [A]	40.85 [B]	39.36 [C]	42.28	0.64	<0.001

[A–C] The same letter indicates non-significant differences ($p > 0.05$) and different letters indicate significant differences ($p < 0.05$). Capital letters are significant differences between treatment groups. a, the rapidly degrading part; b, the slowly descending part; a + b, potentially degraded part; c, degradation rate constant; ED, effective degradation rate.

Table 16. Degradation rate of the content of ADL of mixed silage with different ratios of SS-LC in rumen of sheep.

Items	Treatments					Mean	SEM	*p*-Value
	0%SS	25%SS	50%SS	75%SS	100%SS			
ADL								
4 (h)	9.46 C	10.09 B	11.24 A	11.85 A	12.65 A	11.05	0.84	<0.001
8 (h)	10.25 C	12.02 B	12.68 B	12.34 B	13.12 A	12.08	0.93	<0.001
12 (h)	11.65 D	13.11 C	13.51 B	13.26 B	14.58 A	13.22	0.93	<0.001
24 (h)	12.24 C	14.84 B	15.68 A	15.37 A	15.26 A	14.67	0.93	<0.001
48 (h)	13.69 C	15.31 B	16.24 A	16.08 AB	15.59 A	15.38	0.51	<0.001
72 (h)	14.22 C	15.54 B	16.56 A	16.24 A	16.61 A	15.83	0.72	<0.001
Degradability parameters of ADL								
a (mL)	8.51 B	7.29 B	9.20 A	10.24 A	11.5 A	9.34	0.56	<0.001
b (mL)	5.75 B	8.16 A	7.34 A	6.17 B	4.76 C	6.43	0.54	<0.001
a + b (mL)	14.26 B	15.45 B	16.54 A	16.41 A	16.26 A	15.78	1.09	<0.001
c (mL/h)	0.05 B	0.11 A	0.08 A	0.06 B	0.07 B	0.07	0.01	<0.001
ED (%)	12.06 C	13.66 B	14.49 A	14.31 A	14.80 A	13.86	0.96	<0.001

A–D The same letter indicates non-significant differences ($p > 0.05$) and different letters indicate significant differences ($p < 0.05$). Capital letters are significant differences between treatment groups. a, the rapidly degrading part; b, the slowly descending part; a + b, potentially degraded part; c, degradation rate constant; ED, effective degradation rate.

4. Discussion

4.1. Chemical Compositions of SS-LC Mixed Silages

Except that the WSC contents of the ensiled silages were lower than their pre-ensiled counterparts, there were no obvious differences in the nutritional composition among the treatment groups. Similar results have been reported by Ni et al. (2018) [17], who found that the CP, EE, NDF, and ADF contents of ensiled silages were not significantly different from the pre-ensiled silages. Similar results have been reported by Ren et al. (2021) [21], who observed that the WSC contents in SS silage after 60 days were significantly lower than the pre-ensiled SS, and the WSC contents continued to reduce with increasing ensiling time.

After 150 days of ensiling, the CP, DM, Ash, and ADL contents in the mixed silages gradually decreased with the increasing amount of SS. While the WSC contents in mixed silages gradually increased, other nutrients did not change significantly. Interestingly, this is in agreement with Tássia et al. (2021) [39], Wang et al. (2020) [40], and Pedram et al. (2022) [41]. They reported reduced CP but increased WSC contents with increased proportions of gramineous plants in the mixed silages of legumes and gramineous plants.

The possible reason is that the grasses contain higher WSC but lower CP than legumes. The presence of the two forages in mixed silages helped to make the nutrition more comprehensive by improving the digestion nutritive value of the mixed silages [17,20,42].

4.2. Fermentation Characteristics of SS-LC Mixed Silages

Pre-ensiled forages with high WSC content (>5% DM) and sufficient LAB population (>10^5 cfu/g FM) were found in high-quality silage production by Weinberg et al. (2008) [43]. In this study, all mixtures of forages were able to be ensiled successfully. High-quality silage was reported to have high LA, low pH, low NH_3-N, and negligible BA contents by Catchpoole and Henzell (1971) [44]. Hence, as the proportion of SS increased, the fermentation quality of the mixed silages gradually improved. Interestingly, this is in agreement with Wang et al. (2021) [20], Zeng et al. (2020) [42], and Ni et al. (2017) [45], who reported that the fermentation quality of mixed silages would be gradually improved with the increase in the proportion of grasses and WSC content.

Dai et al. (2022) [46] found that the propagation of LAB was suppressed when the pH was below 4.00–4.20. Hence, the pH of mixed silages was below 4.20, which indicated that they tended to ensile successfully [46]. Auerbach and Nadeau (2020) [47] reported that better fermentation is indicated by a lower pH.

In this study, the pH of the 50%SS, 75%SS, and 100%SS was 4.15, 4.05, and 4.01, respectively. Therefore, 50%SS, 75%SS, and 100%SS were conducive to making high-quality silage because the 50%SS, 75%SS, and 100%SS silage had a sufficient LAB population, which produced a higher amount of lactic acid.

However, in all mixed silages, probably due to the low WSC and high DM contents, the pH gradually increased with increasing LC. Morgan et al. (1980) [48] reported that a high DM content may retard the proliferation of undesirable microorganisms and LAB in silage. The level of LC decreased the level of WSC in the mixed silages, which explains the reduction in LA, perhaps because there was less substrate available for LAB.

Tatulli et al. (2023) reported that when the WSC content was limited, heterofermentative LAB tended to be active, and several homofermentative strains such as *Lactobacillus plantarum* could carry out the lactic lactate/acetate conversion [49]. Keles and Demirci (2011) found that inoculation with a heterofermentative LAB (*Lactobacillus buchneri*) resulted in silage with higher ($p < 0.05$) concentrations of acetic and propionic acids and lower concentrations of LA and WSC. Therefore, in this study, the LC silage may be more inclined to be homofermentative, while the SS silage may be more inclined to be heterofermentative [50].

The NH_3-N content in silage is always indicative of protein breakdown and often defined as a product of amino acid deamination during silage fermentation [51,52]. In this study, as the proportion of LC increased, the NH_3-N content increased gradually. Similar results have been reported by Pedram et al. (2020) and Wu et al. (2022), who reported that when mixed silages were made of legume and Gramineae forages, the concentration of ammonia nitrogen increased gradually with an increase in the legume proportion in mixed forage-based silages [41,53].

4.3. Aerobic Exposure to LA and AA and Microbial Changes of SS-LC Mixed Silages

In tropical and subtropical regions, the high temperatures and humid climate during the summer months are major factors in the deterioration of silage mixes. Once a silo is opened, the aerobic microorganisms begin to multiply. Wilkinson and Davies (2012) showed good preservation of mixed silage, although this was subject to inherent instability following exposure to aerobes during the feed-out stage [32]. In fact, well-fermented silage is more unstable in the presence of oxygen than poorly fermented silage. Li et al. (2022) observed a negative correlation between the levels of lactic acid and WSC and the aerobic stability of silages [54]. In this study, similar observations showed that with an increase in the proportion of SS in mixed silages, the WSC and lactic acid contents in the mixed silages were gradually increased, and the aerobic stability was also improved. Johnson et al. (2002) showed that undesirable microbes can use LA and WSC as substrates to release carbon dioxide and heat, increasing the pH and increasing the nutrient loss when the silage is exposed to the air [55].

In this study, after aerobic exposure, with an increase in the SS proportion in mixed silages, the DM content gradually decreased, while aerobic bacteria and yeast gradually increased. As the ratio of SS increased, the number of aerobic bacteria and yeast increased, but the aerobic stability also gradually decreased. Interestingly, this is in agreement with Blajman et al. (2018) and Borreani et al. (2018), who reported that these samples were characterized by a lower DM that favored the activity of aerobic microorganisms, including yeasts, which, in the presence of air, depleted the available nutrient sources and contributed to the silage instability [56,57].

In this study, after 25 days of aerobic exposure, 50%SS, 75%SS, and 100%SS silages showed a strong increase in pH and a decrease in LA. The high levels of yeast, LA, and WSC in 50%SS, 75%SS, and 100%SS silages may explain this result. In general, yeasts are the main factor responsible for the aerobic deterioration of silage. When the yeast community exceeds 10^5 cfu/g FM, silage deterioration tends to occur [58]. Aerobic bacteria were also found to be responsible for aerobic deterioration by Courtin and Spoelstra (1990) [59]. Similarly, after 150 days of aerobic exposure, more than 10^5 cfu/g FM of yeasts and 10^7 cfu/g FM of aerobic bacteria were found in 100%SS silage.

Acetic acid can effectively inhibit the proliferation of fungi, mold, and yeast during aerobic exposure and is known to be an important indicator for predicting the aerobic stability of silages [32]. Therefore, as reflected by the lower LAB and yeast populations, the 0%SS and 25%SS silages with higher AA contents were more stable during aerobic exposure than the control. Furthermore, the increased aerobic stability of the SS-LC mixture silages could be explained by the LC material used in this study.

4.4. In Vitro Parameters of SS-LC Mixed Silages

Digestibility has become widely accepted in the evaluation of the nutritional value and intake of feed [60]. In the meantime, the increase in the in vitro GP is commonly used as an indicator of the efficiency of the rumen digestibility and the predicted metabolizable energy of animal feed [34]. In this study, all GP and digestibility parameters, as well as the estimated ME, were improved by mixed silages of 50%SS and 100%SS. This is consistent with results from Bender et al. (2016) and Lucas et al. (2020) [61,62]. Mixed silage effectively improved digestibility and provided more complete nutrition. This indicates that SS-LC-based mixed silages do not have unfavorable effects on the rumen utilization of mixed silages, although no significant difference was observed between 75%SS and 25%SS silages.

4.5. Degradation Rates of DM, CP, ADF, NDF, and ADL in the Rumen of Sheep

The nylon bag method is a commonly used method for assessing the degradation characteristics of feeds and is suitable for studying the extent of feed degradation within the rumen of ruminants and its effect on rumen microbial activity. Cone et al. (2004) reported the efficiency of protein degradation and its effective utilization by employing a nylon pouch approach in the rumen and the use of Streptomyces staphylococcus protein-hydrolyzing enzyme preparations for the detection of rumen escape proteins in grass and grass silage [63]. The nylon bag method allows the rate of feed degradation to be assessed by monitoring the degree of degradation of the residue in the bag.

In this experiment, with an increase in the sweet sorghum ratio, the degradation rates of DE, ME, NEg, and ADL in the rumen of sheep showed an increasing trend, indicating that sweet sorghum was easier to digest, degrade, utilize, and absorb than licorice. Due to the high carbohydrate content in sweet sorghum, it can provide energy for rumen microorganisms. Ma et al. (2023) have shown that adding lactic acid bacteria additive to mixed silage of amaranth and corn straw can effectively improve the degradation of dry matter [64]. Microbial fermentation of sugars and organic acids in crops releases gases such as carbon dioxide and hydrogen, which leads to dry matter loss [57]. The results of this study are consistent with previous research indicating that an increase in sweet sorghum leads to a gradual increase in the rate of degradation of dry matter.

Lignin is deposited in the space between cellulose, hemicellulose, and pectin molecules in the secondary cell wall, where it forms a cross-linked network with other components of the cell wall [65,66]. This cross-linking helps to increase the strength and stiffness of the cell wall, allowing the plant to stand upright and resist external pressures [67]. Lignin is a complex structural polymer resistant to microbial degradation, and its deposition in plant cell walls makes the cellulose and hemicellulose components of cell walls less accessible to rumen microbial enzymes, thus reducing rumen fermentation efficiency and nutrient availability to animals [68]. In addition, lignin inhibits the activity of digestive enzymes in the small intestine, further reducing the digestibility of the feed [69]. The results are consistent with the present study. In this study, the ADL content of mixed silage gradually decreased as the proportion of sweet sorghum gradually increased, so that the mixed silage with lower lignification was more easily degraded. In terms of digestibility, the main difference between grasses and legumes is that while legumes with thicker lignified walls can only be lightly digested by rumen microbes, grass tissues with the same thick lignified walls degrade extensively, albeit slowly [70]. Therefore, in rumen degradation experiments, the ADL degradation rates of 75%SS and 100%SS were higher than those of other groups.

In this experiment, as the proportion of sweet sorghum increased, the degradation rates of CP, ADF, and NDF showed a trend of first increasing and then decreasing (Tables 13–15), indicating that the ratio of sweet sorghum and licorice mixed silage was the key factor, and only at the appropriate ratio could the effective components in the feed be effectively absorbed. At 25%SS and 50%SS, the bacterial community in the rumen is more conducive to the absorption of CP, ADF, and NDF nutrients, and a large number of fibrinolytic bacteria may be produced in the rumen of animals, such as *Ruminococcus albus*, *Fibrobacter succinogenes*, *Ruminococcus flavefaciens*, *Butyrivibrio fibrisolvens*, and *Bacillus*, etc. [71]. *Bacillus* can secrete many enzymes, such as amylase, xylanase, chitinase, β-1, 3-glucanase, β-glucosidase, lipase, protease, cellulase, etc. [72]. The rate of degradation of CP is influenced by the true protein concentration and amino acid composition of CP in the feed [73]. This result is consistent with that of the experiment. In this experiment, the crude protein content of 25%SS and 50%SS was relatively high, so that the rumen degradation rate of CP was high, which may be due to the large number of microorganisms that can secrete proteases during rumen degradation of the feed. However, although the crude protein content of 0%SS was the highest in this experiment, the degradation rate of the CP of 0%SS in the rumen was not the fastest, possibly because the feed composition of 0%SS was all leguminous licorice, there was no sweet sorghum in the gramineous family, and there was no material to provide carbohydrates for the microorganisms secreting protease to provide energy, which was not good for digestion and absorption.

Cellulase treatment breaks the connection between polyester and cellulose, which can be degraded and utilized by the microbiota in the rumen [74]. Rumen fibro-degrading bacteria degrade plant cellulose and hemicellulose mainly by producing a series of cellulases and hemicellulose enzymes. These enzymes, often referred to as cellulase complexes and hemi-cellulase complexes, are composed of a number of different enzymes, including β-glucosidase, xylanase, galacturonase, etc. [75]. These enzymes are able to break the chemical bonds between cellulose and hemicellulose, breaking it down into smaller soluble carbohydrates [76].

Rumen fibro-degrading bacteria produce volatile fatty acids (VFAs) in the process of degradation of cellulose and hemicellulose [77]. Volatile fatty acids include short-chain fatty acids such as acetic acid, propionic acid, and butyric acid, which are metabolites of cellulose and hemicellulose fermentation. These VFAs are absorbed by the animal as an energy source in the rumen and play an important role in the physiological function and nutritional status of the host animal [78]. However, the limitation of this paper is that the microorganisms and metabolites in the rumen fluid of different treatment groups have not been detected.

5. Conclusions

In conclusion, research was conducted on the fermentation quality, aerobic stability, degradation characteristics of nylon bags, and in vitro degradation characteristics of mixed silage of whole sweet sorghum and licorice with stems and leaves in different proportions, which showed that the mixed silage of whole sweet sorghum and licorice with stems and leaves in a 50:50 ratio (50%SS) effectively improved fermentation quality and rumen degradation rate. The addition of licorice with stems and leaves could effectively improve the aerobic stability and CP content of the mixed silages.

Author Contributions: Conceptualization, S.Z.; methodology, J.W. and F.C.; software, F.C.; validation, S.Z., J.W.; formal analysis, F.C.; investigation, J.W.; resources, F.C.; data curation, S.Z.; writing—original draft, F.C.; preparation, F.C.; writing—review and editing, J.W., A.S.C., H.K. and S.Z.; visualization, F.C.; supervision, S.Z. and J.W.; funding acquisition, S.Z. All authors have read and agreed to the published version of the manuscript.

Funding: This present work was funded by the National Natural Science Foundation of China (No. 32260858) and Key Laboratory of Tarim Animal Husbandry Science and Technology, Xinjiang Production & Construction Group (No. HS202101).

Institutional Review Board Statement: Not applicable.

Informed Consent Statement: Not applicable.

Data Availability Statement: All data are presented in this article in the form of figures and tables.

Acknowledgments: The authors also acknowledge the assistance of Instrumental Analysis Center of Tarim University.

Conflicts of Interest: The authors declare no conflicts of interest.

Abbreviations

AA, acetic acid; AB, aerobic bacteria; ADF, acid detergent fiber; ADL, acid detergent lignin; Ash, crude ash; BA, butyric acid; CP, crude protein; DE, digestible energy; DM, dry matter; ED, effective degradation rate; EE, ether extract; FM, fresh matter; GP, gas production; IVDMD, in vitro dry matter digestibility; LA, high lactic acid; LAB, lactic acid bacteria; LC, aerial parts of licorice; ME, metabolizable energy; MOD, digestible organic matter; NDF, neutral detergent fiber; Neg, net energy for gain; NH_3-N, ammonia nitrogen; PA, propionic acid; SS, sweet sorghum; TN, total nitrogen; WSC, water-soluble carbohydrate.

References

1. Heng, T.; Liao, R.K.; Wang, Z.H.; Wu, W.Y.; Li, W.H.; Zhang, J.Z. Effects of combined drip irrigation and sub-surface pipe drainage on water and salt transport of saline-alkali soil in Xinjiang, China. *J. Arid Land* **2018**, *10*, 932–945. [CrossRef]
2. Qu, H.; Liu, X.B.; Dong, C.F.; Lu, X.Y.; Shen, Y.X. Field performance and nutritive value of sweet sorghum in eastern China. *Field Crops Res.* **2014**, *157*, 84–88. [CrossRef]
3. Rooney, W.L.; Blumenthal, J.; Bean, B.; Mullet, J.E. Designing sorghum as a dedicated bioenergy feedstock. *Biofuels Bioprod. Biorefining* **2007**, *1*, 147–157. [CrossRef]
4. Amer, S.; Seguin, P.; Mustafa, A.F. Short communication: Effects of feeding sweet sorghum silage on milk production of lactating dairy cows. *J. Dairy Sci.* **2012**, *95*, 859–863. [CrossRef]
5. Campanili, P.R.B.; Sarturi, J.O.; Ballou, M.A.; Trojan, S.J.; Sugg, J.D.; Ovinge, L.A.; Alrumaih, A.U.; Pellarin, L.A.; Hoffman, A.A. Effects of silage type and inclusion level on ruminal characteristics and feeding behavior of steers fed finishing diets. *J. Anim. Sci.* **2017**, *95*, 4623–4637. [CrossRef]
6. Colombini, S.; Galassi, G.; Crovetto, G.M.; Rapetti, L. Milk production, nitrogen balance, and fber digestibility prediction of corn, whole plant grain sorghum, and forage sorghum silages in the dairy cow. *J. Dairy Sci.* **2012**, *95*, 4457–4467. [CrossRef]
7. Zhang, S.J.; Chaudhry, A.S.; Osman, A.; Shi, C.Q.; Edwards, G.R.; Dewhurst, R.J.; Cheng, L. Associative effects of ensiling mixtures of sweet sorghum and alfalfa on nutritive value, fermentation and methane characteristics. *Anim. Feed Sci. Technol.* **2015**, *206*, 29–38. [CrossRef]
8. National Research Council (NRC). *Nutrient Requirements of Small Ruminants: Sheep, Goats, Cervids and New World Camelids*; National Academies Press: Washington, DC, USA, 2007.
9. Jiang, M.Y.; Fu, C.M.; He, X.G.; Li, R.; Lin, X. An "essential herbal medicine" licorice: A review of phytochemicals and its effects in combination preparations. *J. Ethnopharmacol.* **2020**, *249*, 112439. [CrossRef]
10. Kondo, K.; Shiba, M.; Yamaji, H.; Morota, T.; Zhengmin, C.; Huixia, P.; Shoyama, Y. Species identification of licorice using nrDNA and cpDNA genetic markers. *Biol. Pharm. Bull.* **2007**, *30*, 1497–1502. [CrossRef] [PubMed]
11. Jiang, D.; Li, P.; Liu, C.; Yin, Y.; Ren, G. Molecular cloning and functional characterization of UGTs from Glycyrrhiza uralensis flavonoid pathway. *Int. J. Biol. Macromol.* **2021**, *192*, 1108–1116. [CrossRef] [PubMed]
12. Cui, X.; Lou, L.; Zhang, Y.; Yan, B. Study of the distribution of Glycyrrhiza uralensis production areas as well as the factors affecting yield and quality. *Sci. Rep.* **2023**, *13*, 5160. [CrossRef]
13. Zhang, Q.; Ye, M. Chemical analysis of the Chinese herbal medicine Gan-Cao (licorice). *J. Chromatogr. A* **2009**, *1216*, 1954–1969. [CrossRef]
14. Abarghuei, M.J.; Salem, A.Z.M. Sustainable impact of pulp and leaves of Glycyrrhiza glabra to enhance ruminal biofermentability, protozoa population, and biogas production in sheep. *Environ. Sci. Pollut. Res.* **2021**, *28*, 33371–33381. [CrossRef]
15. Tahereh, M.H.; Mohammad, J.S.; Gholamreza, K.; Abolfazl, J. Monitoring amino acid profile and protein quality of Licorice (*Glycyrrhiza glabra* L.) under drought stress, silicon nutrition and mycorrhiza inoculation. *Sci. Hortic.* **2022**, *295*, 110808.
16. Liu, X.; Li, D.; Ge, Q.; Yang, B.; Li, S.H. Effects of harvest period and mixed ratio on the characteristic and quality of mixed silage of alfalfa and maize. *Anim. Feed Sci. Technol.* **2023**, *306*, 115796. [CrossRef]
17. Ni, K.K.; Zhao, J.Y.; Zhu, B.G.; Su, R.N.; Pan, Y.; Ma, J.K.; Zhou, G.A.; Tao, Y.; Liu, X.R.; Zhong, J. Assessing the fermentation quality and microbial community of the mixed silage of forage soybean with crop corn or sorghum. *Bioresour. Technol.* **2018**, *265*, 563–567. [CrossRef]

18. Larsen, S.U.; Hiort-Gregersen, K.; Vazifehkhoran, A.H.; Triolo, J.M. Co-ensiling of straw with sugar beet leaves increases the methane yield from straw. *Bioresour. Technol.* **2017**, *245*, 106–115. [CrossRef] [PubMed]
19. Jiang, Y.; Dennehy, C.; Lawlor, P.G.; Hu, Z.; Zhan, X.; Gardiner, G. Inactivation of enteric indicator bacteria and system stability during dry co-digestion of food waste and pig manure. *Sci. Total Environ.* **2018**, *612*, 293–302. [CrossRef] [PubMed]
20. Wang, J.; Yang, B.Y.; Zhang, S.J.; Amar, A.; Chaudhry, A.S.; Cheng, L.; Abbasi, I.H.R.; Al-Mamun, M.; Guo, X.F.; Shan, A.S. Using mixed silages of sweet sorghum and alfalfa in total mixed rations to improve growth performance, nutrient digestibility, carcass traits and meat quality of sheep. *Animal* **2021**, *15*, 100246. [CrossRef] [PubMed]
21. Ren, H.W.; Sun, W.L.; Yan, Z.H.; Zhang, Y.C.; Wang, Z.Y.; Song, B.; Zheng, Y.; Li, J.P. Bioaugmentation of sweet sorghum ensiling with rumen fluid: Fermentation characteristics, chemical composition, microbial community, and enzymatic digestibility of silages. *J. Clean. Prod.* **2021**, *294*, 126308. [CrossRef]
22. Orrico-Junior, M.A.P.; Vendramini, J.M.B.; Erickson, J.; Moriel, P.; Silveira, M.L.A.; Aguiar, A.D.; Sanchez, J.M.D.; Silva, W.L.; Silva, H.M. Nutritive value and fermentation characteristics of silages produced from different sweet sorghum plant components with or without microbial inoculation. *Appl. Anim. Sci.* **2020**, *36*, 777–783. [CrossRef]
23. Ran, T.; Tang, S.X.; Yu, X.; Hou, Z.P.; Hou, F.J.; Beauchemin, K.A.; Yang, W.Z.; Wu, D.Q. Diets varying in ratio of sweet sorghum silage to corn silage for lactating dairy cows: Feed intake, milk production, blood biochemistry, ruminal fermentation, and ruminal microbial community. *J. Dairy Sci.* **2021**, *104*, 12600–12615. [CrossRef]
24. AOAC. *Official Methods of Analysis*, 15th ed.; Association of Official Analytical Chemists: Arlington, VA, USA, 1990.
25. Thomas, T.A. An automated procedure for the determination of soluble carbohydrates in herbage. *J. Sci. Food Agric.* **1977**, *28*, 639–642. [CrossRef]
26. Van Soest, P.J.; Robertson, J.B.; Lewis, B.A. Methods for dietary Fiber, neutral detergent Fiber, and nonstarch polysaccharides in relation to animal nutrition. *J. Dairy Sci.* **1991**, *74*, 3583–3597. [CrossRef] [PubMed]
27. Broderick, G.A.; Kang, J.H. Automated simultaneous determination of ammonia and total amino acids in ruminal fluid and in vitro media. *J. Dairy Sci.* **1980**, *63*, 64–75. [CrossRef] [PubMed]
28. Desta, S.T.; Yuan, X.; Li, J.; Shao, T. Ensiling characteristics, structural and nonstructural carbohydrate composition and enzymatic digestibility of Napier grass ensiled with additives. *Bioresour. Technol.* **2016**, *221*, 447–454. [CrossRef] [PubMed]
29. Zhao, J.; Dong, Z.H.; Chen, L.; Wang, S.R.; Shao, T. The replacement of whole-plant corn with bamboo shoot shell on the fer-mentation quality, chemical composition, aerobic stability and in vitro digestibility of total mixed ration silage. *Anim. Feed Sci. Technol.* **2020**, *259*, 114348. [CrossRef]
30. Seppälä, A.; Heikkilä, T.; Mäki, M.; Miettinen, H.; Rinne, M. Controlling aerobic stability of grass silage-based total mixed rations. *Anim. Feed Sci. Technol.* **2013**, *179*, 54–60. [CrossRef]
31. Takahashi, T.; Horiguchi, K.; Goto, M. Effect of crushing unhulled rice and the addition of fermented juice of epipHytic lactic acid bacteria on the fermentation quality of whole crop rice silage, and its digestibility and rumen fermentation status in sheep. *J. Anim. Sci.* **2005**, *76*, 353–358. [CrossRef]
32. Wilkinson, J.M.; Davies, D.R. The aerobic stability of silage: Key findings and recent developments. *Grass Forage Sci.* **2013**, *68*, 1–19. [CrossRef]
33. Menke, K.H.; Steingass, H. Estimation of the energetic feed value obtained from chemical analysis and in vitro gas production using rumen fluid. *Anim. Res. Develop.* **1988**, *28*, 7–55.
34. Contreras-Govea, F.E.; Muck, R.E.; Mertens, D.R.; Weimer, P.J. Microbial inoculant effects on silage and in vitro ruminal fermentation, and microbial biomass estimation for alfalfa, bmr corn, and corn silages. *Anim. Feed Sci. Technol.* **2011**, *163*, 2–10. [CrossRef]
35. Ørskov, E.R.; McDonald, I. The estimation of protein degradation in the rumen from incubation measure ments weighted according to the rate of passage. *J. Agric. Sci.* **1979**, *92*, 499–503. [CrossRef]
36. Menke, K.H.; Raab, L.; Salewski, A.; Steingass, H.; Fritz, D.; Schneider, W. The estimation of the digestibility and metabolizable energy content of ruminant feeding stuffs from the gas production when they are incubated with rumen liquor in vitro. *J. Agric. Sci.* **1979**, *93*, 217–222. [CrossRef]
37. Mirzaci-Aghsaghali, A.; Maheri-sis, N.; Mirza-Aghazadeh, A.; Safaei, A.R. Use of nylon bag technique to determine nutritive value and degradation kinetics of Iranian alfalfa varieties. *Asian J. Anim. Vet. Adv.* **2008**, *3*, 214–221.
38. Sehu, A.; Çakir, S.; Sahin, T. Determination of rumen degradability of some oilseeds and meals using nylon bag technique. *Vet. Fak. Derg.* **2010**, *57*, 173–178.
39. Tássia, B.P.S.; Tiago, A.D.V.; Lucas, G.G.; Guilherme, G.S.; Larissa, S.G.; Júlia, A.M.; Mauro, S.S.D.; Alanne, T.N.; Nathália, T.S.G.; Caio, S.T.; et al. Partial replacement of corn silage with whole-plant soybean and black oat silages for dairy cows. *J. Dairy Sci.* **2021**, *104*, 9842–9852.
40. Wang, M.S.; Franco, M.; Cai, Y.M.; Zhu, Y. Dynamics of fermentation profile and bacterial community of silage prepared with alfalfa, whole-plant corn and their mixture. *Anim. Feed Sci. Technol.* **2020**, *270*, 114702. [CrossRef]
41. Pedram, P.; Hamidreza, M.A.; Mehdi, K.B.; Jörg, R.A. Growth performance, nutrient digestibility, and ruminal fermentation of dairy calves fed starter diets with alfalfa hay versus corn silage as forage and soybean oil versus palm fatty acids as fat source. *J. Dairy Sci.* **2022**, *105*, 9597–9609.

42. Zeng, T.R.; Li, X.L.; Guan, H.; Yang, W.Y.; Liu, W.G.; Liu, J.; Du, Z.C.; Li, X.M.; Xiao, Q.Y.; Wang, X.C.; et al. Dynamic microbial diversity and fermentation quality of the mixed silage of corn and soybean grown in strip intercropping system. *Bioresour. Technol.* **2020**, *313*, 123655. [CrossRef] [PubMed]
43. Weinberg, Z.G.; Chen, Y.; Weinberg, P. Ensiling olive cake with and without molasses for ruminant feeding. *Bioresour. Technol.* **2008**, *99*, 1526–1529. [CrossRef]
44. Catchpoole, V.R.; Henzell, E.F. Silage and silage making from tropical herbage species. *Herb. Abstr.* **1971**, *41*, 213–221.
45. Ni, K.K.; Wang, F.F.; Zhu, B.G.; Yang, J.X.; Zhou, G.A.; Pan, Y.; Tao, Y.; Zhong, J. Effects of lactic acid bacteria and molasses additives on the microbial community and fermentation quality of soybean silage. *Bioresour. Technol.* **2017**, *238*, 706–715. [CrossRef]
46. Dai, T.; Wang, J.; Dong, D.; Yin, X.; Cheng, Z. Effects of brewers' spent grains on fermentation quality, chemical composition and in vitro digestibility of mixed silage prepared with corn stalk, dried apple pomace and sweet potato vine. *Ital. J. Anim. Sci.* **2022**, *21*, 198–207. [CrossRef]
47. Auerbach, H.; Nadeau, E. Effects of additive type on fermentation and aerobic stability and its interaction with air exposure on silage nutritive value. *Agron. J.* **2020**, *10*, 1229. [CrossRef]
48. Morgan, C.A.; Edwards, R.A.; McDonald, P. Intake and metabolism studies with fresh and wilted silages. *J. Agric. Sci.* **1980**, *94*, 287–298. [CrossRef]
49. Tatulli, G.; Cagliani, L.R.; Sparvoli, F.; Brasca, M.; Consonni, R. NMR-Based Metabolomic Study on Phaseolus vulgaris Flour Fermented by Lactic Acid Bacteria and Yeasts. *Molecules* **2023**, *28*, 4864. [CrossRef] [PubMed]
50. Keles, G.; Demirci, U. The effect of homofermentative and heterofermentative lactic acid bacteria on conservation characteristics of baled triticale–Hungarian vetch silage and lamb performance. *Anim. Feed Sci. Technol.* **2011**, *164*, 21–28. [CrossRef]
51. He, L.W.; Zhou, W.; Wang, C.; Chen, X.Y.; Zhang, Q. Effect of applying lactic acid bacteria and cellulase on the fermentation quality, nutritive value, tannins profile and in vitro digestibility of Neolamarckia cadamba leaves silage. *J. Anim. Physiol. Anim. Nutr.* **2018**, *102*, 1429–1436. [CrossRef]
52. Ohshima, M.; McDonald, P. A review of the changes in nitrogenous compounds of herbage during ensilage. *J. Sci. Food Agric.* **1978**, *29*, 497–505. [CrossRef]
53. Wu, Q.C.; Li, W.J.; Wang, W.K.; Wang, Y.L.; Zhang, F.; Liu, L.K.; Yang, H.J. Foxtail millet (*Setaria italica* L.) silage compared peanut vine hay (*Arachis hypogaea* L.) exhibits greater feed efficiency via enhancing nutrient digestion and promoting rumen fermentation more efficiently in feedlotting lambs. *Small Rumin. Res.* **2022**, *215*, 0921–4488. [CrossRef]
54. Li, Y.Y.; Du, S.; Sun, L.; Cheng, Q.M.; Hao, J.F.; Lu, Q.; Ge, G.T.; Wang, Z.J.; Jia, Y.S. Effects of Lactic Acid Bacteria and Molasses Additives on Dynamic Fermentation Quality and Microbial Community of Native Grass Silage. *Front. Microbiol.* **2022**, *13*, 830121. [CrossRef]
55. Johnson, L.M.; Harrison, J.H.; Davidson, D.; Mahanna, W.C.; Shinners, K.; Linder, D. Corn silage management: Effects of maturity, inoculation, and mechanical processing on pack density and aerobic stability. *J. Dairy Sci.* **2002**, *85*, 434–444. [CrossRef] [PubMed]
56. Blajman, J.E.; Paez, R.B.; Vinderola, C.G.; Lingua, M.S.; Signorini, M.L. A meta-analysis on the effectiveness of homofermentative and heterofermentative lactic acid bacteria for corn silage. *J. Appl. Microbiol.* **2018**, *125*, 1655–1669. [CrossRef] [PubMed]
57. Borreani, G.; Tabacco, E.; Schmidt, R.J.; Holmes, B.J.; Muck, R.E. Factors Affecting Dry Matter and Quality Losses in Silages. *J. Dairy Sci.* **2018**, *101*, 3952–3979. [CrossRef] [PubMed]
58. Filya, I.; Sucu, E. The effects of lactic acid bacteria on the fermentation, aerobic stability and nutritive value of maize silage. *Grass Forage Sci.* **2010**, *65*, 446–455. [CrossRef]
59. Courtin, M.G.; Spoelstra, S.F. A simulation model of the microbiological and chemical changes accompanying the initial stage of aerobic deterioration of silage. *Grass Forage Sci.* **1990**, *45*, 153–165. [CrossRef]
60. Huhtanen, P.; Rinne, M.; Nousiainen, J. Evaluation of the factors affecting silage intake of dairy cows: A revision of the relative silage dry-matter intake index. *Animal* **2007**, *1*, 758–770. [CrossRef] [PubMed]
61. Bender, R.W.; Lopes, F.; Cook, D.E.; Combs, D.K. Effects of partial replacement of corn and alfalfa silage with tall fescue hay on total-tract digestibility and lactation performance in dairy cows. *J. Dairy Sci.* **2016**, *99*, 5436–5444. [CrossRef] [PubMed]
62. Lucas, G.; Tiago, A.; Del, V.; Elissandra, M.C.; Zilio, L.Y.; Sakamoto, J.A.; Marques, M.S.S.; Dias, A.T.; Nunes, L.S.; Gheller, T.B.; et al. Partial replacement of corn silage with soybean silage on nutrient digestibility, ruminal fermentation, and milk fatty acid profile of dairy cows. *Anim. Feed Sci. Technol.* **2020**, *266*, 114526.
63. Cone, J.W.; van Gelder, A.H.; Mathijssen-Kamman, A.A.; Hindle, V.A. Rumen escape protein in grass and grass silage determined with a nylon bag and an enzymatic technique. *Anim. Feed Sci. Technol.* **2004**, *111*, 1–9. [CrossRef]
64. Ma, J.; Fan, X.; Ma, Z.; Huang, X.W.; Tang, M.H.; Yin, F.Q.; Zhao, Z.H.; Gan, S.Q. Silage Additives Improve Fermentation Quality, Aerobic Stability and Rumen Degradation in Mixed Silage Composed of Amaranth and Corn Straw. *Front. Plant Sci.* **2023**, *14*, 1189747. [CrossRef] [PubMed]
65. Inoue, K.; Sewalt, V.J.H.; Ballance, G.M.; Ni, W.T.; Stürzer, C.; Dixon, R.A. Developmental Expression and Substrate Specifici-ties of Alfalfa Caffeic Acid 3-Methyltransferase and Caffeoyl Coenzyme a 3-Methyltransferase in Relation to Lignification. *Plant Physiol.* **1998**, *117*, 761–770. [CrossRef] [PubMed]
66. Casler, M.D.; Vogel, K.P. Accomplishments and Impact from Breeding for Increased Forage Nutritional Value. *Crop Sci.* **1999**, *39*, 12–20. [CrossRef]

67. Guo, D.J.; Chen, F.; Inoue, K.; Blount, J.W.; Dixon, R.A. Downregulation of Caffeic Acid 3-Methyltransferase and Caffeoyl Coa Methyltransferase in Transgenic Alfalfa: Impacts on Lignin Structure and Implications for the Biosynthesis of G and S Lignin. *Plant Cell* **2001**, *13*, 73–88. [CrossRef] [PubMed]
68. Damiran, D.; Biligetu, B.; Lardner, H. Evaluation of Rumen Degradation Kinetics of Low-Lignin Alfalfa 'Hi-Gest® 360' in Sas-katchewan Canada. *Animals* **2023**, *13*, 1047. [CrossRef] [PubMed]
69. Wang, Y.L.; Wang, W.K.; Wu, Q.C.; Yang, H.J. The Release and Catabolism of Ferulic Acid in Plant Cell Wall by Rumen Mi-crobes: A Review. *Anim. Nutr.* **2022**, *9*, 335–344. [CrossRef] [PubMed]
70. Jung, H.G.; Engels, F.M. Alfalfa Stem Tissues: Rate and Extent of Cell-Wall Thinning During Ruminal Degradation. *NJAS Wagening. J. Life Sci.* **2001**, *49*, 3–13. [CrossRef]
71. Nyonyo, T.; Shinkai, T.; Mitsumori, M. Improved Culturability of Cellulolytic Rumen Bacteria and Phylogenetic Diversity of Culturable Cellulolytic and Xylanolytic Bacteria Newly Isolated from the Bovine Rumen. *FEMS Microbiol. Ecol.* **2014**, *88*, 528–537. [CrossRef]
72. Su, Y.; Liu, C.; Fang, H.; Zhang, D.W. A Universal Cell Factory for Industry, Agriculture, Biomaterials and Medicine. *Microb. Cell Factories* **2020**, *19*, 173. [CrossRef]
73. Cardozo, P.W.; Calsamiglia, S.; Ferret, A.; Kamel, C. Effects of Natural Plant Extracts on Ruminal Protein Degradation and fermentation Profiles in Continuous Culture. *J. Anim. Sci.* **2004**, *82*, 3230–3236. [CrossRef] [PubMed]
74. Li, J.F.; Yuan, X.J.; Dong, Z.H.; Mugabe, W.; Shao, T. The Effects of Fibrolytic Enzymes, Cellulolytic Fungi and Bacteria on the Fermentation Characteristics, Structural Carbohydrates Degradation, and Enzymatic Conversion Yields of Pennisetum Sinese Silage. *Bioresour. Technol.* **2018**, *264*, 123–130. [CrossRef] [PubMed]
75. Bai, T.T.; Pu, X.X.; Guo, X.F.; Liu, J.F.; Zhao, L.B.; Zhang, X.P.; Zhang, S.J.; Cheng, L. Effects of Dietary Nonfibrous Carbohydrate/Neutral Detergent Fiber Ratio on Methanogenic Archaea and Cellulose-Degrading Bacteria in the Rumen of Karakul Sheep: A 16s Rrna Gene Sequencing Study. *Appl. Environ. Microbiol.* **2023**, *89*, 1. [CrossRef] [PubMed]
76. Huang, X.Y.; Xu, Y.Y.; Wu, X.Y.; Ding, Y.H.; Fan, C.Y.; Xue, Y.F.; Zhuo, Z.; Cheng, J.B. Mixed Fermentation of *Lactiplantibacillus plantarum* and *Bacillus licheniformis* Changed the Chemical Composition, Bacterial Community, and Rumen Degradation Rate of Tea Residue. *Fermentation* **2022**, *8*, 380. [CrossRef]
77. Xu, Y.; Aung, M.; Sun, Z.Y.; Zhou, Y.Q.; Cheng, Y.F.; Hao, L.Z.; Padmakumar, V.; Zhu, W.Y. Bio-Fermentation Improved rumen Fermentation and Decreased Methane Concentration of Rice Straw by Altering the Particle-Attached Microbial Community. *Fermentation* **2022**, *8*, 72. [CrossRef]
78. Vasta, V.; Daghio, M.; Cappucci, A.; Buccioni, A.; Serra, A.; Viti, C.; Mele, M. Plant Polyphenols and Rumen Microbiota responsible for Fatty Acid Biohydrogenation, Fiber Digestion, and Methane Emission: Experimental Evidence and Methodological Approaches. *J. Dairy Sci.* **2019**, *102*, 3781–3804. [CrossRef]

agriculture

MDPI

Article

Isolation of *Bacillus velezensis* from Silage and Its Effect on Aerobic Stability and In Vitro Methane Production of Whole-Plant Corn Silage

Chen Zhang [1,2,†], Zimo Zhang [1,†], Mengfan Zhu [1,2], Yongliang Wang [2], Tangtang Zhou [1], Fachun Wan [2], Yunhua Zhang [3,*] and Lijuan Chen [1,*]

[1] The Biological Feedstuff Laboratory, College of Animal Science and Technology, Anhui Agricultural University, Hefei 230036, China; zhchnadp@163.com (C.Z.)
[2] Animal Nutritional Genome and Germplasm Innovation Research Center, College of Animal Science and Technology, Hunan Agricultural University, Changsha 410128, China; wanfc@sina.com (F.W.)
[3] Anhui Province Key Laboratory of Farmland Ecological Conservation and Pollution Prevention, College of Resources and Environment, Anhui Agricultural University, Hefei 230036, China
* Correspondence: yunhua9681@163.com (Y.Z.); zhangchenlijuan@163.com (L.C.)
† These authors contributed equally to this work.

Abstract: Once a silo has been opened, the silage inside will face challenges such as aerobic deterioration, rot, and contamination. Biocontrol bacteria, as a kind of biological antiseptic, are highly effective and natural and are gaining increasing attention. This study aimed to screen a strain with anti-microbial activity against silage spoilage microorganisms and examine its effects on the fermentation quality, aerobic stability, in vitro digestion, and methane production of silage. Lactic acid bacteria, pathogenic and rot-causing microorganisms, were used as indicators to screen the strains for putrefactive silage. The bacteriostatic spectrum, growth performance, and tolerance to the silage environment of the strain were tested. A strain named D-2 was screened from rotten whole-plant corn silage and identified as *Bacillus velezensis* through physiological and biochemical tests as well as 16S rDNA sequencing. This study found that D-2 exhibits antibacterial effects on several microorganisms, including *Escherichia coli*, *Staphylococcus aureus*, *Salmonella enteritidis*, *Aspergillus niger*, *Saccharomyces cerevisiae*, *Fusarium oxysporum*, and *Fusarium graminearum*. However, it has no adverse effect on *Lactobacillus reuteri*, *Enterococcus faecium*, or *Lactobacillus casei*. D-2 can attain a stable stage within 10 h and withstand temperatures of up to 70 °C. Moreover, this study found that D-2 had a high survival rate of over 97% after 48 h in a lactic acid environment with pH 4. Freshly chopped whole-plant corn was inoculated without or with D-2 and ensiled for 60 days. The results show that D-2 inoculations increase the content of water-soluble carbohydrates, acetic acid, and propionic acid in the silage and decrease the number of yeasts and molds, the NH_4^+-N/TN ratio, and the pH. We also found that fermenting whole-plant corn with D-2 significantly increased the in vitro digestibility and the propionic acid content, while also significantly inhibiting methane production. After being exposed to air for 10 days, D-2 can still effectively reduce the total number of yeasts and molds, prevent the decrease in lactic acid bacteria, and inhibit the increase in the pH and NH_4^+-N/TN ratio of silage products. Overall, D-2 is resistant to pathogenic and rot-causing microorganisms, allowing for easy adaptation to silage production conditions. D-2 can effectively improve aerobic stability and reduce losses in the nutritional value of silage, indicating possible applications for the prevention of silage rot and methane production.

Keywords: silage; antibacterial activity; *Bacillus velezensis*; isolation and identification; aerobic stability; methane production

Citation: Zhang, C.; Zhang, Z.; Zhu, M.; Wang, Y.; Zhou, T.; Wan, F.; Zhang, Y.; Chen, L. Isolation of *Bacillus velezensis* from Silage and Its Effect on Aerobic Stability and In Vitro Methane Production of Whole-Plant Corn Silage. *Agriculture* **2024**, *14*, 830. https://doi.org/10.3390/agriculture14060830

Academic Editor: Joshua D. Klein

Received: 19 March 2024
Revised: 21 May 2024
Accepted: 22 May 2024
Published: 25 May 2024

1. Introduction

Silage is a type of moist, stored, and nutrient-rich feed formed by fermenting fresh greenfeed with lactic acid bacteria (LAB) under anaerobic conditions [1]. It breaks the seasonal limitations and is an important guarantee for ruminant overwintering feed [2]. During the ensiling process, LAB (such as *Lactobacillus plantarum*, *Lactobacillus casei*, *Lactobacillus reuteri*, etc.) produces a large amount of lactic and acetic acids to form an acidic environment. This phenomenon quickly reduces the pH of the silage system to around 4.0, inhibits the activities of undesirable microbes [3], and produces antagonistic metabolites, such as nisin, ε-polylysine, and phenyllactic acid, which inhibit the growth of pathogenic and rot-causing microorganisms [4] and improve the nutritional properties and safety of silage [5]. To some extent, the ensiling of forage grass alleviates the problem of a lack of high-quality forage materials in winter in some farming areas in China. However, during actual production and application, silage faces the risk of pollution and deterioration because forage itself carries impurities, such as rot-causing microorganisms (such as *Aspergillus niger*, *Fusarium oxysporum*, *Fusarium graminearum*, etc.), soil, and uncompacted raw materials. After the silo has been opened, the silage comes into contact with oxygen. This results in the recovery and rapid multiplication of rot-causing microorganisms, increased silage pH, reduced aerobic stability and quality, nutrient loss, decreased animal feed utilization, and the production of multiple mycotoxins [6,7], thus compromising the animals' health.

In addition to the large number of LAB, aerobic microorganisms (such as yeasts, coliforms, and molds) are present in silage and seriously affect its stability during aerobic exposure [8,9]. Responsible for the spoilage of silage are *Fusarium*, *Aspergillus niger*, *Aspergillus flavus*, and *Penicillium* [10,11]. It is common to add citric, fumaric, and formic acids and other acidity regulators into silage feed as preservatives to inhibit the proliferation of these spoilage microorganisms and reduce losses [12,13]. However, this process requires repeated spraying, is costly, and negatively affects the palatability of feed. Whole-plant corn silage has become the predominant forage used in dairy cattle diets worldwide [14]. The spoilage of corn silage caused by pathogenic and rot-causing microorganisms can affect the individual herd as well as impact the overall profitability of the industry [15].

Based on this, we hypothesize the existence of microorganisms in silage that can inhibit the growth of pathogenic and spoilage microorganisms without affecting the growth of some LAB. The aim of this study is to screen the microorganism and investigate its impact on fermentation quality, aerobic stability, and in vitro ruminal methane production.

2. Materials and Methods

2.1. Materials

Samples: Rotten silage (whole-plant corn silage) was collected from Huangshan Junfeng ecological cattle farm (Huangshan, China). Whole-plant corn (Qingzhu No. 6) was grown in an experimental field at Anhui Agricultural University (latitude: 31°58′ N, 117°24′ E). Culture media: nutrient, Luria-Bertani (LB), and deMan, Rogosa, and Sharpe (MRS) broths, and potato dextrose agar (PDA) (HB0108, HB0128, HB0384-1, and HB0233; Qingdao Hi-Tech Industrial Park Hope Bio-Technology Co., Ltd., Qingdao, China). About 20 g/L of agar powder (HB8274, Qingdao Hi-Tech Industrial Park Hope Bio-Technology Co., Ltd., Qingdao, China) was added to each broth to prepare solid culture media. Indicator microorganisms: *Escherichia coli* (*E. coli*), *Staphylococcus aureus* (*S. aureus*), and *Salmonella enteritidis* (*S. enteritidis*) were used as pathogenic indicator microorganisms. *Aspergillus niger* (*A. niger*), *Saccharomyces cerevisiae* (*S. cerevisiae*), *Fusarium oxysporum* (*F. oxysporum*), and *Fusarium graminearum* (*F. graminearum*) were used as rot-causing indicator microorganisms. *Lactobacillus casei* (*L. casei*), *Lactobacillus reuteri* (*L. reuteri*), and *Enterococcus faecium* (*E. faecium*) were used as LAB (probiotic) indicator microorganisms. *E. coli*, *S. aureus*, *S. enteritidis*, *L. casei*, *L. reuteri*, and *E. faecium* were obtained from the Anhui Province Engineering Laboratory for Animal Food Quality and Bio-safety (Hefei, China); *A. niger*, *S. cerevisiae*, *F. oxysporum*, and *F. graminearum* were obtained from the Anhui Academy of Agricultural Sciences (Hefei, China).

2.2. Isolation Procedure

(1) Primary screening: Plant tissues with intact structure and no fungal colonies from rotten silage were collected as test samples. The samples were immersed in sterilized

physiological saline (sample/physiological saline = 1 g: 9 mL), vortexed for 5 min, and incubated at 28 °C for 24 h. An asepsis mop was used to collect the sample solution, which was spread onto nutrient agar (NA) and cultured at 37 °C with an O_2 concentration $\leq$ 0.5% for five days. After the microorganisms had grown, the strains with different colony morphologies were picked up separately and subjected to isolation and purification, and the purified strains were inoculated into nutrient broth and incubated statically at 37 °C for 24 h.

(2) Second screening: To assess the bacteriostatic properties of purified strains using the Kirby–Bauer (K-B) method, a disk diffusion test was performed on the tested bacteria [16]. The fermentation fluid of the tested bacteria, LABs, pathogenic microorganisms, and yeasts in a stationary phase was collected separately, and the concentration of each microbe was adjusted with phosphate buffered saline (PBS) to $\times 10^6$ CFU/mL. The culture mediums inoculated with rot-causing molds were taken, the spores were rinsed with sterile normal saline containing 0.05% Polysorbate-80, and the mycelia were scraped to prepare a spore suspension or mycelium suspension. Then, the indicator microorganism suspensions were spread using sterilized swabs from the plates. The pathogenic microorganisms were inoculated on an LB medium, the rot-causing microorganisms were inoculated on a PDA medium, and the LAB were inoculated on an MRS medium. After allowing the surface of each medium to dry for 3–5 min, dry sterile ADVANTEC paper (Toyo Roshi Kaisha, Ltd., Tokyo, Japan) disks (Φ = 6 mm, δ = 0.7 mm) were placed on the plates and pressed with forceps to ensure complete contact with each agar surface. A total of 10 μL of the tested bacteria solution ($\times 10^6$ CFU/mL) was loaded onto the paper disks on the agar plates, spread, and cultivated for 24 h–72 h at 37 °C (bacteria) or 28 °C (fungi) to observe and measure the inhibition zones. All the experiments were conducted 3 times. The microorganisms that could inhibit the growth of pathogenic and rot-causing microorganisms were screened but had no effect on LAB growth.

2.3. Identification

(1) Morphological tests: after Gram staining, the morphological characteristics of a single colony after purification were observed using oil immersion microscopy.

(2) Physiological and biochemical tests: the tests were conducted in accordance with Bergey's Manual of Determinative Bacteriology [17].

(3) Genetic tests: The genomic DNA from stable culture fluid was extracted using a Bacteria Genomic DNA Kit (Trelief®, Tsingke Biotechnology Co., Ltd., Beijing, China), and its 16S rDNA gene fragment was amplified using two universal bacterial primers (namely, 27F: 5′-AGAGTTTGATCCTGGCTCAG-3′ as the forward primer and 1492R: 5′-GGTTACCTTGTTACGACTT-3′ as the reverse primer). Totals of 27 μL PCR mix (GoldenStar® T6 Super PCR Mix Ver.2, Tsingke Biotechnology Co., Ltd., Beijing, China), 1 μL forward primer, 1 μL reverse primer, and 1 μL DNA template were used for the PCR reaction. The PCR reaction conditions are as follows: initial denaturation at 98 °C for 180 s, followed by 35 cycles of denaturation at 98 °C for 10 s, annealing at 55 °C for 10 s, and extension at 72 °C for 10 s. The final extension was made at 72 °C for 120 s. The PCR products were sent to Nanjing Tsingke Biotechnology Co., Ltd. (Nanjing, China) for sequencing. The 16S rDNA sequence obtained was explored using BLAST (https://blast.ncbi.nlm.nih.gov/Blast.cgi) (accessed on 28 June 2022) to query for highly similar sequences, and the relative phylogenetic position of the isolate was determined using ML tree analysis with MEGA7.0 ver.

2.4. Growth Characteristics

About 2 mL of bacterial solution (OD_{600} = 0.6) was inoculated into 100 mL of nutrient broth and cultivated at 140 rpm and 37 °C. The OD at 600 nm wavelength was measured every 2 h, and a growth curve was drawn [18]. The fermentation fluid in the stationary phase was inoculated into the nutrient broth to obtain a 1% concentration and cultured at different temperatures (30, 40, 50, 60, and 70 °C) and pH values (2, 4, 6, 8, and 10) for 12 h. Viable counts were determined. The measurements were repeated three times.

2.5. Lactic Acid Tolerance

The pH conditions of silage were simulated. The pH of the nutrient broth was adjusted to 4 by using lactic acid [3], and the nutrient broth was filtered through a Millipore filter (0.2 μm). The tested bacterial solution ($\times 10^6$ CFU/mL) was inoculated into the nutrient broth at 2% inoculum volume and cultured at 37 °C. The samples were collected at 0, 12, 24, 36, and 48 h for viable bacterial counting. The experiment was repeated three times.

2.6. Antibiotic Resistance

The minimum inhibitory concentration (MIC) of 8 antibiotics for *Bacillus* spp., regulated by the European Food Safety Agency (EFSA) [19] against D-2 (tested bacteria), was tested with the broth microdilution method according to the guidelines of the Clinical and Laboratory Standards Institute (CLSI) [20]. The concentration (mg/L) range of 8 antibiotics (Shanghai Yuanye Bio-Technology Co., Ltd., Shanghai, China) used in the test is as follows [21]: Vancomycin (0.25–16.00); Gentamicin (0.25–32.00); Erythromycin (0.25–16.00); Clindamycin (0.25–8.00); Tetracycline (0.5–32.00); Chloramphenicol (1.00–64.00); Kanamycin (1.00–64.00); Streptomycin (0.25–32.00). The MIC values were determined after being placed at 35 °C and incubated for 20 h, and the results were judged against antibiotic susceptibility standards based on the live bacterial components used as feed additives as specified by EFSA.

2.7. Silage Preparation and Oxygen Exposure

Whole-plant corn (dough stage) was chopped into approximately 1–2 cm segments and then left to wilt via air drying at an ambient temperature to a DM concentration of 31.40% fresh weight (FW). The chopped forage was randomly divided into 10 sub-samples (2 treatments × 5 replicates, about 500 g for each sub-sample) and then randomly assigned to one of the following treatments: (1) D-2, the dosage of inoculation was 1×10^6 CFU/g fresh forage, which was mixed thoroughly in a disinfected plastic container; (2) Control group, in which an equal volume of distilled water was applied under the same conditions. The treated or untreated forages were packed into polyethylene plastic bags (290 mm × 230 mm; fermented feed bag with a one-way valve; Wenzhou Wangting Packaging Co., Ltd., Wenzhou, China) and ensiled individually, vacuum-sealed, and stored in an air-conditioned room (25 ± 2 °C) for 60 d.

The silage samples in the D-2 and control groups at 60 days after fermentation were held separately in a sterile beaker (200 g each), covered with plastic film, incubated at 28 °C and 75% RH, exposed to air at intervals of 12 h, and then thoroughly stirred for 5 min per day with a sterile glass rod; this operation was repeated for 10 d. The samples were collected 60 days after fermentation and 10 days after aerobic exposure for the analysis of fermentation quality, microbial population, and chemical composition.

2.8. Analysis of Aerobic Stability, Chemical Composition, and In Vitro Gas Production of Silage

Each group of silage was evaluated in accordance with the scoring method of the Deutsche Landwirtschafts-Gesellschaft [22]. The samples (10 g) from each group were diluted in 90 mL of sterilized physiological saline, and the mixture was thoroughly mixed. A portion (50 mL) was further diluted to 10^{-2} with sterilized physiological saline, and 100 μL diluted samples were spread onto MRS or Rose Bengal agar to determine the total LAB, yeast, and mold counts [23]. The MRS plates were incubated anaerobically for 72 h at 37 °C, and the Rose Bengal plates were incubated aerobically at 28 °C for 48 h. The pH of another silage extract aliquot was measured using a KASVI potentiometer [24]. The organic acid and water-soluble carbohydrate (WSC) contents were measured using high-performance liquid chromatography methods [25].

The Kjeldahl method [26] was used to determine the total nitrogen (TN) content, and the phenol-sodium hypochlorite method [27] was used to determine the ammoniacal nitrogen (NH_4^+-N) concentration of silage. The neutral detergent fiber (NDF) and acid detergent fiber (ADF) contents of the silage were presented based on their dry matter (DM) bases. The determination methods of DM, NDF, and ADF contents were referenced according to Bai et al. [28] and Van Soest et al. [29].

The ruminal fermentation of whole-plant corn silage samples was conducted using an in vitro gas production technique; the in vitro dry matter digestibility (IVDMD) and methane production of silage were determined using the method of Khota et al. [30].

2.9. Statistical Analysis

Data on growth characteristics (temperature and pH), bacteriostatic spectrum, and survival rate of D-2 were analyzed using a one-way ANOVA on SPSS 17.0 (SPSS, Inc., Chicago, IL, USA), and data on fermentation quality, aerobic stability, digestion, and methane production of whole-plant corn silage were analyzed using unpaired *t*-tests. All the values are expressed as mean ± SD, and the minimum and maximum levels of significance were set to $p < 0.05$ and $p < 0.01$, respectively. Origin7.5 (OriginLab, Northampton, MA, USA) was used to draw the figures.

3. Results

3.1. Identification of D-2

The D-2 strain was isolated through preliminary screening of the plates and a second bacteriostatic evaluation. D-2 grew well on the NA plate and had a milky white color, a nearly round shape, irregular edges, rough and opaque surfaces, bulges in the middle, and wrinkles (Figure 1A). The strain is a Gram-positive, endospore-forming, and rod-shaped microorganism (Figure 1B). The examination of morphological, microscopic, physiological, and biochemical characteristics (Table 1) preliminarily identified D-2 as *Bacillus* spp. After the PCR amplification of D-2, a 16S rDNA gene fragment of 1438 bp in size was obtained (Figures 1C and A1), and the sequencing results were analyzed through BLAST alignment at NCBI, showing that the isolates had 99.51%–100% sequence homology with the *Bacillus* spp. gene. The above sequences were collected, and the 16S rDNA of 11 strains with the closest similarity was used to construct a phylogenetic tree using MEGA 7.0 software. For statistical testing, >1000 bootstraps were used. The results are shown in Figure 1D. In accordance with morphological, physiological, biochemical, and molecular observations, the D-2 strain was named *Bacillus velezensis* D-2.

Table 1. Physiological and biochemical identification results of D-2.

Physiological and Biochemical Test	Reaction	Physiological and Biochemical Test	Reaction
Anaerobic fermentation test	+	Esculin hydrate test	+
MR/VP test	MR (−)	Xylose fermentation test	+
	VP (+)	Arabinose fermentation test	−
Citrate test	−	Raffinose test	−
Propionate test	−	Starch hydrolysis test	+
Catalase test	−	Indole test	−
Mannitol fermentation test	−	Nitrate reduction test	+

Note: MR/VP test, Methyl Red/Voges-Proskauer test; "+" positive; "−" negative.

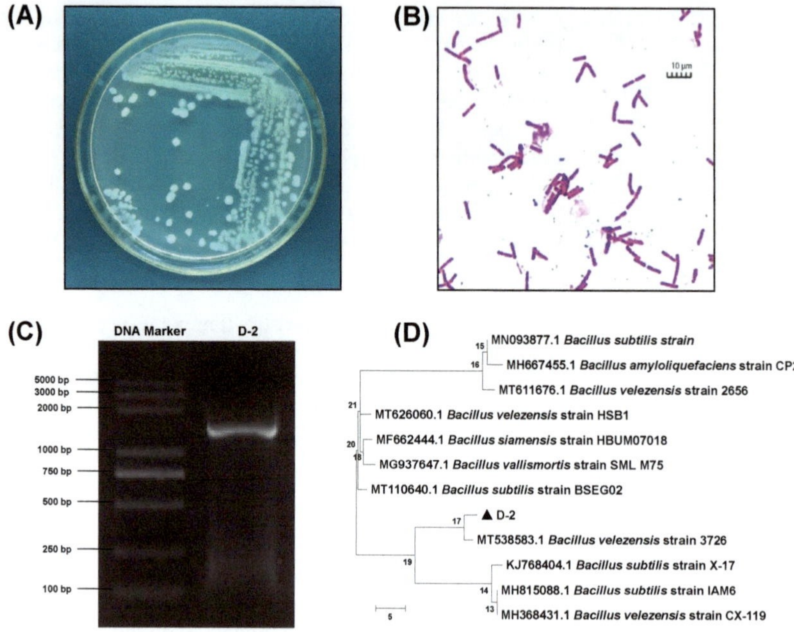

Figure 1. Identification of D-2. (**A**) Morphology. (**B**) Gram staining and optical micrographs. (**C**) PCR amplification electrophoresis result of D-2 16S rDNA. (**D**) Evolutionary relationships of strain D-2 based on the 16S rRNA gene sequences.

3.2. Growth Characteristics of D-2

In accordance with the test results, the average value was determined. The time and OD_{600} represent the abscissa and ordinate, respectively, and a growth curve was drawn. The results are shown in Figure 2A. At 0–2 h, D-2 was in a lag phase, and no significant increase in cell quantity was observed. At 2–10 h, the strain entered the exponential phase, and the cell quantity showed geometric growth. At 10–16 h, the increase in cell quantity slowed down, entering the stationary phase. After 16 h, the strain began to deteriorate.

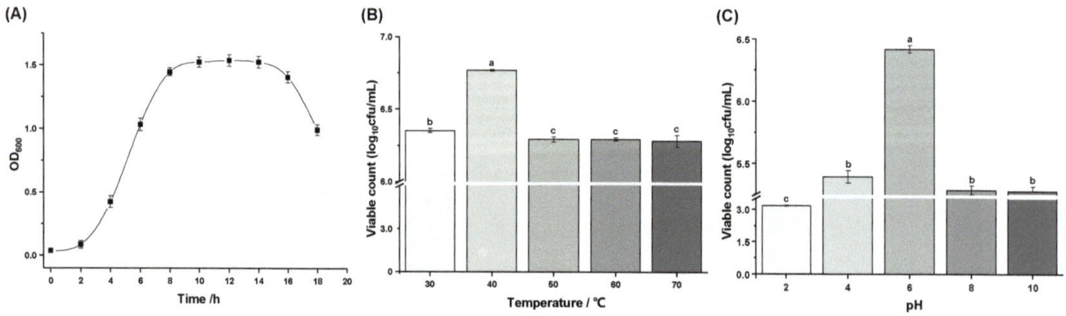

Figure 2. Growth characteristics of D-2. (**A**) Growth curve. (**B**) Living D-2 concentration under different temperatures. (**C**) Living D-2 concentration at different pH values. Different lowercase letters indicate significant differences among treatment groups ($p < 0.05$).

According to the growth curve, the bacterial liquid in the stationary phase and the equivalent live bacteria ($\times 10^5$ CFU/mL) were treated at different temperatures and pH conditions for 12 h. D-2 could survive at different temperatures, and the pH (Figure 2B,C) and viable cell count significantly increased at 40 °C ($p < 0.01$), but at pH = 2, the viable cell quantity was 1.45 × 10^3 CFU/mL, representing a significant reduction ($p < 0.01$), and at pH = 6, the viable cell quantity was 2.63 × 10^6 CFU/mL, representing a significant increase ($p < 0.01$). The results showed that D-2 grows rapidly and is very adaptable at different pHs and temperatures.

3.3. Bacteriostatic Effect of D-2

The bacteriostatic tests showed that D-2 inhibited the pathogenic and rot-causing microorganisms to different degrees but had no inhibitory effect against the LAB, as represented by *L. reuteri*, *E. faecium*, and *L. casei* (Table 2). The inhibition ability of D-2 on the above microorganisms followed this order: *F. graminearum* > *S. aureus* > *S. enteritidis* > *F. oxysporum* > *E. coli* > *A. niger* ≥ *S. cerevisiae*. The bacteriostatic effect is shown in Figure 3.

Table 2. Bacteriostatic spectrum of D-2.

Indicator	Inhibitory Circle Radius (mm)	Indicator	Inhibitory Circle Radius (mm)
E. coli	10.66 ± 0.27 [e]	*F. oxysporum*	11.74 ± 0.26 [d]
S. aureus	17.56 ± 0.54 [b]	*F. graminearum*	19.60 ± 0.15 [a]
S. enteritidis	12.22 ± 0.50 [c]	*L. reuteri*	−
A. niger	9.58 ± 0.28 [f]	*E. faecium*	−
S. cerevisiae	9.08 ± 0.52 [f]	*L. casei*	−

Note: "−" negative; no bacteriostasis. Different lowercase letters indicate significant differences among treatment groups ($p < 0.05$).

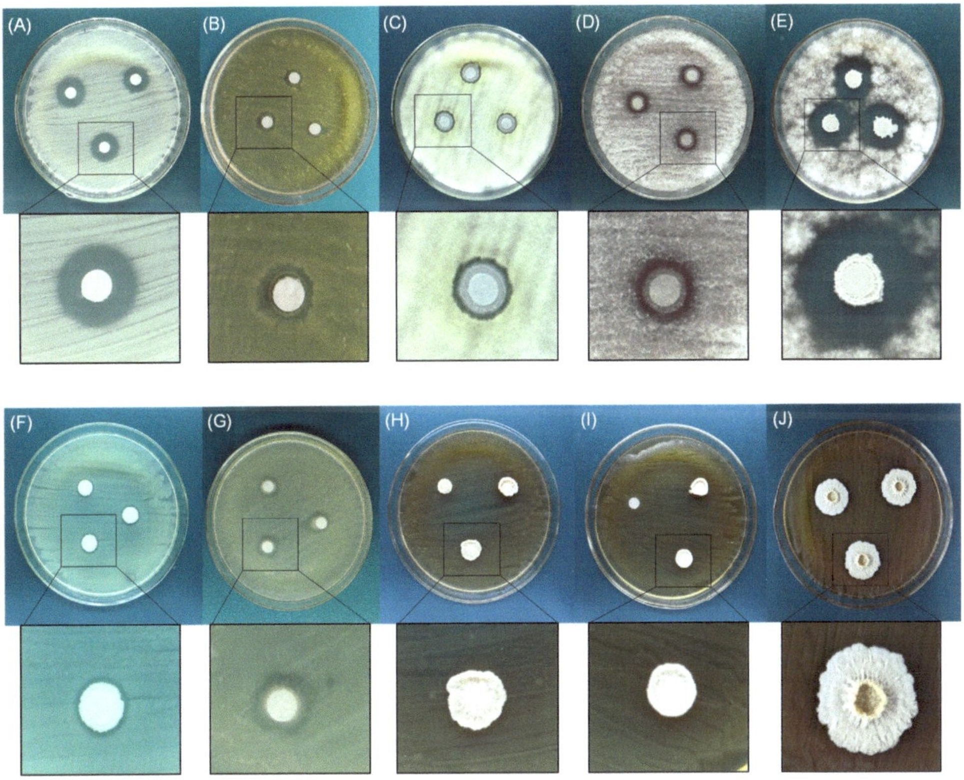

Figure 3. Bacteriostatic effect of D-2. (**A**) *S. aureus*; (**B**) *S. cerevisiae*; (**C**) *A. niger*; (**D**) *F. oxysporum*; (**E**) *F. graminearum*; (**F**) *E. coli*; (**G**) *S. enteritidis*; (**H**) *L. reuteri*; (**I**) *E. faecium*; (**J**) *L. casei*.

3.4. Tolerance of D-2 to an Acidic Environment Where pH = 4

The D-2 survival rates are shown in Table 3. At a pH of four, the survival rate of D-2 within 24 h was 100%, and the viable counts increased significantly ($p < 0.05$). At 36–48 h, the viable counts decreased, but the results were not significantly different from those observed at 0 h ($p > 0.05$). The survival rates were over 97%, indicating that the strain exhibited strong lactic acid tolerance. Combined with the growth curve, D-2 entered its decline phase after 16 h. However, at pH 4, the viable counts were still increasing at 24 h, and the difference between the viable counts at 24 and 12 h was not significant ($p > 0.05$). Thus, the stationary phase of D-2 was speculated to be extended under this condition. After 36 h, the nutrients were depleted in the medium, the bacteria began to decline in number, and a decreased viable count was observed. These values slightly increased at 48 h, but with no significant difference ($p > 0.05$).

Table 3. Survival rate of D-2 at pH 4.

Items	Time				
	0 h	**12 h**	**24 h**	**36 h**	**48 h**
Viable count (log_{10} CFU/mL)	5.49 ± 0.02 ^{b}	5.52 ± 0.03 ^{a}	5.53 ± 0.01 ^{a}	5.46 ± 0.01 ^{b}	5.44 ± 0.07 ^{b}
Survival rate (%)	100	100	100	97.28	98.29
Increasing rate (%)	0	+12.93	+16.33	−2.72	−1.70

Note: Different lowercase letters indicate significant differences among treatment groups ($p < 0.05$).

3.5. D-2 Resistance to Eight Antibiotics

The results of the antibiotic sensitivity test of D-2 are shown in Table 4. EFSA specifies the eight antibiotics and MIC values that must be tested for *Bacillus* spp. It can be seen that for D-2, all of the MIC values are lower than the limited values. D-2 is sensitive to eight antibiotics and meets EFSA's standards for live bacteria intended to be used as a feed additive.

Table 4. Antibiotic sensitivity test of D-2.

Antibiotics	MIC (mg/L)	Criterion (mg/L)	Susceptibility
Vancomycin	1	≤4	S
Gentamicin	1	≤4	S
Erythromycin	0.5	≤4	S
Clindamycin	0.5	≤4	S
Tetracycline	2	≤8	S
Chloramphenicol	4	≤8	S
Kanamycin	2	≤8	S
Streptomycin	4	≤8	S

Note: S, susceptible.

3.6. Effect of D-2 on Fermentation Quality of Whole-Plant Corn

Using D-2 as the silage strain, the whole-plant corn was ensiled for 60 days, and the fermentative results are shown in Table 5. Compared with the control group, the pH, the number of yeasts and molds, and the NH_4^+-N/TN ratio of the D-2 group were lower. And the contents of WSC and acetic acid in the D-2 group were significantly increased ($p < 0.05$). There were no effects on the LAB, TN, DM, NDF, ADF, lactic acid, and propionic acid of whole-plant maize silage when using D-2 or not ($p > 0.05$).

Table 5. Microorganism counts, chemical composition, fermentative characteristics, and quality of whole-plant corn at 60 days after fermentation.

Parameter	Control	D-2
Number of LABs (log_{10} CFU/g FW)	8.24 ± 0.01	8.20 ± 0.03
Number of yeasts and molds (log_{10} CFU/g FW)	4.80 ± 0.02 [a]	2.65 ± 0.01 [b]
NH_4^+-N (% FW)	0.32 ± 0.02	0.28 ± 0.03
TN (% FW)	2.30 ± 0.05	2.33 ± 0.02
NH_4^+-N/TN (%)	13.64 ± 0.13 [a]	12.22 ± 0.20 [b]
DM (% FW)	30.50 ± 0.50	31.05 ± 0.62
NDF (% DM)	34.22 ± 0.07	34.24 ± 0.10
ADF (% DM)	24.32 ± 0.02	24.27 ± 0.06
WSC (% DM)	3.38 ± 0.03 [b]	4.61 ± 0.02 [a]
pH	3.65 ± 0.01 [a]	3.50 ± 0.02 [b]
Lactic acid (% DM)	8.36 ± 0.03	8.21 ± 0.03
Acetic acid (% DM)	1.74 ± 0.01 [b]	1.96 ± 0.03 [a]
Propionic acid ($\times 10^{-2}$% DM)	0.82 ± 0.01	0.87 ± 0.01
Butyric acid (% DM)	ND	ND

Note: FW, fresh weight; DM, dry matter; TN, total nitrogen; NDF, neutral detergent fiber assayed with a heat-stable amylase expressed inclusion of residual ash; ADF, acid detergent fiber expressed inclusion of residual ash; WSC, water-soluble carbohydrate. Different lowercase letters indicate significant differences among treatment groups ($p < 0.05$). ND means not detected.

3.7. Effect of D-2 on the Aerobic Stability of Whole-Plant Corn Silage

The effect of D-2 on the aerobic stability and quality of silage is shown in Table 6. The results showed that, after 10 days of oxygen exposure, the numbers of LAB, TN, and DM in the D-2 group were significantly higher than those in the control group ($p < 0.05$). The number of yeasts and molds, the NH_4^+-N quantity, the NH_4^+-N/TN ratio, and the pH were significantly lower than those in the control group ($p < 0.05$). The results showed that D-2 could effectively prolong the aerobic stability of silage ($p < 0.05$).

Table 6. Aerobic stability of whole-plant corn silage after 10 days of oxygen exposure.

Parameter	Control	D-2
Number of LABs (log_{10} CFU/g FW)	6.25 ± 0.27 [b]	7.82 ± 0.02 [a]
Number of yeasts and molds (log_{10} CFU/g FW)	6.18 ± 0.03 [a]	4.60 ± 0.11 [b]
NH_4^+-N (% FW)	0.45 ± 0.05 [a]	0.30 ± 0.06 [b]
TN (% FW)	2.22 ± 0.06 [b]	2.28 ± 0.02 [a]
NH_4^+-N/TN (%)	20.36 ± 0.22 [a]	13.23 ± 0.19 [b]
DM (% FW)	27.48 ± 0.80 [b]	29.53 ± 0.72 [a]
NDF (% DM)	34.55 ± 0.03	33.87 ± 0.16
ADF (% DM)	25.08 ± 0.15	24.79 ± 0.04
pH	6.60 ± 0.01 [a]	4.26 ± 0.04 [b]

Note: FW, fresh weight; DM, dry matter; TN, total nitrogen; NDF, neutral detergent fiber assayed with heat-stable amylase expressed inclusion of residual ash; ADF, acid detergent fiber expressed inclusion of residual ash. Different lowercase letters indicate significant differences among treatment groups ($p < 0.05$).

3.8. Effect of D-2 on In Vitro Digestion and Methane Production of Whole-Plant Corn Silage

After 24 h of in vitro incubation, the sum of acetic and butyric acid did not differ significantly (Table 7). However, compared with the control group, methane production in the D-2 group was significantly decreased ($p < 0.05$), and the IVDMD and propionic acid totals for the D-2 group were higher than those of the control group.

Table 7. IVDMD, methane production, and volatile fatty acid concentration at 24 h incubation of whole-plant corn silage.

Parameter	Control	D-2
IVDMD (%)	61.33 ±1.52 [b]	63.15 ± 1.70 [a]
CH_4 (L/kg IVDMD)	12.88 ± 0.57 [a]	10.13 ± 0.62 [b]
Acetic acid (mmol/L)	23.13 ± 0.30	22.61 ± 0.52
Propionic acid (mmol/L)	18.42 ± 0.34 [b]	22.56 ± 0.03 [a]
Butyric acid (mmol/L)	5.85 ± 0.15	5.60 ± 0.20

Note: IVDMD, in vitro dry matter digestibility. Different lowercase letters indicate significant differences among treatment groups ($p < 0.05$).

4. Discussion

The D-2 strain, which was isolated from rotten silage samples, was revealed to be *Bacillus velezensis* via morphological, physiological, biochemical, and molecular characterization. *Bacillus* produces many metabolites with antibiotic activity during growth and multiplication [31]. Other studies showed that *B. velezensis* has a good antagonistic effect against pathogenic and rot-causing microorganisms, such as *E. coli, S. aureus*, and *F. graminearum*, and has the potential to promote animal growth and disease resistance [32,33]. These previous findings agree with the results of this study. In addition, the D-2 strain screened in our study shows some inhibitory effects against *A. niger, S. cerevisiae*, and *S. enteritidis*, but not against *L. reuteri, E. faecium*, and *L. casei*, providing a foundation for the use of this strain during ensiling.

During silage fermentation, the silage central temperature can reach up to 70 °C [34], and the pH of silage systems can reduce to four during the rapid fermentation stage [3]. In this study, D-2 was screened under anoxic conditions, and it can survive in an environment with a pH ranging from two to ten and tolerate 70 °C. D-2 can tolerate different pH values, temperatures, and lactic acid contents, has good antibacterial spectrum characteristics, and is sensitive to the eight tested antibiotics. It also provides a favorable environment for the proliferation of LABs by inhibiting the growth of pathogenic and rot-causing microorganisms consuming oxygen and improving the fermentation quality. D-2 is expected to be successfully applied to corn silage fermentation. When silage is exposed to air, oxygen triggers the activity of microorganisms such as yeasts and molds. In addition, the loss of volatile fatty acids such as acetic and butyric acids in silage results in a decrease in organic acid concentration, diminishing the inhibitory effect of organic acids on yeasts and molds. With prolonged exposure, yeasts, molds, and other spoilage bacteria gradually become the dominant microbes [6,7]. D-2 can inhibit yeasts and molds, but not LAB. Thus, the D-2 group has a significantly lower number of yeasts and molds than the control group does.

NH_4^+-N exists in silage as free ammonia (NH_3) or ammonium salt (NH_4^+), and their composition ratio depends on the pH; they are present in a free state when the pH is high, or conversely, in an ammonium salt form. In high-quality silage, NH_4^+-N is present in the form of ammonia salt, but when the silage is seriously decayed, it has a pungent ammonia smell. NH_4^+-N in silage is produced by the decomposition of nitrogen-containing organic substances by plant enzymes and aerobic bacteria [35]. Therefore, the content of NH_4^+-N can be used as an indicator of the degree of protein decomposition. The NH_4^+-N/TN ratio reflects the degree of protein and amino acid decomposition in silage, and a large ratio indicates increased protein decomposition and the poor quality of silage. In this study, D-2 inhibited the proliferation of spoilage microorganisms, resulting in a lower NH_4^+-N content and NH_4^+-N/TN ratio in the D-2 group compared to the control group. This finding shows that D-2 has a protective effect on the aerobic stability of silage and can delay proteolysis via aerobic exposure and extend the shelf-life of silage under aerobic exposure. Water activity (WA) is an important indicator of bacterial activity [36]. In this study, the effect of WA is excluded as a plastic film covered the samples, which were placed in an environment with a constant temperature and humidity. Given the decomposition microorganisms, the DM of silage is reduced, which increases the relative proportion of

NDF and ADF in the treatment group, but no difference is observed among these groups ($p > 0.05$). The main feature of silage spoilage is protein degradation, with a minimal effect on the NDF and ADF.

The low pH of silage can inhibit the activity of undesirable microbes and stop the loss of organic matter in silage fermentation [37]. In this study, compared with the control group, the D-2 group had a lower pH in the whole-plant corn silage, and D-2 can also inhibit the activity of undesirable microbes in silage, so it is speculated that the D-2 group loses less organic matter and more DM. In the rumen, a small amount of hydrogen (1 $\mu mol \cdot L^{-1}$) is produced during metabolite conversion, and both propionic acid and methane production need to consume hydrogen, causing competition between the two [38]. The content of propionic acid in the D-2 group was higher than that in the control group. It is speculated that after fermentation with D-2, the propionic acid production pathway was enhanced, more hydrogen was utilized by this pathway, the concentration of hydrogen in the rumen was reduced, and methane production was inhibited [39].

D-2 demonstrated many application advantages. However, it was conducted under laboratory conditions, which minimized external interference to the greatest extent possible. In actual production, the natural temperature and larger silos may influence the results. In addition, climatic variations, the chemical composition of plants, and bacterial flora can affect the results of inoculants. Therefore, to effectively apply these findings of D-2 in practical production settings, further verification is required through repeated experiments conducted for two consecutive years or in two different locations to improve the reliability of the data obtained.

5. Conclusions

In this study, a strain of *Bacillus velezensis* (D-2) with anti-microbial activity was successfully isolated from silage. D-2 could selectively inhibit pathogenic and rot-causing microorganisms that cause silage spoilage and could adapt to the silage production conditions. Inoculation with D-2 could increase aerobic stability, maintain the nutritional quality of silage, and increase methane production. Thus, D-2, which is a potential microbial additive that slows down the spoilage of silage, may effectively protect the aerobic stability of silage and, when fed to ruminants, may reduce their emission of methane gas compared to uninoculated corn silage as feed.

Author Contributions: C.Z.: conceptualization; methodology; writing—original draft. Z.Z.: investigation; writing—original draft. M.Z. and T.Z.: software; validation; formal analysis. Y.W.: formal analysis; data curation. F.W.: writing—review and editing. Y.Z.: resources; writing—review and editing. L.C.: conceptualization; funding acquisition; supervision; writing—review and editing. All authors have read and agreed to the published version of the manuscript.

Funding: This study was supported by the National Natural Science Foundation of China (grant number: 32172769), the National Key R&D Program (2021YFD1301004), and the Science and Technology Major Project of Yunnan Province (grant number: 202202AE090032).

Institutional Review Board Statement: Not applicable.

Data Availability Statement: All data generated or analyzed during this study are included in this published article.

Acknowledgments: We would like to thank Zuojun Liu, Dafeng Xv, and Qiang Yin (Anhui Academy of Agricultural Sciences, CN) and the Anhui Province Engineering Laboratory for Animal Food Quality and Bio-Safety for providing support and assistance.

Conflicts of Interest: The authors declare the following financial interests/personal relationships, which may be considered potential competing interests: Lijuan Chen, Chen Zhang, Yunhua Zhang, Gangqing Ying, Gangqin Shu, Yili Wang, and Xi Li are the inventors of a Chinese patent (CN CN115161234B, 5 September 2023) related to the procedure for the isolation of *Bacillus velezensis* from rotten silage described in this manuscript.

Abbreviations

ADF: acid detergent fiber; *A. niger, Aspergillus niger;* CH_4, methane; CFU, colony-forming unit; CLSI, Clinical and Laboratory Standards Institute; DM, dry matter; *E. coli, Escherichia coli;* E. faecium, Enterococcus faecium; EFSA, European Food Safety Agency; *F. oxysporum, Fusarium oxysporum; F. graminearum,* Fusarium graminearum; FW, fresh weight; IVDMD, in vitro dry matter digestibility; LAB, lactic acid bacteria; LB, Luria-Bertani (medium); *L. casei, Lactobacillus casei; L. reuteri, Lactobacillus reuteri;* MIC, minimum inhibitory concentration; MRS, DeMan, Rogosa, and Sharpe (Medium); MR/VP test, Methyl Red/Voges-Proskaurer test; NA, nutrient agar; ND, not detected; NDF, neutral detergent fiber; NH_4^+-N, ammoniacal nitrogen; PBS, phosphate buffered saline; PCR, polymerase chain reaction; PDA, potato dextrose agar (medium); pH, hydrogen ion concentration; rDNA, ribosomal deoxyribonucleic acid; rRNA, ribosomal ribonucleic acid; *S. aureus, Staphylococcus aureus; S. cerevisiae, Saccharomyces cerevisiae; S. enteritidis, Salmonella enteritidis;* TN, total nitrogen; WA, water activity; WSC, water-soluble carbohydrate.

Appendix A

```
1    TATGTAGAGC ATGCTAGTCG AGCGGACAGA TGGGAGCTTG CTCCCTGATG TTAGCGGCGG
61   ACGGGTGAGT AACACGTGGG TAACCTGCCT GTAAGACTGG GATAACTCCG GGAAACCGGG
121  GCTAATACCG GATGGTTGTT TGAACCGCAT GGTTCAGACA TAAAAGGTGG CTTCGGCTAC
181  CACTTACAGA TGGACCCGCG GCGCATTAGC TAGTTGGTGA GGTAACGGCT CACCAAGGCG
241  ACGATGCGTA GCCGACCTGA GAGGGTGATC GGCCACACTG GGACTGAGAC ACGGCCCAGA
301  CTCCTACGGG AGGCAGCAGT AGGGAATCTT CCGCAATGGA CGAAAGTCTG ACGGAGCAAC
361  GCCGCGTGAG TGATGAAGGT TTTCGGATCG TAAAGCTCTG TTGTTAGGGA AGAACAAGTG
421  CCGTTCAAAT AGGGCGGCAC CTTGACGGTA CCTAACCAGA AAGCCACGGC TAACTACGTG
481  CCAGCAGCCG CGGTAATACG TAGGTGGCAA GCGTTGTCCG GAATTATTGG GCGTAAAGGG
541  CTCGCAGGCG GTTTCTTAAG TCTGATGTGA AAGCCCCCGG CTCAACCGGG GAGGGTCATT
601  GGAAACTGGG GAACTTGAGT GCAGAAGAGG AGAGTGGAAT TCCACGTGTA GCGGTGAAAT
661  GCGTAGAGAT GTGGAGGAAC ACCAGTGGCG AAGGCGACTC TCTGGTCTGT AACTGACGCT
721  GAGGAGCGAA AGCGTGGGGA GCGAACAGGA TTAGATACCC TGGTAGTCCA CGCCGTAAAC
781  GATGAGTGCT AAGTGTTAGG GGGTTTCCGC CCCTTAGTGC TGCAGCTAAC GCATTAAGCA
841  CTCCGCCTGG GGAGTACGGT CGCAAGACTG AAACTCAAAG GAATTGACGG GGGCCCGCAC
901  AAGCGGTGGA GCATGTGGTT TAATTCGAAG CAACGCGAAG AACCTTACCA GGTCTTGACA
961  TCCTCTGACA ATCCTAGAGA TAGGACGTCC CCTTCGGGGG CAGAGTGACA GGTGGTGCAT
1021 GGTTGTCGTC AGCTCGTGTC GTGAGATGTT GGGTTAAGTC CCGCAACGAG CGCAACCCTT
1081 GATCTTAGTT GCCAGCATTC AGTTGGGCAC TCTAAGGTGA CTGCCGGTGA CAAACCGGAG
1141 GAAGGTGGGG ATGACGTCAA ATCATCATGC CCCTTATGAC CTGGGCTACA CACGTGCTAC
1201 AATGGACAGA ACAAAGGGCA GCGAAACCGC GAGGTTAAGC CAATCCCACA AATCTGTTCT
1261 CAGTTCGGAT CGCAGTCTGC AACTCGACTG CGTGAAGCTG GAATCGCTAG TAATCGCGGA
1321 TCAGCATGCC GCGGTGAATA CGTTCCCGGG CCTTGTACAC ACCGCCCGTC ACACCACGAG
1381 AGTTTGTAAC ACCCGAAGTC GGTGAGGTAA CCTAGAGGAG CCAACCACCG CGTGAGTG
```

Figure A1. Sequencing results of 16S rDNA from the strain.

Table A1. The chemical composition and count of microorganisms in maize green forage before ensilage.

Whole-Plant Crop Maize	
Number of LAB (log_{10} CFU/g FW)	7.84 ± 0.01
Number of yeasts and molds (log_{10} CFU/g FW)	6.80 ± 0.02
DM (% FW)	31.42 ± 0.38
TN (% FW)	2.65 ± 0.05
NDF (% DM)	33.51 ± 0.37
ADF (% DM)	23.81 ± 0.14
pH	6.4 ± 0.01

Note: FW, fresh weight; DM, dry matter; TN, total nitrogen; NDF, neutral detergent fiber assayed with heat-stable amylase expressed inclusion of residual ash; ADF, acid detergent fiber expressed inclusion of residual ash.

References

1. Okoye, C.O.; Wang, Y.; Gao, L.; Wu, Y.; Li, X.; Sun, J.; Jiang, J. The performance of lactic acid bacteria in silage production: A review of modern biotechnology for silage improvement. *Microbiol. Res.* **2023**, *266*, 127212. [CrossRef] [PubMed]
2. Guo, X.; Xu, D.; Li, F.; Bai, J.; Su, R. Current approaches on the roles of lactic acid bacteria in crop silage. *Microb. Biotechnol.* **2023**, *16*, 67–87. [CrossRef] [PubMed]
3. Weinberg, Z.G. Preservation of forage crops by solid-state lactic acid fermentation-ensiling. In *Current Developments in Solid-State Fermentation*; Springer: New York, NY, USA, 2008; pp. 443–467.
4. Dhundale, V.; Hemke, V.; Desai, D.; Dhundale, P. Evaluation and exploration of lactic acid bacteria for preservation and extending the shelf life of fruit. *Int. J. Fruit. Sci.* **2018**, *4*, 355–368. [CrossRef]
5. Huang, Q.L.; Zhou, X.; Wang, J.; Zhang, J.; Cao, Y. Effects of lactic acid bacteria, molasses and inorganic acid supplementations on fermentation quality and in vitro dry matter disappearance rate of leymus chinensis silage. *Chin. J. Anim. Nutr.* **2021**, *33*, 420–427.
6. Mcdonald, P.; Henderson, A.R.; Heron, S.J.E. The Biochemistry of Silage, 2nd ed.Chalcombe Publications: London, UK, 1991.
7. Muck, R.E. Silage microbiology and its control through additives. *Rev. Bras. Zootec.* **2010**, *39*, 183–191. [CrossRef]
8. Xu, D.M.; Zhang, P.; Ke, W.C.; Guo, X.S. Research process in silage microorganism and its effects on silage quality. *Acta Agrestia Sin.* **2017**, *25*, 460–465.
9. Huang, F.; Zhang, L.; Zhou, B.; Zou, C. Research process in silage microorganism and its effect on silage aerobic stability. *Chin. J. Anim. Nutr.* **2019**, *31*, 82–89.
10. Cavallarin, L.; Tabacco, E.; Antoniazzi, S.; Borreani, G. Aflatoxin accumulation in whole crop maize silage as a result of aerobic exposure. *J. Sci. Food Agric.* **2011**, *91*, 2419–2425. [CrossRef] [PubMed]
11. Ogunade, I.M.; Martinez-Tuppia, C.; Queiroz, O.C.M.; Jiang, Y.; Drouin, P.; Wu, F.; Vyas, D.; Adesogan, A.T. Silage review: Mycotoxins in silage: Occurrence, effects, prevention, and mitigation. *J. Dairy. Sci.* **2018**, *101*, 4034–4059. [CrossRef]
12. Muck, R.E.; Nadeau, E.M.G.; Mcallister, T.A.; Contreras-Govea, F.E.; Santos, M.C.; Jr, L.K. Silage review: Recent advances and future uses of silage additives. *J. Dairy. Sci.* **2018**, *101*, 3980–4000. [CrossRef]
13. Queiro, O.C.M.; Ogunade, I.M.; Weinberg, Z.; Adesogan, A.T. Silage review: Foodborne pathogens in silage and their mitigation by silage additives. *J. Dairy. Sci.* **2018**, *101*, 4132–4142. [CrossRef]
14. Wang, Y.; Ying, G.; Zhang, Z.; Tang, Y.; Zhang, Y.; Chen, L. Bacillus velezensis promotes the proliferation of lactic acid bacteria and influences the fermentation quality of whole-plant corn silage. *Front. Plant Sci.* **2024**, *15*, 1285582. [CrossRef] [PubMed]
15. Kristensen, N.B.; Sloth, K.H.; Højberg, O.; Spliid, N.H.; Jensen, C.; Thøgersen, R. Effects of microbial inoculants on corn silage fermentation, microbial contents, aerobic stability, and milk production under field conditions. *J. Dairy. Sci.* **2010**, *8*, 3764–3774. [CrossRef]
16. Soares, A.; Pestel, C.M.; Leysour, D.R.F.; Bourgoin, G.; Boyer, S.; Caron, F. Area of technical uncertainty for susceptibility testing of amoxicillin/clavulanate against *Escherichia coli*: Analysis of automated system, etest and disk diffusion methods compared to the broth microdilution reference. *Clin. Microbiol. Infect.* **2020**, *26*, 1685.e1–1685.e6.
17. Buchanan, R.E.; Gibbons, N.E. *Bergey's Manual of Determinative Bacteriology*, 9th ed.; Science Press: Beijing, China, 1995.
18. Zhang, C.; Yu, X.; Xu, H.; Cui, G.; Chen, L. Action of *Bacillus natto* 16 on deoxynivalenol (DON) from wheat flour. *J. Appl. Microbiol.* **2021**, *131*, 2317–2324. [CrossRef]
19. EFSA Panel on Additives and Products or Substances used in Animal Feed (FEEDAP). Guidance on the as-sessment of bacterial susceptibility to antimicrobials of human and veterinary importance. *EFSA J.* **2012**, *10*, 2740.
20. Clinical and Laboratory Standards Institute (CLSI). *Methods for Dilution Antimicrobial Susceptibility Tests for Bacteria that Grow Aerobically*, 11th ed.; CLSI Standard M07; Clinical and Laboratory Standards Institute: Wayne, PA, USA, 2018.
21. Clinical and Laboratory Standard Institute. *Methods for Antimicrobial Dilution and Disk Susceptibility Testing of Infrequently Isolated or Fastidious Bacteria*, 3rd ed.; CLSI Guideline M45; Clinical and Laboratory Standards Institute: Wayne, PA, USA, 2016.
22. Wang, J.; Zhang, Y.D.; Zheng, N.; Wang, J.; Zhang, P. Research rrogress in sensory evaluation of sliage. *China Dairy Cattle* **2019**, *1*, 1–3.
23. Wang, Y.; Wang, C.; Zhou, W.; Yang, F.; Chen, X.; Zhang, Q. Effects of wilting and *Lactobacillus plantarum* addition on the fermentation quality and microbial community of *Moringa oleifera* leaf silage. *Front. Microbiol.* **2018**, *9*, 1817. [CrossRef]
24. Carlos, E.F.T.; Renato, S.F. Sensory evaluation of winter cereal silage. *J. Chem. Chem. Eng.* **2017**, *11*, 102–106.
25. Cai, Y.; Benno, Y.; Ogawa, M.; Kumai, S. Effect of applying lactic acid bacteria isolated from forage crops on fermentation characteristics and aerobic deterioration of silage. *J. Dairy. Sci.* **1999**, *82*, 520–526. [CrossRef]
26. Association of Official Analytical Chemists (AOAC). *Official Methods of Analysis*; Association of Official Analytical Chemists: Gaithersburg, MD, USA, 2002.
27. Broderick, G.A.; Kang, J.H. Automated simultaneous determination of ammonia and total amino acids in ruminal fluid and in vitro media. *J. Dairy. Sci.* **1980**, *63*, 64–75. [CrossRef] [PubMed]
28. Bai, J.; Franco, M.; Ding, Z.; Hao, L.; Ke, W.; Wang, M.; Xie, D.; Li, Z.; Zhang, Y.; Ai, L.; et al. Effect of *Bacillus amyloliquefaciens* and *Bacillus subtilis* on fermentation, dynamics of bacterial community and their functional shifts of whole-plant corn silage. *J. Anim. Sci. Biotechnol.* **2022**, *13*, 864–877. [CrossRef] [PubMed]
29. Van, S.P.J.; Robertson, J.B.; Lewis, B.A. Methods for dietary fiber, neutral detergent fiber, and nonstarch polysaccharides in relation to animal nutrition. *J. Dairy. Sci.* **1991**, *74*, 3583–3597.

30. Khota, W.; Pholsen, S.; Higgs, D.; Cai, Y. Fermentation quality and in vitro methane production of sorghum silage prepared with cellulase and lactic acid bacteria. *Asian-Australas. J. Anim. Sci.* **2017**, *30*, 1568–1574. [CrossRef] [PubMed]
31. Zhang, D.F.; Gao, Y.X.; Wang, Y.J.; Liu, C.; Shi, C.B. Advances in taxonomy, antagonistic function and application of *Bacillus velezensis*. *Microbiol. China* **2020**, *47*, 3634–3649.
32. Li, X.; Wang, X.; Shi, X.; Wang, B.; Li, M.; Wang, Q.; Zhang, S. Antifungal effect of volatile organic compounds from *Bacillus velezensis* CT32 against *Verticillium dahliae* and *Fusarium oxysporum*. *Processes* **2020**, *8*, 1674. [CrossRef]
33. Liu, S.; Zhang, B.; Xiang, D.; Zhao, Z.; Chang, Y.; Sha, Q.; Zhao, Y. Effects of *Bacillus velezensis* on growth performance, fecal microbiota and metabolites in pigs. *Chin. J. Anim. Nutr.* **2020**, *32*, 5622–5635.
34. Zhang, L.; Liu, D.; Shao, T. The application of plastic sheet in silage and silage making. *China Herbiv. Sci.* **2007**, *1*, 55–57.
35. Yan, J.; Wang, W.; Gao, Y.; Pan, Z.; Mu, S. Rapid determintion of ammoniacal nitrogen in silage by automatic kjeldahl apparatus. *Cereal Feed Ind.* **2009**, *4*, 47–48.
36. Lv, C.; Jin, J.; Wang, P.; Dai, X.; Liu, Y.; Zheng, M.; Xing, F. Interaction of water activity and temperature on the growth, gene expression and aflatoxin production by *Aspergillus flavus* on paddy and polished rice. *Food Chem.* **2019**, *293*, 472–478. [CrossRef]
37. Jiang, F.G.; Cheng, H.J.; Liu, D.; Wei, C.; An, W.J.; Wang, Y.F.; Sun, H.T.; Song, E.L. Treatment of whole-plant corn silage with lactic acid bacteria and organic acid enhances quality by elevating acid content, reducing pH, and inhibiting undesirable microorganisms. *Front. Microbiol.* **2020**, *11*, 593088. [CrossRef] [PubMed]
38. Guo, Y.Q.; Hu, W.L.; Liu, J.X. Methanogens and manipulation of methane production in the rumen. *Acta Microbiol. Sin.* **2005**, *1*, 145–148.
39. Asanuma, N.; Hino, T. Prevention of rumen acidosis and suppression of ruminal methanogenesis by augmentation of lactate utilization. *Nihon Chikusan Gakkaiho* **2004**, *75*, 543–550. [CrossRef]

agriculture

MDPI

Article

Nutritional Value Evaluation of Corn Silage from Different Mesoregions of Southern Brazil

Mikael Neumann [1], Ellen Baldissera [1], Livia Alessi Ienke [1], André Martins de Souza [2,*], Paulo Eduardo Piemontez de Oliveira [3] and Valter Harry Bumbieris Junior [2]

1 Department of Veterinary Medicine, Midwestern Parana State University, Guarapuava 85040-167, Brazil; neumann.mikael@hotmail.com (M.N.); ellen_baldissera@outlook.com (E.B.); liviaienke@gmail.com (L.A.I.)

2 Departament of Zootechnics, State University of Londrina, Londrina 86057-970, Brazil; jrbumbieris@uel.br

3 Veterinarian, Collaborator of the NUPRAN Group, Guarapuava 85040-167, Brazil; pauloep@agraria.com.br

* Correspondence: andre.martins.souza@uel.br

Abstract: Corn silage is widely used in livestock farming; however, its quality is easily altered, and one of the factors that has a high influence in this regard is the region of production. The objective was to evaluate the chemical–bromatological composition of 498 samples of corn silage from mesoregions in Southern Brazil during the 2022/2023 summer harvest. The following were studied in relation to our objective: nutritional composition, dry matter, mineral matter, ether extract, starch, crude protein, neutral detergent fiber, acid detergent fiber, acid detergent lignin, total digestible nutrients, total carbohydrates, and fractions of carbohydrates. The silages from Central South-PR had higher levels of starch and ether extract (30.68% ± 6.24% and 3.41% ± 0.92%, respectively), whereas in West-SC, the silages had higher levels in the A + B1 fraction of carbohydrates (49.59% ± 6.34%). Silages in North-PR had higher concentrations of neutral detergent fiber and acid detergent fiber (49.86% ± 5.92% and 29.70% ± 4.38%, respectively), while in Northwest-RS and West-PR, silages had higher levels of the B2 carbohydrate fractions (46.25% ± 1.98% and 44.55% ± 3.84%, respectively). The nutritional composition differences presented were due to the variables of each mesoregion, interfering in the scenario of formulating diets and animal nutrition.

Keywords: chemical analysis; climate variations; dry matter

Citation: Neumann, M.; Baldissera, E.; Alessi Ienke, L.; Martins de Souza, A.; Piemontez de Oliveira, P.E.; Harry Bumbieris Junior, V. Nutritional Value Evaluation of Corn Silage from Different Mesoregions of Southern Brazil. *Agriculture* **2024**, *14*, 1055. https://doi.org/10.3390/agriculture14071055

Academic Editor: Siran Wang

Received: 21 May 2024
Revised: 21 June 2024
Accepted: 28 June 2024
Published: 30 June 2024

1. Introduction

Extensive farming is still characterized as the main form of cattle breeding in Brazil, as it is still thought to be an economical form due to the low investment made. However, this system has low production rates, increases the production cycle, is unprofitable, and has management problems [1].

In contrast, intensive farming practices have increased considerably, aiming to obtain greater profitability, produce better-quality animals, produce more milk, and improve production rates [2]. With this growth, the demand for preserved feeds has also increased. According to a survey conducted by Silvestre and Millen [3], corn silage is the most commonly used roughage in Brazilian feedlots (beef and dairy livestock).

Due to its high productivity, good nutritional value, high energy content, ease of fermentation inside the silo and good acceptance of most species of ruminants, corn (*Zea mays* L.) is the most used crop for making silage in Brazil [4–6]. However, environmental conditions differ considerably between cultivation regions and directly influence the productive and qualitative behavior of this crop, resulting in changes in the yield and quality of the silage produced [7].

The production of corn silage with favorable characteristics depends on several controllable and non-controllable factors; therefore, each step of the process must be carried out precisely to produce high-quality forage [6]. For precise nutrition, diet formulation must be based on accurate and reliable data regarding the forage in question, the time of ensiling and supply to animals, and fundamental factors for improving diet formulation and animal production [8].

The southern region, for example, is characterized by a subtropical climate or Cfa (Köppen classification) [9], which has well-defined seasons and a hot and humid climate that is beneficial for the development of the corn crop [10].

Therefore, the present study aimed to evaluate the chemical composition of corn silage from different mesoregions of Southern Brazil, namely the Central South, North, West, and Southwest of Paraná, West of Santa Catarina, and Northwest of Rio Grande do Sul.

2. Materials and Methods

2.1. Collection Locations

The silages used in this study were collected from different mesoregions of Southern Brazil, divided between the states of Paraná, Santa Catarina, and Rio Grande do Sul, as shown in Figure 1.

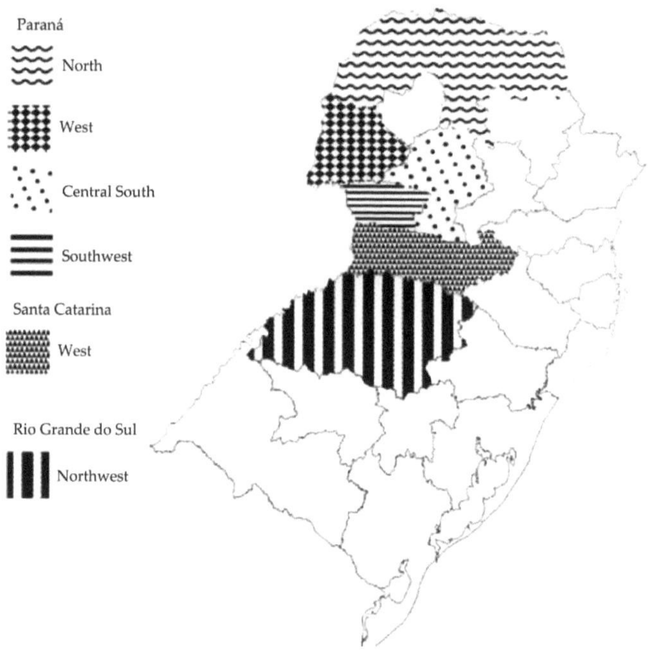

Figure 1. Mesoregions of the southern region of Brazil used for data collection.

These collection areas are of great importance in livestock farming and are characterized by the large presence of milk producers with designated areas for silage production, varying from 2 to 20 ha, and beef producers who have areas of 7 to 50 ha designated for corn silage production.

The climatic description of all the evaluated mesoregions was performed according to the Köppen classification. The climate of the North, West, and Southwest mesoregions of Paraná is of Cfa type—humid subtropical (mesothermal), with no defined dry season, hot summers, and less frequent frosts. The collection areas in these mesoregions have average altitudes of 538, 502, and 565 m, with average maximum temperatures of 27.4, 27.0, and 23.0 °C, and average minimum temperatures of 17.0, 16.9, and 12.6 °C. The annual rainfall averages vary from 1300 to 1500 mm in the North, from 1500 to 1700 mm in the West, and from 1900 to 2100 mm in the Southwest. The Central South mesoregion of Paraná has a Cfb-type Humid Temperate Climate (mesothermal) without a dry season, with a mild summer and a winter with severe and frequent frosts. The collection areas in the Central

South mesoregion have an average altitude of 971 m, with average maximum temperatures of 20.7 °C and average minimum temperatures of 12.4 °C, and the average annual rainfall varies from 1700 to 1900 mm.

The west of Santa Catarina has a Cfb humid temperate (mesothermal) climate, without a dry season and with a cool summer. The average altitude of the areas used for data collection is 630 m, the average annual maximum temperature is 23.0 °C, the average minimum temperature is 13.5 °C, and the average annual rainfall varies from 1460 to 1820 mm.

The Northwest of Rio Grande do Sul has a Cfa-type subtropical (mesothermal) climate, with hot summers, infrequent frosts, and a tendency for rain to concentrate in the summer months, but without a defined dry season. The collection areas in the Northwest of Rio Grande do Sul have an average altitude of 475 m; the average annual maximum temperature is 24.0 °C, the average minimum temperature is 13.7 °C, and the average annual rainfall varies from 1800 to 2000 mm.

2.2. Data Collection

The bromatological analysis reports on corn silages from the 2022/23 harvest were provided by the Chemical Analysis and Ruminant Nutrition Laboratory (NUPRAN) of the State University of Central West of Paraná (UNICENTRO), by the Cooperativa Agrária Agroindústria de Entre Rios (Guarapuava, Brazil) and by Soluções Nutricionais Ltd.a. (NRC; Francisco Beltrão, Brazil).

In total, 517 reports from the states of Paraná, Santa Catarina, and Rio Grande do Sul were obtained and divided into mesoregions, with 62 reports in the Central South mesoregion, 14 in the North mesoregion, 65 in the West mesoregion, 139 in the Southwest mesoregion of Paraná, 225 in the West mesoregion of Santa Catarina, and 12 in the Northwest mesoregion of Rio Grande do Sul.

Based on these reports, a descriptive statistical analysis was carried out in a box plot using the dry matter (DM) of the silages as a measure of variation, with discrepant data, i.e., "outliers", removed from the samples studied (Figure 2).

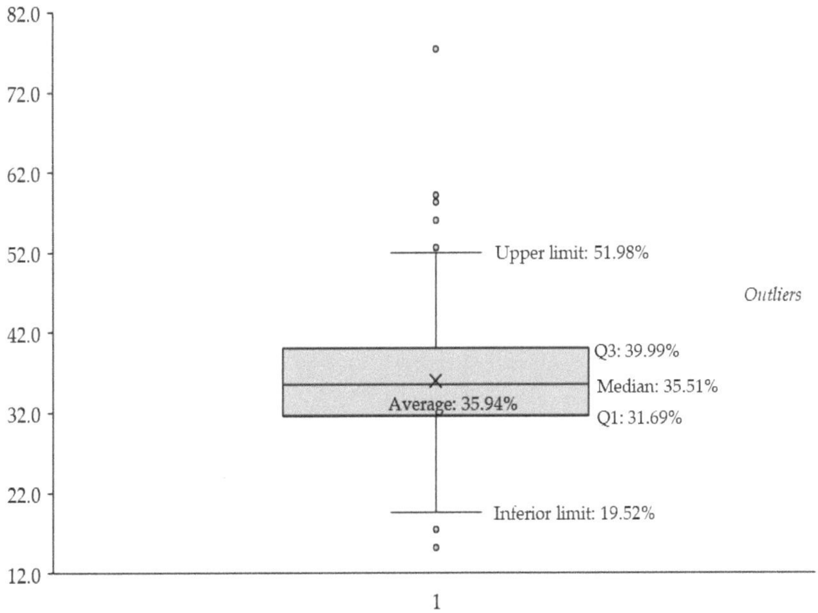

Figure 2. Box plot of the DM content of corn silage harvest 2022/23 from different mesoregions of Southern Brazil.

In total, 498 reports remained after removing the outliers. These were distributed as follows: 60 reports in the Central South mesoregion, 12 in the North mesoregion, 61 in the West mesoregion, 136 in the Southwest mesoregion of Paraná, 218 in the West mesoregion of Santa Catarina, and 11 in the Northwest mesoregion of Rio Grande do Sul.

2.3. Nutritional Value

With the aim of evaluating the nutritional value of the silages studied, and comparing their results between regions, the values of dry matter (DM), mineral matter (MM), ether extract (EE), crude protein (CP), starch, neutral detergent fiber (NDF), acid detergent fiber (ADF), acid detergent lignin (ADL) are available in the reports. Total digestible nutrients (TDNs) was obtained using the equation TDNs = 87.84 − (0.70 × ADF), as suggested by Bolsen et al. [11].

The carbohydrate fractionation was performed using the equation proposed by Sniffen et al. [12]. Total carbohydrates (TCs): TCs = 100 − (CP + EE + MM); Fraction B2, that is, the fraction slowly degraded in the rumen: B2 = NDF − fraction C; fraction C, which represents the indigestible fraction of the cell wall: C = (ADL × 2.4); and fraction A + B1, which corresponds to the fractions of rapid and medium rumen degradation: A + B1 = 100 − (C + B2).

2.4. Statistical Analysis

Descriptive statistics were used for all data using the SAS program [13]. The main component analyses and non-metric multidimensional scale were performed using the Vegan package from the R statistical program, version 4.3.0, which also enabled the evaluation of the correlations between the variables.

3. Results

Table 1 presents the bromatological–chemical composition of 498 corn silage samples collected from six mesoregions of Southern Brazil, represented by the variables MS, MM, EE, crude protein (CP), starch, NDF, ADF, ADL, TDNs, total CHO, and the percentage of fractions A + B1, B2, and C of carbohydrates.

Table 1. Nutritional value (overall average ± SD/min–max) of corn silages sampled in Southern Brazil (summer harvest 2022/23).

Variables	*n* Mesoregions	*n* Sample	$\overline{X} \pm$ SD	min–max
DM, %	6	498	35.82 ± 5.91	19.96–51.98
MM, % DM	6	498	4.66 ± 0.91	2.20–9.98
EE, % DM	6	498	3.29 ± 0.57	1.19–5.64
CP, % DM	6	498	8.11 ± 0.98	4.31–10.84
Starch, % DM	6	498	28.52 ± 6.96	9.63–46.36
NDF, % DM	6	498	43.09 ± 5.17	29.49–61.33
ADF, % DM	6	498	25.33 ± 3.56	14.76–36.98
ADL, % DM	6	498	3.34 ± 1.47	0.60–11.43
TDNs, % DM	6	498	69.87 ± 3.33	56.01–78.00
Total CHO, % DM	6	498	83.92 ± 1.40	79.88–90.51
A+B1 (% Total CHO)	6	498	48.63 ± 6.21	27.11–65.98
B2 (% Total CHO)	6	498	41.83 ± 5.60	23.49–58.51
C (%Total CHO)	6	498	9.54 ± 4.15	1.70–32.61

DM: dry matter; MM: mineral matter; EE: ether extract; CP: crude protein; NDF: neutral detergent fiber; ADF: acid detergent fiber; ADL: acid detergent lignin; TDNs: total digestible nutrients; CHO: carbohydrates; A + B1: rapid degradation fraction; B2: slow degradation fraction; C: non-degraded fraction; *n*: sample number; $\overline{X}$: average; SD: standard deviation; min: minimum; max: maximum.

Greater variations were observed in DM, starch, NDF, ADF, fraction A + B1, fraction B2, and fraction C of silage carbohydrates, with variations between the minimum and maximum values of 32.02, 36.73, 31.84, 22.22, 38.87, 35.02, and 30.91%, respectively. The variations were smaller for EE, CP, MM, ADL, TDNs, and total CHO (4.45, 6.53, 7.78, 10.83, 21.99, and 10.63%, respectively).

The same assessments presented in Table 1 were also carried out separately between the mesoregions in question and are presented in Tables 2–4.

Table 2. DM, MM, EE, CP, and starch (average ± SD/min–max) of corn silages sampled in different mesoregions of Southern Brazil (2022/23 harvest).

Mesoregions	n	DM, % DM		MM, % DM		EE, % DM		CP, % DM		Starch, % DM	
		$\overline{X} \pm SD$	min–max	$\overline{X} \pm SD$	min–max	$\overline{X} \pm SD$	min–max	$\overline{X} \pm SD$	min–max	$\overline{X} \pm SD$	min–max
Central South-PR	60	36.50 ± 5.17	26.89–49.68	4.29 ± 1.17	2.35–6.72	3.41 ± 0.92	1.19–5.64	7.06 ± 0.91	5.14–8.62	30.68 ± 6.24	12.64–39.97
North-PR	12	35.83 ± 6.19	24.73–45.70	5.93 ± 1.51	4.36–9.98	2.91 ± 0.58	2.14–3.84	6.32 ± 1.31	4.55–8.35	25.02 ± 5.92	11.49–32.02
West-PR	61	36.92 ± 6.17	23.48–50.57	4.91 ± 0.52	3.64–6.75	2.90 ± 0.41	2.02–4.11	8.56 ± 0.79	6.60–10.65	25.25 ± 8.11	9.63–39.81
Southwest-PR	136	36.27 ± 5.88	22.90–51.63	4.69 ± 0.83	2.53–8.83	3.21 ± 0.49	2.25–5.05	8.33 ± 0.75	4.31–9.70	28.15 ± 6.49	12.02–46.36
West-SC	218	35.24 ± 6.01	19.96–51.98	4.59 ± 0.85	2.20–6.69	3.46 ± 0.44	2.20–4.51	8.19 ± 0.86	5.62–10.84	29.49 ± 6.73	12.28–46.15
Northwest-RS	11	32.07 ± 5.11	24.91–40.04	4.83 ± 0.28	4.36–5.28	2.90 ± 0.22	2.59–3.23	8.77 ± 0.76	7.32–9.95	24.40 ± 6.44	12.97–33.02

PR: Paraná; SC: Santa Catarina; RS: Rio Grande do Sul; %: percentage; DM: dry matter; MM: mineral matter; EE: ether extract; CP: crude protein; n: sample number; X: average; SD: standard deviation; min: minimum; max: maximum.

Table 3. NDF, ADF, ADL, and TDNs (average ± SD/min–max) of corn silages sampled in different mesoregions of Southern Brazil (2022/23 harvest).

Mesoregions	n	NDF, (% DM)		ADF, (% DM)		ADL, (% DM)		TDNs, (% DM)	
		$\overline{X} \pm SD$	min–max	$\overline{X} \pm SD$	min–max	$\overline{X} \pm SD$	min–max	$\overline{X} \pm SD$	min–max
Central South-PR	60	45.56 ± 5.32	35.33–57.59	25.68 ± 3.15	19.94–32.66	4.97 ± 1.21	2.03–8.82	70.07 ± 2.79	64.61–76.19
North-PR	12	49.86 ± 5.92	41.45–61.33	29.70 ± 4.38	23.68–36.59	5.53 ± 2.00	3.93–11.43	66.37 ± 4.42	56.01–72.06
West-PR	61	43.53 ± 4.34	35.39–56.97	25.75 ± 3.63	19.25–35.12	2.62 ± 1.20	1.14–6.01	69.03 ± 4.26	60.43–76.81
Southwest-PR	136	42.66 ± 4.78	31.30–58.82	25.26 ± 3.17	17.66–33.16	2.85 ± 1.19	1.31–6.53	69.95 ± 3.31	58.53–76.00
West-SC	218	42.20 ± 5.19	29.49–58.81	24.95 ± 3.72	14.76–36.98	3.34 ± 1.29	0.60–8.24	70.22 ± 2.93	61.73–78.00
Northwest-RS	11	43.01 ± 2.23	40.33–47.26	25.00 ± 2.27	21.79–28.61	1.83 ± 0.33	1.28–2.41	69.57 ± 4.16	62.15–75.21

PR: Paraná; SC: Santa Catarina; RS: Rio Grande do Sul; %: percentage; NDF: neutral detergent fiber; ADF: acid detergent fiber; ADL: acid detergent lignin; TDNs: total digestible nutrients; DM: dry matter; n: sample number; $\overline{X}$: average; SD: standard deviation; min: minimum; max: maximum.

Table 4. Carbohydrate fractionation (average ± SD/min–max) of corn silages sampled in different mesoregions of Southern Brazil (2022/23 harvest).

Mesoregions	*n*	Total CHO, % DM		A+B1 (% Total CHO)		B2 (% Total CHO)		C (% Total CHO)	
		$\overline{X} \pm SD$	min–max	$\overline{X} \pm SD$	min–max	$\overline{X} \pm SD$	min–max	$\overline{X} \pm SD$	min–max
Central South-PR	60	85.09 ± 2.10	79.88–90.51	46.41 ± 6.48	33.01–58.07	39.57 ± 6.51	27.11–52.54	14.02 ± 3.30	5.68–23.83
North-PR	12	84.84 ± 1.77	80.23–86.67	41.18 ± 7.49	27.11–51.14	43.14 ± 5.04	35.34–51.38	15.68 ± 5.82	11.14–32.61
West-PR	61	83.63 ± 1.04	81.04–86.69	47.95 ± 5.21	30.81–57.73	44.55 ± 3.84	35.25–51.67	7.51 ± 3.42	3.24–17.52
Southwest-PR	136	83.76 ± 1.06	81.06–87.09	49.06 ± 5.76	31.68–62.65	42.77 ± 4.66	27.37–53.30	8.17 ± 3.39	3.79–18.80
West-SC	218	83.76 ± 1.26	79.25–87.47	49.59 ± 6.34	31.07–65.98	40.83 ± 5.94	23.49–58.51	9.58 ± 3.67	1.70–23.47
Northwest-RS	11	83.50 ± 0.93	82.41–85.45	48.50 ± 2.51	44.61–51.87	46.25 ± 1.98	43.37–49.66	5.25 ± 0.94	3.69–6.77

PR: Paraná; SC: Santa Catarina; RS: Rio Grande do Sul; %: percentage; DM: dry matter; CHO: carbohydrates; A + B1: rapid degradation fraction; B2: slow degradation fraction; C: non-degraded fraction; *n*: sample number; $\overline{X}$: average; SD: standard deviation; min: minimum; max: maximum.

The DM, MM, EE, CP, and starch contents of silages in the Northwest-RS mesoregion had smaller variations, with a difference between minimum and maximum values of 15.13, 0.92, 0.64, 2.63, and 20.05%, respectively, while the silages evaluated in the West-PR, North-PR, and Central South-PR mesoregions, respectively, showed greater variations for DM, EE, and starch contents (27.09, 2.09, and 30.18%); (20.97, 1.70, and 20.53%); and (22.79, 4.45, and 27.33%) (Table 2).

The West-SC and Southwest-PR mesoregions contained silages with similar variations in DM (32.02 and 28.73%), MM (4.49 and 6.30%), EE (2.31 and 2.80%), CP (5.22 and 5.39%), and starch (33.87 and 34.34%, respectively).

Table 3 presents the NDF, ADF, ADL, and TDN contents, and the NDF and ADL contents of silages belonging to the Northwest-RS mesoregion showed the smallest variations (6.93 and 1.13%, respectively).

The West-PR, Southwest-PR, and West-SC mesoregions produced silages with greater variation in ADF levels (15.87, 15.50, and 22.22%, respectively). Silages collected from Central South-PR, North-PR, and Northwest-RS showed smaller variations in this variable (12.72, 12.91, and 6.82%, respectively).

The TDNs of silages collected in the West- and Southwest-PR mesoregions showed greater variation (16.38 and 17.47%, respectively). The smallest variations were observed in the Central South-PR, North-PR, West-SC, and Northwest-RS mesoregions, at 11.58, 16.05, 16.27, and 13.06%, respectively.

Evaluating the fractionation of carbohydrates present in silages revealed that silages collected in the Central South-PR and West-SC mesoregions had greater variations in total CHO content (10.63 and 7.52%, respectively) (Table 4). Silages collected from the North-PR, West-PR, Southwest-PR, and Northwest-RS mesoregions showed smaller variations of 6.44, 5.65, 6.03, and 3.04%, respectively. Silages collected in Northwest-RS showed less variation for fractions A + B1, B2, and C (7.26, 6.29, and 3.08%, respectively).

Figure 3 presents the principal component analysis (PCA), which aims to reduce the amount of data used, eliminate overlaps, and promote a linear transformation in the data so that the data resulting from this transformation has its most relevant components in the first dimensions on the so-called main axes.

The first two main components, PC 1 (57.90%) and PC 2 (34.60%), explained 92.50% of the total variation in the data, which allowed for a safe interpretation of the variables and mesoregions studied. The contribution of the analyzed variables to the formation of principal components is defined by the length of the arrows and their proximity to the axis. The variables NDF and ADF had a greater influence on the formation of PC 1, whereas fractions A + B1 and CP had less influence. The formation of PC 2 had a greater influence on EE and starch and a lesser influence on the B2 fraction.

Silages collected in the North-PR mesoregion had higher concentrations of MM, ADF, and NDF, but lower concentrations of TDNs and fractions A + B1. In contrast, silages from the Central South-PR mesoregion showed higher concentrations of starch and EE and lower concentrations of the B2 fraction.

Silages collected from the West-PR and Northwest-RS mesoregions contained the highest content of fraction B2 and the lowest content of starch, EE, ADL, total CHO, and fraction C of carbohydrates. Silages from the West-SC and Southwest-PR mesoregions showed higher levels of TDNs and fraction A + B1, and lower concentrations of NDF and ADF.

Correlations between the variables were defined according to the angles formed between them. If the variables form an angle close to 0° or acute, the correlation is high and positive; if it is close to 180°, the correlation is considered high but negative; and if the angle is 90° or close, the variables have little or no correlation [14].

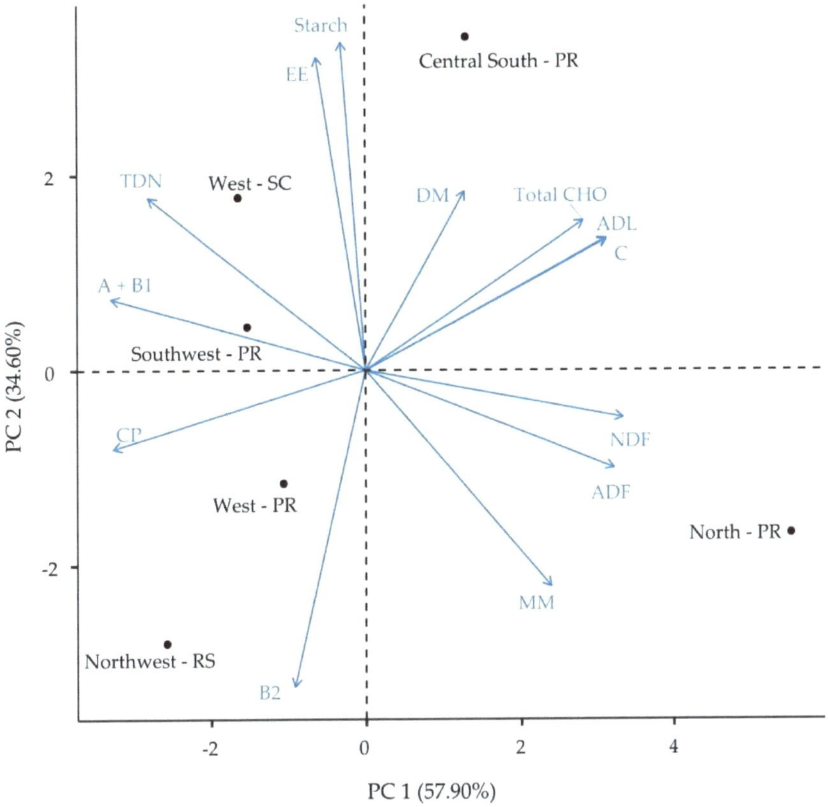

Figure 3. Analysis of main components of corn silages for the mesoregions of Southern Brazil and bromatological–chemical composition variables, referring to the 2022/23 summer harvest.

In view of this, there was a high correlation between ADL and the C fraction with total CHO, NDF, ADF, starch, and EE. High but inversely proportional correlations were observed between MM, ADF, and NDF with TDNs; between NDF and ADF with the A + B1 fraction; between the C, total CHO, and ADL fractions with CP; and between DM and the B2 fraction.

Figure 4 shows the grouping of silages, which is generated based on measures of similarities and dissimilarities; the variables within the group have great homogeneity, whereas there is great heterogeneity between the groups.

The silages from the West-SC and Southwest-PR mesoregions showed similarities, with the silages having higher levels of TDNs and fraction A + B1 and lower levels of NDF, ADF, and MM. The silages from West-PR and Northwest-RS also showed similarities, having a higher content of the B2 fraction and lower levels of starch, EE, ADL, total CHO, and fraction C. Silages from the North-PR and Central South-PR mesoregions differed from the others, establishing separate groups.

This section may be divided by subheadings. It should provide a concise and precise description of the experimental results, their interpretation, as well as the experimental conclusions that can be drawn.

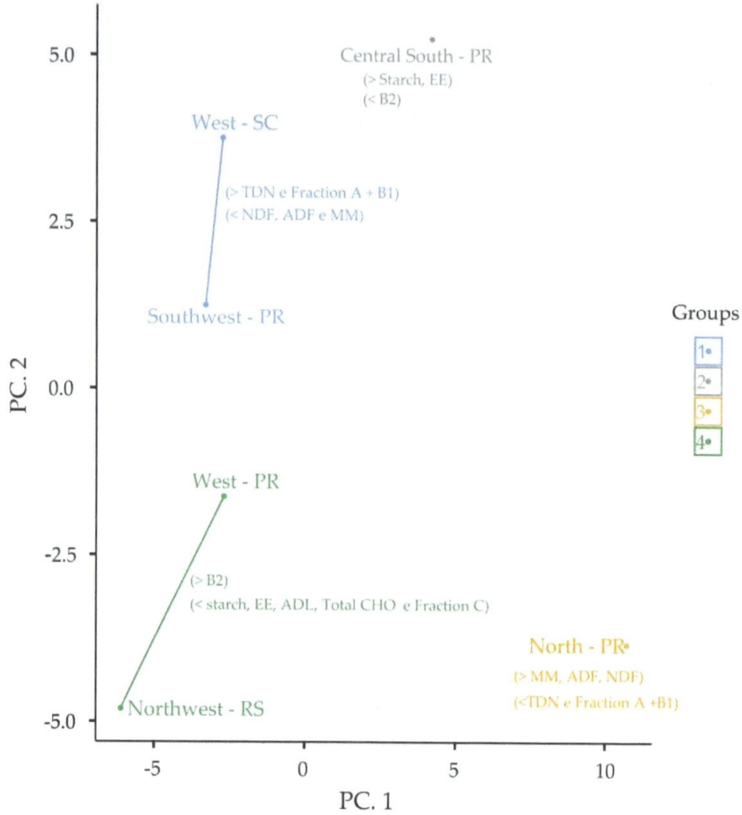

Figure 4. Non-metric multidimensional scale (NMDS) showing the grouping of different mesoregions in Southern Brazil according to the bromatological–chemical composition of corn silages from the 2022/23 harvest. Blue: Group 1; Gray: Group 2; Orange: Group 3; Green: Group 4.

4. Discussion

Generally, the DM content of the evaluated silages was within the recommended range (28–37% DM) as cited in the literature, regardless of the mesoregion (Tables 1 and 2). A variation in DM content may be due to stresses occurring during corn cultivation, such as climate variability (excess or lack of rain), occurrence of insects, especially the corn leafhopper, level of fertilization of crops, origin of samples, climatic conditions on the day of harvest, cutting point, particle processing, storage conditions, and sending the material to the laboratory [8] to which each corn silage was submitted. It is also worth noting that these conditions interfere, whether positively or negatively, in the chemical parameters of silages [15].

In addition to the points mentioned above, the plastic film used to seal the silo also influences the DM content of the silage, whether or not it is associated with the specific mass of the material. For example, films that allow for high oxygen permeability will alter the fermentation process, producing more heat and water inside the silo, in addition to increasing DM loss and damaging the bromatological quality of the silage [16].

Among the evaluated parameters, MM was the most stable. When evaluating silages from different corn hybrids, Neumann et al. [6] found no differences in the MM content. Neumann et al. [17] evaluated different corn hybrids in different cultivation locations and observed a difference in MM content between the locations, suggesting that the type and fertility of the soil in which the hybrid is cultivated may interfere with the concentration of minerals in the plant. The MM content is easily increased by contamination during production (soil), which can increase the concentration of undesirable microorganisms inside the silo, such as those of the Clostridium genus.

The EE content, and especially starch, is directly related to the maturation stage at which the plant is harvested for ensiling, and its content can be affected for this reason [15].

By evaluating the different harvest stages of the corn plant, Buso et al. [18] observed that when the plant grains were in the mealy stage (R4 stage), there was a greater participation of grains in the plant structure, which increased the starch concentration. Upon evaluating corn silage made at different maturation stages, Souza et al. [15] observed a greater share of grains in the plant structure, a higher EE content, and a lower share of fibrous compounds at the time of harvest than at the R4 stage. When the corn plant is ensiled with a DM content below 25%, for example, when it is a young plant, the NDF and starch contents are reduced because of the low deposition of starch in the grains, generating losses in consumption and animal performance [19].

The CP can also be affected by the harvest stage, but to a lesser extent. However, when it comes to preserved feeds such as silage, the fermentation process must be monitored. Silage fermentation inevitably generates proteolysis, which reduces the protein percentage in food. This action can be minimized by ensuring efficiency in the silage-making process, promoting adequate compaction in the ensiled mass, sealing the silo in the shortest possible time with quality plastic film, and avoiding cross-contamination. Even with these possible causes of change in CP, it is important to point out that the values of this nutrient obtained in the present study are within the values reported by Faria et al. [8] and are very close to those of corn silage with good nutritional quality.

The nutritional index of a food product can be measured using its TDN value. When corn silage has low TDN levels, it may have come from an older ensiled plant or one that has undergone environmental stress (water or pest attack) during its cycle, which can lead to less grain filling, reduced starch content, greater stalk participation in older plants, and a greater percentage of leaves in the senescent stage [20].

In addition to the parameters discussed above, the carbohydrates (concentration and absorption) present in corn silage play a fundamental role in animal production, as they directly affect the performance of animals, whether intended for milk or meat production. The ingestion and degradation of silage, in addition to sanitary aspects, are influenced by the concentration of carbohydrates (NDF, ADF, and ADL)—which can be increased by a thickening of the stems and a lignification of the leaves (older plants)—intrinsic characteristics of the plant itself, and the health status of the plant [21].

When silage carbohydrates are divided into A + B1, B2, and C, this indicates the stage the plant was at the time of harvest or whether it may have gone through some stress in addition to indicating nutritional quality. A higher content of fraction A + B1 represents a higher concentration of easily degraded fractions, which is found in younger and healthier plants. In contrast, silages with a higher proportion of fraction C represent silages from more mature plants; that is, they have a higher concentration of lignin [22,23].

In a compilation of 323 corn silage samples evaluated from 2004 to 2015, Faria et al. [8] found average levels of 53.30, 32.02, and 4.25% of NDF, ADF, and ADL, respectively, which were higher than those found in the present study. These inferences indicate that the carbohydrate content obtained in the present study was within acceptable standards without causing losses to animal production.

By evaluating the behavior of the bromatological components of silages within the mesoregions (Figure 3), it was possible to obtain a more accurate understanding of the harvest stage and the occurrence of possible stress arising from the environment.

In the northern mesoregion of Paraná, silages had the highest levels of NDF and ADF but lower levels of TDNs and fraction A + B1, a behavior which may be suggestive of silages made with plants at a more advanced stage, or a reflection of climatic particularities such as greater irregularity in higher rainfall and nighttime temperatures, than the other regions studied, which may have resulted in a higher prevalence of foliar diseases. Plants at a more advanced stage of maturity tend to generate silage with an average NDF content of 45.6% or higher [9]. Foliar diseases result in an increase in these carbohydrates because of the plant's defense mechanism, which increases the production of lignin to prevent the entry of pathogens into the cells, consequently increasing the NDF and ADF contents of the silage [24]. However, fungicides control the incidence of foliar diseases in corn plants and reduce NDF and ADF levels [25]. When there is an increase in fibrous components owing to dilution, it is expected that there will be a decrease in the TDNs and A + B1 fractions.

The opposite was observed in silages made in the Southwest-PR, West-SC, Central South-PR, West-PR, and Northwest-RS mesoregions (Figure 3), which had higher levels of TDNs, fraction A + B1, starch, EE, and fraction B2. These results tend to occur in silages made with plants at younger stages, or very close to the optimum harvest point, and/or are associated with conditions of less irregular precipitation, lower night temperatures than other regions studied, and an even lower prevalence of foliar diseases.

Using silages from the Southwest-PR, West-SC, and Central South-PR mesoregions might result in greater animal performance or, in some way, reduce costs with the inclusion of a concentrate such as TDNs, starch, EE, and fraction A + B1 in the diet. In addition to being highly digested in the rumen, the energy density of these roughages is high as they are compounds made up mainly of non-fibrous carbohydrates [15,26]. The silages from West-PR and Northwest-RS probably had a lower energy input, as they had a greater share of the B2 fraction. The ruminal fermentation of this fraction is extremely dependent on the food passage rate owing to its slow degradation, and this characteristic is related to the levels of cellulose and hemicellulose, which are potentially digestible carbohydrates [12,27].

The positive correlations observed in the present study occurred because these variables were formed by shared compounds. The NDF and ADF contents of the silage are linked to the cellulose, hemicellulose, and lignin concentrations of the plant, which are closely related to the size of the plant, percentage of stem in the plant structure, number of senescent leaves, percentage of bracts plus cob, and phytosanitary status of the plant [25,28].

The total CHO content depends on the concentration of fractions A + B1, B2, and C of carbohydrates, indicating that a higher concentration of total CHO does not always indicate good-quality silage. When the content of fraction A + B1 is higher, it is suggested that it is a silage with greater energy input, which has a fibrous fraction with good digestibility, and with lower proportions of straw and cob, which are portions of the plant that have low degradation, resulting in better-quality silage that will certainly be more useful to animals.

Starch and EE contents are linked to the maturity stage at which the plant is harvested, and both are related to the energy input of the silage; therefore, their relationships are directly proportional [15].

The negative correlations obtained were due to the dilution of the compounds. According to Gralak et al. [29], an increase in the proportion of ADF, which comprises ADL and other fibrous plant compounds, reduces the fractions that represent non-fibrous carbohydrates, negatively interfering with silage quality. The increase in NDF content is related to the simultaneous increase in MM and ADF content, which reduces the TDN content of the silage [24].

The similarity or dissimilarity between the chemical and bromatological compositions of the evaluated silages that originated in the groups (Figure 4) may involve several factors. Evaluating the effects of climatic variables on corn cultivation, Nied et al. [30] presented the results of several studies, demonstrating that precipitation is the main climatic variable that interferes with corn cultivation in Southern Brazil.

In the present study, precipitation and other climatic conditions were not correlated with the bromatological quality of the silages. However, these conditions were possibly similar in the regions where the silages belonged to the same group, and different (both positively and negatively) in the regions where the silages belonged to different groups (Figure 4) in this harvest. Perhaps this is not the reality in all harvests, but these results showed that it is necessary to go beyond the trivial when evaluating the bromatological quality of silages and pay attention to other variables; this applies to all regions of Brazil.

These studies to monitor the nutritional value of corn silages produced at each harvest must be routine in such a way that, over time, it is possible to identify the factors that generate the variations found between the different regions studied, thus aiming to promote technical correction recommendations in the corn ensiling process.

5. Conclusions

In the Southwest-PR and West-SC, the silages presented better nutritional value and a higher proportion of rapidly degrading carbohydrates; in the Central South-PR, the silages presented higher energy content; in the West-PR, Northwest-RS, and North-PR, the silages had a higher content of low-digestible fibers and worse nutritional value.

Author Contributions: M.N.: Conceptualization, methodology, validation, data curation, visualization, project administration, writing—review. E.B.: investigation, writing—original draft preparation, writing—editing. L.A.I.: Writing—original draft preparation, writing—editing. A.M.d.S.: Methodology, software, data curation, investigation. P.E.P.d.O.: Investigation. V.H.B.J.: Investigation, visualization, writing—review. All authors have read and agreed to the published version of the manuscript.

Funding: This research received no external funding.

Data Availability Statement: The datasets used and/or analyzed during the current study are available from the corresponding author on reasonable request.

Conflicts of Interest: The authors declare no conflicts of interest.

References

1. Carvalho, T.B.; Zen, S. The Beef Cattle Farming chain in Brazil: Evolution and trends. *Ipecege Mag.* **2017**, *3*, 85–99. [CrossRef]
2. Pinto, A.C.J.; Millen, D.D. Nutritional recommendations and management practices adopted by feedlot cattle nutritionists: The 2016 Brazilian survey. *Can. J. Anim. Sci.* **2019**, *99*, 392–407. [CrossRef]
3. Silvestre, A.M.; Millen, D.D. The 2019 Brazilian survey on nutritional practices provided by feedlot cattle consulting nutritionists. *Braz. J. Zootech.* **2021**, *50*, e20200189. [CrossRef]
4. Daniel, J.L.P.; Bernardes, T.F.; Jobim, C.C.; Schmidt, P.; Nussio, L.G. Production and utilization of silages in tropical areas with focus on Brazil. *Grass Forage Sci.* **2019**, *74*, 188–200. [CrossRef]
5. Silveira, E.S.; Carvalho, M.N.; Lima, B.B.; Oliveira, T.R.A.; Oliveira, G.H.F. Characterization of different genetic classes of corn grown in a semi-arid region regarding forage potential. *Matter* **2021**, *26*, e13084. [CrossRef]
6. Neumann, M.; Leão, G.F.M.; Coelho, M.G.; Figueira, D.N.; Spada, C.A.; Perussolo, L.F. Productive, nutritional and bioeconomic aspects of corn hybrids for silage production. *Zootech. Arch.* **2017**, *66*, 51–57. [CrossRef]
7. Vieira, A.F.; Nunes, R.L.C.; Torres, R.A.; Dias, N.S.; Oliveira, A.B. Agronomic evaluation of corn hybrids for silage in Baraúna, Northeastern semi-arid region. *Braz. Corn Sorghum Mag.* **2016**, *4*, 283–290. [CrossRef]
8. Faria, T.F.R.; Pinese, F.; Gimene, F.M.A.; Demarchi, J.J.A.A.; Campos, F.P.; Premazzi, L.M.; Gerdes, L. Bromatological composition of commercial corn silages produced in Brazil. *Zootech. Arch.* **2021**, *70*, 20–27. [CrossRef]
9. Horst, E.H.; Neumann, M. Assessing crop and corn silage profile in beef cattle farms in Southern Brazil: Ten years' results. *Agriculture* **2022**, *2*, 1200. [CrossRef]
10. Alvarenga, R.C.; Novotny, E.H.; Pereira-Filho, I.A.; Santana, D.P.; Pereira, F.T.F.; Hermani, L.C.; Cruz, J.C.C. *Cultivo do Milho*, 4th ed.; Embrapa Milho and Sorg: Sete Lagoas, Brazil, 2010; pp. 5226–5232, ISSN 1679-012X.
11. Bolsen, K.K.; Ashbell, G.; Weinberg, Z.G. Silage fermentation and silage additives—Review. *Asian-Australas. J. Anim. Sci.* **1996**, *9*, 83–494. [CrossRef]
12. Sniffen, C.J.; O'Connor, J.D.; Van Soest, P.J.; Fox, D.G.; Russell, J.B. A net carbohydrate and protein system for evaluating cattle diets: II. Carbohydrate and protein availability. *J. Anim. Sci.* **1992**, *70*, 3562–3577. [CrossRef] [PubMed]
13. SAS Institute. *Statistical Analysis System*; SAS Institute Inc.: Cary, NC, USA, 1993.
14. Hongyu, K.; Sandanielo, V.L.M.; Oliveira Junior, G.J. Principal component analysis: Theory, interpretations and applications. *Eng. Sci.* **2016**, *1*, 83–90.
15. Souza, A.M.; Neumann, M.; Rampim, L.; Almeida, E.; Matchula, A.; Cristo, F.B. Effect of storage time on the chemical composition of whole and grainless corn plant silage harvested at different maturity stages. *Braz. J. Anim. Sci.* **2022**, *51*, e20200180. [CrossRef]
16. Gianchini, J.C.; Barreta, D.A.; Danieli, B.; Schogor, A.L.B. Suitability of corn silage for use on family dairy farms. *Anim. Sci. Mag.* **2020**, *30*, 23–35.
17. Neumann, M.; Poczynek, M.; Leão, G.F.M.; Figueira, D.N.; Souza, A.M. Performance of corn hybrids for silage grown in different locations with three population densities. *Braz. Corn Sorghum Mag.* **2018**, *1*, 49–62. [CrossRef]

18. Buso, W.H.D.; Machado, A.S.; Ribeiro, T.B.; Silva, L.O. Production and chemical composition of corn hybrid silage under two cutting heights. *Neotrop. Agric. Mag.* **2018**, *5*, 74–80. [CrossRef]
19. Khan, N.A.; Yu, P.; Ali, M.; Cone, J.W.; Hendriks, W.H. Nutritional value of corn silage in relation to dairy cow performance and milk quality. *Sci. J. J. Food Agric.* **2015**, *95*, 238–252. [CrossRef]
20. Neumann, M.; Nörnberg, J.L.; Leão, G.F.M.; Horst, E.H.; Figueira, D.N. Chemical fractionation of carbohydrate and protein composition of corn silages fertilized with increasing doses of nitrogen. *Rural. Sci.* **2017**, *47*, e20160270. [CrossRef]
21. Garcez, B.S.; Alves, A.A.; Araújo, D.L.C.; Lacerda, M.D.S.B.; Souza, L.G.C.; Carvalho, L.F. Ruminal degradability of colonião grass (*Panicum maximum* jacq. cv. colonião) at three post-regrowth ages. *Acta Vet. Bras.* **2016**, *10*, 130–134. [CrossRef]
22. Sampaio, I.B.M.; Pike, D.J.; Owen, E. Optimal design for studying dry matter degradation in the rumen. *Braz. Arch. Vet. Med. Zootech.* **1995**, *47*, 373–383.
23. Silva, S.P.; Silva, M.M.C. Carbohydrate and protein fractionation according to the CNCPS system. *Vet. News* **2013**, *19*, 95–108.
24. Haerr, K.J.; Pineda, A.; Lopes, N.M.; Weems, J.D.; Bradley, C.A.; Pereira, M.N.; Murphy, M.R.; Fellows, G.M.; Cardoso, F.C. Effects of corn treated with foliar fungicide on in situ corn silage degradability in Holstein cows. *Anim. Feed. Sci. Technol.* **2016**, *222*, 149–157. [CrossRef]
25. Souza, A.M.; Neumann, M.; Bumbieris Junior, V.H.; Manchur, A.D.; Pontarolo, G.B.; Heker Junior, J.C.; Gavlak, T.F.; Plodoviski, D.C. Effect of advancing maturity stages of corn for silage on chemical characterization, digestibility and production costs. Seminar. *Agric. Sci.* **2021**, *42*, 283–300. [CrossRef]
26. Gayer, T.O.; Kasper, N.F.; Tadielo, L.E.; Krolow, R.H.; De Azevedo, E.B.; Oaigen, R.P.; Castagnar, D.D. Different dry matters content used for the conservation of annual ryegrass (*Lolium multiflorum* Lam.) in anaerobic environment. *Afr. J. Agric. Res.* **2019**, *14*, 369–378. [CrossRef]
27. Lupatini, G.C.; Maccari, M.; Zanette, S.; Piacentini, E.; Neumann, M. Assessment of the agronomic performance of corn hybrids (*Zea mays*, L.) for silage production. *Braz. Corn Sorghum Mag.* 2004; 3, 193–203. [CrossRef]
28. Mendes, M.C.; Stadler, A.J.; Oliveira, K.S.; Neumann, M.; Jesus, S.S.; Faria, M.V. Bromatological characteristics and digestibility of forage corn hybrids associated with foliar fungicide application. *Semin. Agric. Sci.* **2020**, *41*, 3285–3298. [CrossRef]
29. Gralak, E.; Faria, M.V.; Figueiredo, A.S.T.; Rizzardi, D.A.; Neumann, M.; Mendes, M.C.; Scapim, C.A.; Galbeiro, S. Genetic divergence among corn hybrids and combining ability for agronomic and bromatological traits of silage. *Genet. Mol. Res.* **2017**, *16*, gmr16029643. [CrossRef]
30. Nied, A.H.; Heldwein, A.B.; Estefanel, V.; Silva, J.C.D.; Alberto, C.M. Sowing dates of corn with lower risk of water deficit in Santa Maria, RS, Brazil. *Rural. Sci.* **2005**, *1*, 995–1002. [CrossRef]

agriculture

Article

Influence of Growth Stages and Additives on the Fermentation Quality and Microbial Profiles of Whole-Plant Millet Silage

Na Zhao [1], Xia Hao [1,2], Maozhe Yin [1], Changqing Li [3], Chao Wang [3,*] and Hongyan Han [1,*]

1 State Key Laboratory of Reproductive Regulation and Breeding of Grassland Livestock, School of Life Sciences, Inner Mongolia University, Hohhot 010070, China; 32208252@mail.imu.edu.cn (N.Z.); 32008037@mail.imu.edu.cn (X.H.); 32208105@mail.imu.edu.cn (M.Y.)
2 Dong Sheng District Experimental Middle School, Ordos 017000, China
3 Inner Mongolia Engineering Research Center of Development and Utilization of Microbial Resources in Silage, Inner Mongolia Academy of Agriculture and Animal Husbandry Science, Hohhot 010031, China; lcq@imaaahs.ac.cn
* Correspondence: wangchao@imaaahs.ac.cn (C.W.); hyhan1212@imu.edu.cn (H.H.); Tel.: +86-471-5298583 (H.H.)

Abstract: This study aimed to determine the optimal growth stage and additives for producing high quality millet silage through two experiments. Experiment 1: Whole-plant millet from the same field and under uniform management was harvested at the heading, sizing, milking, dough, and full-maturity stages. Then, it was chopped into 2–3 cm segments, vacuum-sealed in plastic bags without any further treatment, stored at 20 °C, and opened after 60 days. The results indicated that the dough stage had the highest water-soluble carbohydrate (WSC) and crude protein (CP) contents. The lactic acid (LA) and acetic acid (AA) contents during the dough and maturity stages were significantly higher than other stages, with the lowest pH observed during the dough stage. Experiment 2: The whole-plant millet was harvested at the dough stage. It was then chopped into 2–3 cm segments using a forage chopper, mixed thoroughly, and subjected to different treatments—inoculation with 10^6 CFU/g FM of *Lactiplantibacillus plantarum* (LP), adding of 1% FM sucrose (S), and a combination of *Lactiplantibacillus plantarum* and sucrose (MIX)—with a control group (CK) receiving an equivalent amount of water. The MIX treatment significantly enhanced the WSC content compared to other treatments ($p < 0.05$), and both the LP and MIX treatments showed superior LA and AA contents and lactic acid bacteria counts. These additives significantly altered the bacterial community, shifting dominance from Proteobacteria in the CK and raw materials to Firmicutes. *Klebsiella* dominated the CK group but was significantly reduced in the additive treatments, where *Lentilactobacillus* became the dominant genus. Therefore, we recommend harvesting millet at the dough stage and adding a mixture of *Lactiplantibacillus plantarum* and sugar to improve fermentation quality and aerobic stability.

Keywords: whole-plant millet; growth stages; additives; fermentation quality; bacterial communities

Citation: Zhao, N.; Hao, X.; Yin, M.; Li, C.; Wang, C.; Han, H. Influence of Growth Stages and Additives on the Fermentation Quality and Microbial Profiles of Whole-Plant Millet Silage. *Agriculture* **2024**, *14*, 1323. https://doi.org/10.3390/agriculture14081323

Academic Editor: Gustavo Lascano

Received: 18 June 2024
Revised: 21 July 2024
Accepted: 5 August 2024
Published: 9 August 2024

1. Introduction

Millet (*Setaria italica*) is one of the world's most important and oldest domesticated crops. Noted for its low water requirement among all cereal crops and its adaptability to grow well in poor soils [1], it is also a short-season crop. Millet is rich in crude protein and fiber, making it an excellent feedstock [2]. However, its high moisture content can make storage challenging, and silage is a preferred method for preserving such feedstock. Silage involves chopping fresh plants, compacting them to reduce oxygen, and allowing anaerobic fermentation to convert soluble carbohydrates into organic acids, reducing pH and inhibiting the growth of spoilage organisms, thereby preserving nutrients [3].

Due to its soft texture, good nutritional value, and palatability, whole-plant millet has high potential for animal feed. Research by Costa and others has shown that intercropping millet with *Paiaguas palisadegrass* in different pasture systems enhances the dry matter content of the silage [4]. Amer and others suggest that, compared to corn silage, millet silage contains higher CP and ADF levels. Replacing 67% of diet silage with millet silage in dairy cows' feed did not affect dry matter and crude protein intake, milk production, or lactose concentrations, but did increase milk fat concentration, thereby providing more energy-corrected milk [5].

Achieving high-quality whole-plant millet silage is complex due to the influence of multiple factors. Santos et al. [6] explored the effect of the growth period on the nutritional quality of *Pennisetum glaucum* (L.) R. silage and found that the growth period of pearl millet had a significant effect on the nutritional quality of silage, among which the DM content increased with the extension of growth period, while CP and ADF showed a decreasing trend. Hill et al. [7] demonstrated that adding silage inoculants can improve the fermentation characteristics and dry matter (DM) content of millet silage, also providing additional energy when fed to cattle.

Previous studies have shown that the nutritional composition of whole-plant millet, such as cellulose, carbohydrates, protein, moisture content, and the number of naturally adhering lactic acid bacteria, varies at different growth stages [6]. These factors are crucial in influencing the fermentation quality of silage. However, whole-plant millet has low water-soluble carbohydrate (WSC) content before ensiling, which may not meet the substrate requirements for high-quality silage fermentation without any treatments [8]. Research indicates that *Lactiplantibacillus plantarum* has excellent fermentation capabilities, promoting rapid fermentation of silage, lowering pH levels, inhibiting the growth of harmful microorganisms, and helping to improve silage quality [9]. Muck et al. [10] noted that silage microbial inoculants can significantly alter various aspects of silage fermentation, such as pH, lactic acid (LA) content, acetic acid (AA) content, DM loss, and the digestibility of DM and fiber. Similarly, Rinne, M et al. [11] suggest that the addition of LAB to silage accelerates LA fermentation compared to untreated material. Wang, X et al. [12] emphasized that LAB is a key factor in improving the fermentation quality and nutritional value of silage.

Therefore, we hypothesize that the nutritional composition and the quantity and types of adhering microorganisms vary across the important growth stages of millet. The addition of the widely recognized *Lactiplantibacillus plantarum* and soluble sugars can regulate the microbial diversity of millet silage, thereby controlling the fermentation quality. Based on this, our experiment focuses on whole-plant millet, studying the effects of different growth stages and the addition of *Lactiplantibacillus plantarum* and soluble sugars on the fermentation quality, aerobic stability, and bacterial communities of millet. The aim is to explore how various growth stages and additives influence the nutritional components, fermentation quality, aerobic stability, and bacterial dynamics of the silage. The goal is to provide a theoretical foundation for the development of whole-plant millet silage formulations and their application in feed.

2. Materials and Methods

2.1. *Raw Material and Silage Preparation*

Experiment 1: Whole-plant millet from the same field and under uniform management was harvested at the heading, sizing, milking, dough, and full-maturity stages. It was then chopped into 2–3 cm segments using a forage chopper, mixed thoroughly, and packed into silage bags weighing 300 g each without any further treatment. These bags were vacuum-sealed, with four replicates per treatment, and stored at room temperature (20 °C) for 60 days.

Experiment 2: A randomized block design was employed to evaluate the effects of different treatments on whole-plant millet silage harvested at the dough stage. The factors in this design are the treatments, which included inoculation with 10^6 CFU/g FM of *Lactiplantibacillus plantarum* (LP group), adding 1% FM sucrose (S group), and a combination of *Lactiplantibacillus plantarum* (10^6 CFU/g FM) and sucrose (1% FM) (MIX group), with a control group (CK) receiving an equivalent amount of water. The additive used was a homofermentative bacterium, classified as a biological additive. We dissolved the bacterial powder and sugar in water and added them in liquid form. Each treatment was randomly assigned within each block, with four replicates per treatment. The chopped millet segments were mixed thoroughly and packed into silage bags weighing 300 g each. The silage was stored in the laboratory, with the temperature maintained at around 20 °C throughout the entire storage period. At silo opening and after 7 days of aerobic exposure, the samples were analyzed to evaluate the impact of the various additives on the nutritional components, fermentation quality, aerobic stability, and bacterial communities of the whole-plant millet silage.

2.2. *Silage Extract Preparation*

After opening the silo, the silage was thoroughly mixed. A 20 g sample was then placed in a 200 mL conical flask, to which 180 mL of distilled water was added. The mixture was blended for 1 min. The resultant filtrate was passed through double-layer cheesecloth and subsequently through a No. 12 filter paper (Fushun city Minsheng filter paper factory, Fushun, China). The collected filtrate was used for pH analysis, microbial count, and water-soluble carbohydrate (WSC) content analysis. For organic acid analysis, the filtrate was further filtered through a 0.22 μm filter and stored at −20 °C.

2.3. *Nutritional Component Analysis*

An amount of 100 g of silage sample was dried in an oven (Shanghai YiHeng Scientific, Shanghai, China) at 65 °C for 48 h to determine the DM content. The dried samples were ground using a grinder (FS-6D; Shenzhen Fuchi Machinery Equipment Co., Ltd., Shenzhen, China), and samples were collected for nutritional component analysis. CP content was determined via the Kjeldahl method using an automatic Kjeldahl apparatus (KT8000, Foss Co., Ltd., Hillerød, Denmark). NDF and ADF contents were determined using an ANKOM A2000 fiber analyzer (Ankom Technology, Macedon, NY, USA). WSC content was measured using the anthrone–sulfuric acid method [13].

2.4. *Fermentation Quality Analysis*

The pH was measured using a pH meter (PB-10 Sartorius, Gottingen, Germany). Organic acids were analyzed using a high-performance liquid chromatograph (SHI-MADZE-10A, chromatographic column: Shodex Rspak KC-811 S-DVB gel Column 30 mm × 8 mm, detector: SPD-M10AVP, mobile phase: 3 mmol/L perchloric acid, flow rate: 0.6 mL/min; column temperature 50 °C, detection wavelength 210 nm, sample volume 5 μL).

2.5. *Aerobic Stability Analysis*

After opening the silo, 130 g of the sample was transferred into a 500 mL polyethylene bottle and compacted. A thermometer probe (model AZ88598, Hengxin Technology Co., Ltd., Shenzhen, China) was inserted at the center of the sample without touching the bottle. We then placed paper scraps in an empty bottle and inserted the thermometer probe without it touching the sides to record the temperature, which was considered the room temperature. The temperature of the sample was recorded every 15 min. If the sample's temperature exceeded the ambient temperature by more than 2 °C, indicating the onset of spoilage, the time of this occurrence was noted.

2.6. *Microbial Count*

Microbial counts were assessed using the plate count method. Various media supplied by Guangzhou HuanKai Microbial Sci. & Tech. Co., Ltd., Guangzhou, China, were employed for counting different microbes: lactic acid bacteria were cultured on De Man Rogosa Sharpe medium, yeast on potato dextrose agar (PDA), and general bacteria on nutrient agar (NA). These cultures were incubated at 37 °C for 48 h prior to counting.

2.7. Bacterial Community Analysis

Following the method of Mu et al. [14], DNA was extracted using an EZNA® Stool DNA Kit. The primers 341F (5′-CCTACGGGNGGCWGCAG-3′) and 805R (5′-GACTACHVG GGTATCTAATCC-3′) were used for polymerase chain reaction (PCR) to amplify the V3-V4 region of the bacterial 16S rDNA gene. PCR products were purified using AMPure XT beads (Beckman Coulter Genomics, Danvers, MA, USA), quantified using Qubit (Invitrogen, Waltham, MA, USA) [15], and sequenced on the Illumina NovaSeq PE250 platform. The raw data from the sequencer were assembled using FLASH (v1.2.8), low-quality sequences were removed using fqtrim (v0.94), and chimeras were filtered using Vsearch (v2.3.4). Bacterial community diversity was calculated using QIIME2 (2019.7). Additionally, sequence annotation was performed using BLAST (v2.16.0); the comparison databases were SILVA and NT-16S. Related charts were drawn using R-3.4.4.

2.8. Data Processing

Experimental data were preprocessed using Excel 2016. The effects of growth stage and additives on nutritional component, fermentation quality and microbial count and diversity were analyzed with one-way ANOVA. Tukey's multiple comparison method was used to analyze the difference between the least squares means of each treatment group. The results are presented as mean ± standard deviation (Mean ± SEM).

3. Results

3.1. Effect of Different Growth Stages on Nutritional Components of Whole-Plant Millet Silage

As the growth stage progressed, the DM content gradually increased, with the milk and dough stages around 30% and 35%, respectively, reaching the highest at the full-maturity stage (40%). CP content had a decreasing trend before increasing, with a significant increase at the dough stage compared to the milking and sizing stages ($p < 0.05$). The WSC content during the sizing and dough stages was significantly higher than in the other growth stages ($p < 0.05$), with no significant difference between them. At the full-maturity stage, the WSC content significantly decreased compared to other stages ($p < 0.05$). Throughout maturation, the NDF content exhibited wave-like fluctuations, generally decreasing and reaching its lowest at full maturity of 54.2%. There were no significant differences in the ADF content between growth stages (Table 1).

Table 1. Effects of different growth stages on nutrients of whole-plant millet silage.

Items		Growth Stages				
		Heading Stage	Sizing Stage	Milk Stage	Dough Stage	Maturity Stage
pre-ensiling	DM (g/kg DM)	255 ± 2.25 c	270 ± 1.32 c	296 ± 1.70 b	329 ± 1.65 b	371 ± 1.05 a
	CP (g/kg DM)	78.7 ± 6.45 a	54.7 ± 0.95 b	55.5 ± 1.59 b	49.8 ± 2.45 b	52.4 ± 1.45 b
	WSC (g/kg DM)	111 ± 6.65 ab	125 ± 3.35 a	68.1 ± 4.50 d	92.7 ± 5.20 bc	73.6 ± 7.85 cd
	NDF (g/kg DM)	651 ± 5.40 a	661 ± 9.10 a	672 ± 17.9 a	659 ± 5.75 a	599 ± 7.70 b
	ADF (g/kg DM)	359 ± 17.2	364 ± 8.60	358 ± 6.20	365 ± 12.65	337 ± 2.44
post-ensiling	DM (g/kg DM)	271 ± 1.71 e	285 ± 1.11 d	300 ± 0.65 c	355 ± 1.98 b	402 ± 2.90 a
	CP (g/kg DM)	77.8 ± 0.41 a	51.8 ± 0.36 c	52.0 ± 3.00 c	66.8 ± 2.03 b	77.1 ± 0.10 a
	WSC (g/kg DM)	19.2 ± 0.81 c	36.0 ± 0.60 a	24.0 ± 0.91 b	33.7 ± 0.91 a	16.5 ± 0.50 d
	NDF (g/kg DM)	631 ± 3.06 a	591 ± 3.06 b	622 ± 1.92 a	629 ± 3.11 a	542 ± 5.20 c
	ADF (g/kg DM)	352 ± 13.0	360 ± 6.65	356 ± 3.75	355 ± 7.65	333 ± 11.20

Note: Different lowercase letters in the same row indicate significant differences among various growth stages ($p < 0.05$). DM: dry matter; CP: crude protein; WSC: water soluble carbohydrate; NDF: neutral detergent fiber; ADF: acid detergent fiber.

The number of lactic acid bacteria (LAB) was lowest during the milk stage and highest at full-maturity stage ($p < 0.05$). Yeast counts were lowest in the milk stage and highest in the sizing stage. The general bacteria numbers were highest in the milk stage, significantly higher than in other stages ($p < 0.05$), and notably lower in the dough and full-maturity stages compared to the other stages ($p < 0.05$) (Table 2).

Table 2. Effects of different growth stages on microbial counts of whole-plant millet silage.

Items	Growth Stages				
	Heading Stage	**Sizing Stage**	**Milk Stage**	**Dough Stage**	**Maturity Stage**
LAB (log CFU/g)	4.40 ± 0.05 b	4.26 ± 0.07 b	3.48 ± 0.14 c	4.23 ± 0.05 b	6.80 ± 0.23 a
Yeasts (log CFU/g)	5.32 ± 0.10 b	6.06 ± 0.20 a	3.31 ± 0.08 d	5.26 ± 0.07 b	4.36 ± 0.16 c
General bacteria (log CFU/g)	7.11 ± 0.29 b	7.48 ± 0.05 b	8.26 ± 0.08 a	6.20 ± 0.03 c	6.18 ± 0.08 c

Note: Different lowercase letters in the same row indicate significant differences among various growth stages ($p < 0.05$). LAB: lactic acid bacteria.

The highest pH value was observed in the full-maturity stage (4.20), with the lowest in the milk stage (4.03), with significant differences ($p < 0.05$). Additionally, the lactic acid (LA) content during the heading, dough, and full-maturity stages was significantly higher than during the milking and dough stages ($p < 0.05$), although no significant differences were observed among the heading, dough, and full-maturity stages. As the growth stages progressed, the acetic acid (AA) content exhibited an increasing trend, leading to a decreasing trend in the LA/AA ratio. Specifically, the AA content at the dough stage was significantly higher than at the milk stage ($p < 0.05$). No propionic acid (PA) and butyric acid (BA) were detected in the silage at any growth stage (Table 3).

Table 3. Effect of different growth stages on fermentation quality of whole-plant millet silage.

Items	Growth Stages				
	Heading Stage	**Sizing Stage**	**Milk Stage**	**Dough Stage**	**Maturity Stage**
pH	4.13 ± 0.01 b	4.08 ± 0.00 c	4.03 ± 0.01 d	4.10 ± 0.02 c	4.20 ± 0.01 a
LA (g/kg DM)	120 ± 0.61 a	108 ± 0.46 c	114 ± 0.58 b	119 ± 2.87 a	120 ± 1.81 a
AA (g/kg DM)	11.5 ± 0.26 d	11.3 ± 0.11 d	22.6 ± 0.87 c	30.3 ± 0.88 b	34.1 ± 0.72 a
LA/AA	10.4 ± 0.18 a	9.53 ± 0.05 b	5.06 ± 0.18 c	3.95 ± 0.19 d	3.52 ± 0.09 d

Note: Different lowercase letters in the same row indicate significant differences among various growth stages ($p < 0.05$). LA: lactic acid, AA: acetic acid, LA/AA: lactic acid/acetic acid.

3.2. Effect of Different Additives on the Quality of Whole-Plant Millet Silage

The DM content in the S group was the highest at 38.5%, significantly exceeding that of the CK group ($p < 0.05$). At silo opening, the ADF content in the S group was significantly lower than in the other groups ($p < 0.05$), and the NDF content was also lower than in the other groups, although not significantly. After the aerobic stability test, the NDF and ADF contents in all treated groups were significantly lower than those in the CK group ($p < 0.05$). There were no significant differences in CP content among the treatment groups. The WSC content was higher in all additive groups compared to the CK group, with the MIX group displaying the highest level, significantly outperforming other treatment groups ($p < 0.05$) (Table 4).

LAB counts in all treatment groups were significantly greater than in the CK group, and the counts of general bacteria were significantly lower than in the CK group ($p < 0.05$). The general bacteria counts in the LP and S groups were notably lower than in the MIX group, while the yeast count in the MIX group was significantly higher than in the CK group ($p < 0.05$) (Table 5).

Table 4. Effects of different additives on nutrients at silo opening and after aerobic stability test of whole-plant millet silage.

Items	Additives	g/kg DM	NDF (g/kg DM)	ADF g/kg DM)	CP (g/kg DM)	WSC (g/kg DM)
Pre-ensiling	M	388.9 ± 9.75	621 ± 12.8	516 ± 9.20	68.8 ± 1.11	21.3 ± 1.98
At silo opening	CK	373 ± 0.06 b**	586 ± 6.35	503 ± 4.89 a	70.2 ± 1.07 *	33.8 ± 8.65 b
	LP	376 ± 0.70 ab	576 ± 2.73	496 ± 5.25 ab	70.0 ± 0.70	46.3 ± 2.34 b**
	S	385 ± 0.01 **	536 ± 19.4	458 ± 15.3 b	69.8 ± 1.28	43.6 ± 3.88 b*
	MIX	383 ± 0.80 ab**	550 ± 25.6	480 ± 20.8 ab	69.4 ± 0.58	69.4 ± 5.80 a**
After 7 d aerobic stability test	CK	442 ± 0.10	582 ± 8.20 A	497 ± 9.20 A	66.4 ± 0.47	25.0 ± 4.07 B
	LP	439 ± 0.22	531 ± 21.9 AB	456 ± 18.8 AB	68.4 ± 0.49	25.8 ± 2.68 B
	S	446 ± 0.08	524 ± 24.1 B	453 ± 19.3 AB	66.9 ± 2.18	27.5 ± 1.41 B
	MIX	449 ± 0.01	513 ± 5.15 B	435 ± 10.3 B	61.5 ± 4.38	37.1 ± 1.31 A

Note: CK, LP, S, and MIX represent control, *Lactiplantibacillus plantarum*, sucrose, combination of *Lactiplantibacillus plantarum* and sucrose addition treatments, respectively; M represents raw materials. Different lowercase letters indicate significant differences ($p < 0.05$) between different treatments at silo opening; different uppercase letters indicate significant differences ($p < 0.05$) between different treatments after the aerobic stability test. * Indicates significant difference between unsealing and aerobic stability experiments in the same treatment ($p < 0.05$), ** indicates that the difference between the at silo opening and after aerobic stability test in the same treatment is extremely significant ($p < 0.01$). DM: dry matter; CP: crude protein; WSC: water soluble carbohydrate; NDF: neutral detergent fiber; ADF: acid detergent fiber.

Table 5. Effects of different additives on microbial counts at silo opening and after aerobic stability test of whole-plant millet silage.

Items	Additives	LAB (log CFU/g)	Yeasts (log CFU/g)	General Bacteria (log CFU/g)
At silo opening	CK	7.19 ± 0.18 b	7.18 ± 0.24 b	7.37 ± 0.9 a**
	LP	8.02 ± 0.10 a	7.51 ± 0.08 ab*	6.15 ± 0.08 c**
	S	7.98 ± 0.20 a	7.69 ± 0.28 ab	5.94 ± 0.10 c*
	MIX	8.19 ± 0.13 a	7.92 ± 0.11 a*	6.63 ± 0.15 b
After 7 d aerobic stability test	CK	6.98 ± 0.18 B	6.93 ± 0.18 C	6.77 ± 0.12
	LP	8.14 ± 0.05 A	8.33 ± 0.02 AB	ND
	S	7.90 ± 0.14 A	8.04 ± 0.14 B	7.58 ± 0.50
	MIX	8.18 ± 0.16 A	8.47 ± 0.11 A	7.16 ± 0.26

Note: CK, LP, S, and MIX represent control, *Lactiplantibacillus plantarum*, sucrose, combination of *Lactiplantibacillus plantarum* and sucrose addition treatments, respectively; M represents raw materials. Different lowercase letters indicate significant differences ($p < 0.05$) between different treatments during at silo opening; different uppercase letters indicate significant differences ($p < 0.05$) between different treatments after the aerobic stability experiment. * Indicates significant difference between unsealing and aerobic stability experiments in the same treatment ($p < 0.05$), ** indicates that the difference between at silo opening and after aerobic stability test in the same treatment is extremely significant ($p < 0.01$); ND: not detected. LAB: lactic acid bacteria.

3.3. Effect of Different Additives on the Aerobic Stability of Whole-Plant Millet Silage

During 7 days of aerobic exposure, there was no significant temperature increase in any of the experimental groups, with aerobic stability exceeding 168 h. However, post-aerobic stability testing revealed an increase in pH values across all groups, although all additive groups remained significantly lower than the CK group ($p < 0.05$). Additionally, the LA and AA contents in all treatment groups, except the S group, decreased compared to their levels upon opening, with the most notable decrease observed in the LP group ($p < 0.01$) (Table 6).

Table 6. Effect of different additives on fermentation quality of whole-grain silage before and after aerobic stabilization experiment.

Items	Additives	pH	LA(g/kg DM)	AA(g/kg DM)	LA/AA
At silo opening	CK	4.50 ± 0.03 a	37.3 ± 1.81 b*	47.3 ± 4.96 c	0.80 ± 0.04 c
	LP	4.07 ± 0.03 b*	58.8 ± 2.86 a**	66.7 ± 2.07 a**	0.88 ± 0.03 bc
	S	4.03 ± 0.01 bc	46.4 ± 4.92 b	48.9 ± 4.63 bc	0.95 ± 0.03 b
	MIX	3.96 ± 0.04 c	65.9 ± 2.94 a	60.9 ± 1.84 ab**	1.08 ± 0.70 a
After 7 d aerobic stability test	CK	4.55 ± 0.03 A	28.0 ± 1.01 B	32.1 ± 3.18 B	0.89 ± 0.05
	LP	4.15 ± 0.01 BC	30.9 ± 5.45 B	27.5 ± 1.27 B	0.90 ± 0.03
	S	4.27 ± 0.16 B	56.6 ± 0.62 A	43.9 ± 1.26 A	1.15 ± 0.12
	MIX	4.00 ± 0.02 C	55.6 ± 3.87 A	49.1 ± 0.71 A	1.14 ± 0.09

Note: CK, LP, S, and MIX represent control, *Lactiplantibacillus plantarum*, sucrose, and a mixture of *Lactiplantibacillus plantarum* and sucrose treatments, respectively. Different lowercase letters indicate significant differences ($p < 0.05$) between treatments at silo opening, while uppercase letters denote differences after aerobic stability test. An asterisk (*) indicates a significant difference ($p < 0.05$) and two asterisks (**) an extremely significant difference ($p < 0.01$) between at silo opening and post-aerobic conditions within the same treatment.

3.4. Effect of Different Additives on the Bacterial Communities of Whole-Plant Millet Silage

All treatment groups achieved a Good's coverage index of 1.00. The Chao1, Shannon, and Simpson indices decreased for all treatments, with the additive groups showing significantly lower Shannon diversity compared to the CK group ($p < 0.05$). The MIX group had the lowest Shannon and Simpson indices, followed by the LP group. After the aerobic stability test, all groups except CK showed an increase in these diversity indices (Table 7).

Table 7. Effects of different additives on α diversity before and after aerobic stability test of whole-plant millet silage.

Items	Additives	Shannon	Simpson	Chao1	Goods_Coverage
Raw material	M	3.89 ± 0.05 a	0.88 ± 0.01 a	65.0 ± 6.35	1.00
At silo opening	CK	3.19 ± 0.03 b	0.82 ± 0.01 a	47.0 ± 0.71	1.00
	LP	0.97 ± 0.02 d*	0.23 ± 0.01 c*	51.0 ± 1.42	1.00
	S	2.44 ± 0.24 c	0.65 ± 0.07 b	52.5 ± 1.77	1.00
	MIX	0.73 ± 0.06 d	0.17 ± 0.01 c	34.0 ± 11.3	1.00
After 7 d aerobic stability test	CK	2.97 ± 0.10 A	0.79 ± 0.02 A	46.0 ± 10.6	1.00
	LP	1.21 ± 0.01 B	0.31 ± 0.01 B	48.0 ± 1.41	1.00
	S	3.08 ± 0.08 A	0.80 ± 0.02 A	57.0 ± 0.71	1.00
	MIX	0.97 ± 0.01 B	0.25 ± 0.01 B	43.0 ± 5.65	1.00

Note: CK, LP, S, and MIX represent control, *Lactiplantibacillus plantarum*, sucrose, combination of *Lactiplantibacillus plantarum* and sucrose addition treatments, respectively; M represents raw materials. Different lowercase letters indicate significant differences ($p < 0.05$) between different treatments at silo opening; different uppercase letters indicate significant differences ($p < 0.05$) between different treatments after aerobic stability test. An asterisk (*) indicates a significant difference ($p < 0.05$) between at silo opening and post-aerobic conditions within the same treatment.

PCA analysis revealed that the microbial communities in the raw materials were distinct from those in the silage treatments, indicating significant changes in microbial composition due to the ensiling process. Notably, the CK and MIX groups demonstrated closely related microbial communities both before and after the aerobic stability test, suggesting that these treatments may have similar impacts on microbial dynamics in the silage (Figure 1).

At the phylum level, Proteobacteria and Firmicutes were predominant across all samples. In the raw materials and the CK group, Proteobacteria were dominant, with relative abundances of 91.6% and 88.6% respectively, while Firmicutes were less prevalent at 5.16% and 11.29%. Conversely, in the LP, S, and MIX groups, the abundance of Firmicutes significantly increased to 94.64%, 63.63%, and 95.26%, respectively, effectively suppressing the dominant Proteobacteria observed in the CK group. This shift highlights the effectiveness of the additives in altering the microbial landscape of the silage (Figures 2 and 3a).

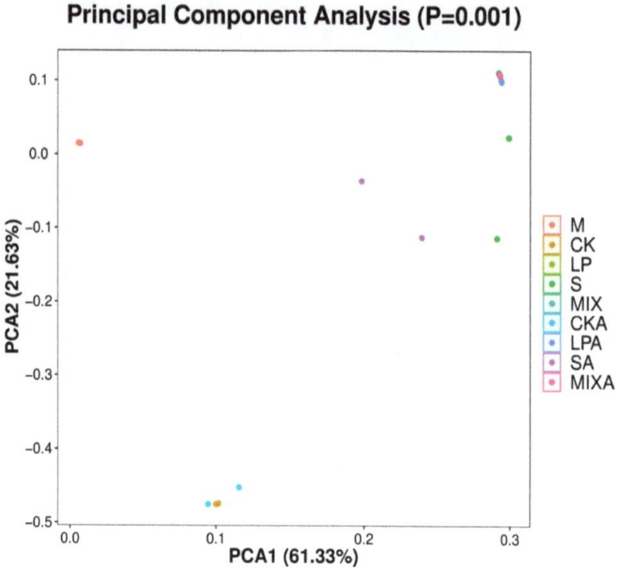

Figure 1. Principal component analysis (PCA) of bacterial communities in whole-plant millet raw material and silage. Note: M: raw material; CK, LP, S, and MIX represent control, *Lactiplantibacillus plantarum*, sucrose, and a mixture of *Lactiplantibacillus plantarum* and sucrose treatments, respectively, at silo opening. Corresponding CKA, LPA, SA, and MIXA denote the groups after the aerobic stability test.

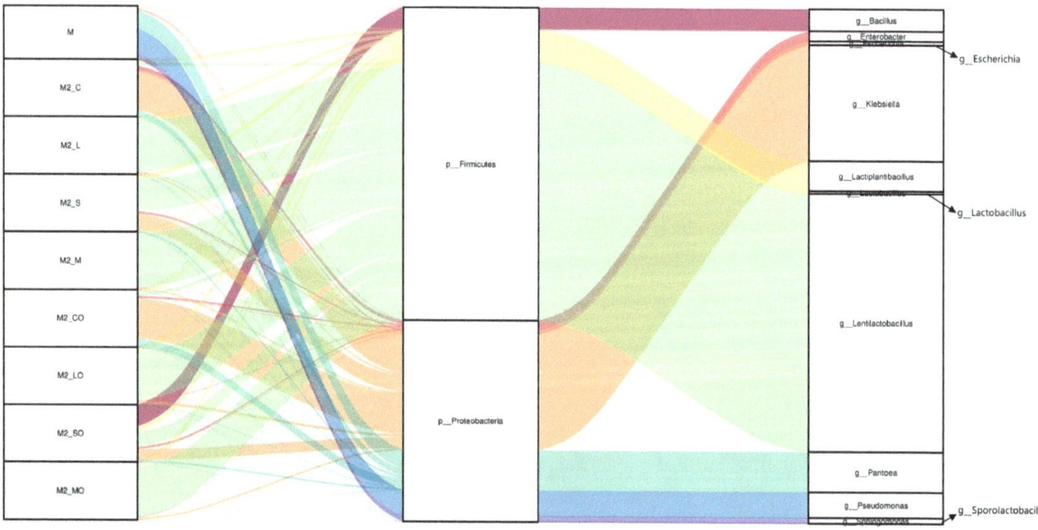

Figure 2. Relative abundance of bacterial communities in whole-plant millet raw material and silage. Note: M: raw material; M2_C, M2_L, M2_S, and M2_M represent control, *Lactiplantibacillus plantarum*, sucrose, and a mixture of *Lactiplantibacillus plantarum* and sucrose treatments, respectively, at silo opening. M2_CO, M2_LO, M2_SO, and M2_MO denote the groups post-aerobic stabilization. CK, LP, S, and MIX represent control, *Lactiplantibacillus plantarum*, sucrose, and a mixture of *Lactiplantibacillus plantarum* and sucrose treatments, respectively. The corresponding CKA, LPA, SA, and MIXA denote the groups after the aerobic stability test.

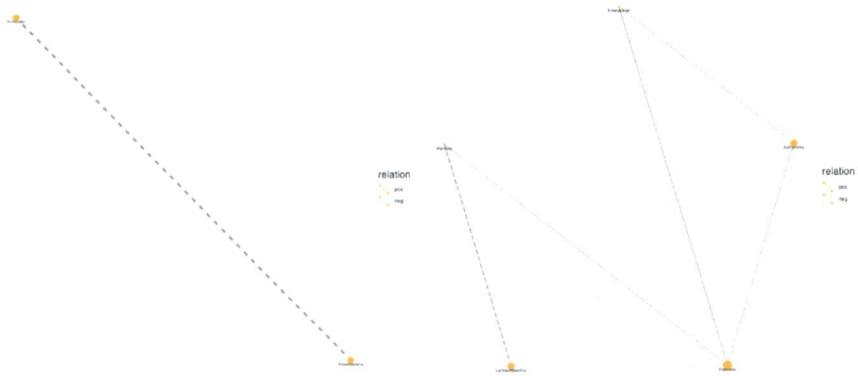

(a) (b)

Figure 3. Relationship between bacterial communities in whole-plant millet silage. (**a**) Phylum level, (**b**) genus level. Note: pos: positive impact; neg: negative impact.

At the genus level, the raw material was dominated by *Pseudomonas*, *Pantoea*, and *Sphingomonas*. However, post-ensiling, the CK group's microbial composition shifted to being dominated by *Klebsiella*, *Pantoea*, and *Enterobacter*. The additive treatments significantly altered these distributions, with *Lentilactobacillus* becoming the dominant genus in the LP, S, and MIX groups, showing a marked ability to suppress undesirable genera such as the *Klebsiella* and *Pantoea* found in the CK group (Figures 2 and 3b). Additionally, *Pantoea* was detected in all treatment groups, with reduced relative abundances compared to the raw material (CK: 11.79%, LP: 0.84%, S: 3.40%, MIX: 0.86%). Notably, after the aerobic stability test, *Bacillus* in the S group increased dramatically (from 0.04% at silo opening to 36.6%), dominating the microbial community (Figure 2).

4. Discussion

4.1. Effect of Different Growth Stages on Nutritional Components of Whole-Plant Millet Silage

The degree of maturity at harvest significantly influences the yield, nutritional content, digestibility, and ensiling potential of crops [16]. As the growth stage advances, changes in some nutritional values of silage are mostly associated with the degree of seed development and vary according to the different types and varieties of forage crops. Among these, the DM content is a crucial factor affecting silage quality. In this experiment, as the growth stage progressed, the DM content in all treatment groups gradually increased. Khan et al. studied the effects of the pre-heading, heading, and milk stages on the quality of millet silage and found that the DM content increased with the maturity of the millet, which is consistent with our findings [17]. By the time of full maturity, the DM content reached about 40%, as the water content in the plants decreased and sugars in the seeds were converted into starch, thereby increasing the DM content in the seed portion and consequently the whole plant [18,19]. A high DM content can complicate the compaction process during ensiling, leaving more air, which promotes the growth of yeasts and molds, thereby increasing the risk of "secondary fermentation" during feeding. Conversely, a low DM content, indicative of high moisture, can lead to excessive propagation of clostridia, which convert lactic acid into butyric acid, degrading silage quality. This process also triggers effluent loss, resulting in the depletion of WSCs and proteins and ultimately reducing intake by livestock such as cattle and sheep.

Hassanat et al. studied millet at the vegetative and heading stages and found that the cell wall components of the plant significantly increased with the extended growth period. The maximum values for NDF, ADF, cellulose, and lignin were found in the stems, followed by the whole plant and then the leaves. In contrast, hemicellulose, cutin, and silica content were higher in the leaves of the plant [9]. Additionally, KILIÇALP et al. studied oat silage at different growth stages and showed that the NDF content increases as the growth stage advances [20]. This difference may be attributed to the varying proportions of stems, leaves, and grains in different varieties, with the ADF and NDF contents being lower in leaves and grains compared to stems [21]. In this study, during the maturation process of forage millet, the NDF content tends to decrease. It is highest at the heading stage, remains relatively stable during the milk and dough stages, and significantly decreases by the mature stage. This trend may be due to the continuous deposition of starch in the grains during the filling process, which leads to a relative decrease in the cellulose content.

As the whole-plant millet matures, the CP content initially decreases until the filling stage, then increases, with the CP content at the dough stage being significantly higher than at the filling and milk stages. Jorge et al. found that the CP content decreases as the growth stage progresses, which is consistent with our results, which showed a decline in the CP content from the heading to the milk ripening stages [22]. This pattern is echoed in studies by Kurniawan [23], which observed a continuous accumulation of CP in whole-plant millet silage from the pre-milk to dough stages. Ferraretto [24] suggests that an increase in starch content within the silage reduces the proportions of both NDF and CP in the whole plant, explaining why the NDF and CP contents are lower during and after the dough stage compared to the heading stage. However, Yin reports a gradual decrease in CP content from the filling to the dough stage, which may be attributed to varietal differences. This complex interaction between growth stages and CP content highlights the dynamic nature of plant biochemistry in response to developmental changes.

WSC is a digestible part of the feed that is directly proportional to the feed's digestibility and nutritional value. In this study, the WSC content gradually increased from the heading to the dough stages, reaching its peak at the dough stage and then is rapidly decreased by full maturity. This pattern is similar to Chu's [25] findings on the impact of irrigation on WSC content and grain yield in different wheat varieties, where the WSC content continuously increased from the jointing to flowering stages and then rapidly decreased during the grain-filling stage. The accumulation of WSCs is closely linked to photosynthesis, which accumulates WSCs through photosynthesis as the plant matures. In the later stages of growth, as light intensity decreases, the rate of photosynthesis drops, leading to a reduction in soluble sugar accumulation [26,27]. Additionally, during the late growth stages, leaf senescence occurs, during which chlorophyll and other cellular components break down, releasing nitrogen and carbohydrates [28] and further contributing to the decline in WSC content as the growth period concludes.

In this study, the numbers of LAB and the contents of LA and AA are significantly higher at the dough stage compared to the milk stage, while the number of general bacteria is significantly lower. LAB efficiently convert WSCs into organic acids such as LA and AA under anaerobic conditions, which helps lower the pH and inhibits the growth of harmful microbes. Lactic acid, in particular, is the most effective organic acid in reducing pH in the silage environment, being 10–12 times more acidic than other organic acids [29]. Additionally, as the growth stage progresses, the number of general bacteria initially increase and then decrease, peaking during the milk stage. These general bacteria compete with LAB for fermentation substrates, leading to energy losses in the silage [30]. Consequently, selecting a growth stage with appropriate DM content is crucial to control moisture and air in the silage, thus reducing the number of general bacteria and enhancing silage quality. The LA content typically increases with the WSC content. However, this correlation was not observed in Experiment 1. This discrepancy could be due to several factors. Variations in the types and activities of the microorganisms present can affect the fermentation process. Different strains of lactic acid bacteria may have varying efficiencies in converting WSCs to lactic acid. Temperature, pH, and moisture content can influence microbial activity and fermentation efficiency. Additionally, other nutrients essential for microbial growth and activity may be limited, affecting the overall fermentation process and the production of LA.

The LA/AA ratio is commonly used as a qualitative indicator of fermentation and the aerobic stability of silage. A higher LA/AA ratio often indicates a more efficient fermentation process, as lactic acid is a stronger acid and more effective at lowering pH, which preserves the silage. A balanced or lower LA/AA ratio can improve the aerobic stability of silage. While lactic acid is important for initial preservation, AA plays a crucial role in inhibiting the growth of spoilage organisms when the silage is exposed to air. It is generally recommended that the concentration of AA should be around 3% DM [31]. Ahtoh et al. found that at low levels of AA (below 4.1% DM), nutritional losses can be highly variable [32]. Dnaiel et al. reported that adding 5% DM of AA to the daily diet of dairy cows resulted in lower feed intake within 4 h of the morning feeding, but no differences in intake were observed over the entire day. Additionally, the average milk yield and milk quality throughout the study period did not differ from the control group [33]. H. Danner et al. indicated that the AA content required for successful silage preservation can be very high. Silage with AA concentrations exceeding 5% DM remained stable for over 100 h, whereas untreated silage with an AA concentration of 1.28% DM remained stable for only 35 h [34]. Therefore, an appropriate amount of acetic acid can help prevent spoilage during feed-out. Additionally, both acids influence the palatability of the silage. Excessive AA might reduce palatability and intake by animals due to its strong vinegar-like smell and taste. However, moderate levels can be beneficial for stability without compromising intake.

4.2. Effect of Different Additives on the Quality of Whole-Plant Millet Silage

At silo open, the pH levels in the LP, S, and MIX additive treatment groups were significantly lower than in the CK group, indicating that the additives positively influenced the preservation of the silage, leading to enhanced fermentation quality [26]. Moreover, the MIX group exhibited superior characteristics, with a pH value of 3.96, an LA content of 6.59% DM and a better LAB count and LA/AA ratio compared to the other two additive groups. In the LP group, homolactic fermentation LAB rapidly produced a significant amount of LA using the WSCs available in the raw materials at the beginning of fermentation, thereby lowering the pH in the silage environment. Adding sucrose provided a fermentation substrate beneficial for LAB proliferation, facilitating further acid production and pH reduction [35]. The S group had the highest DM content (38.5%), which was significantly higher than the CK group, which aligns with similar findings reported by Kang [36]. The addition of sucrose enhanced the early production of LA during ensiling, rapidly reducing the silage pH and thereby inhibiting harmful microbial activity and minimizing nutrient losses [37]. Furthermore, sucrose itself contributes to a high DM content [38]; the MIX group had a lower DM content compared to the S group, likely due to the additional use of LP which consumes nutrients, thus reducing its DM content.

The additive groups also showed lower NDF and ADF contents than the CK group, similar to the findings by You [39], potentially due to the presence of fibrolytic enzyme-producing microbes that reduce fiber content [40]. *Lentilactobbacillus buchneri* has been shown to produce ferulic acid esterase—a part of the hemicellulose-degrading enzyme system that breaks ester bonds between hemicelluloses and lignin—enhancing cellulose utilization [41]. Research by Bao [42] and others indicates that LAB can produce a significant amount of cellulase during fermentation, reducing the fiber content in silage and enhancing the quality. Moreover, the organic acids produced during ensiling can hydrolyze digestible parts of the cell walls, further reducing fiber content [43]. Additionally, Zhang et al. found that *Lactiplantibacillus plantarum* also has the ability to degrade cellulose [44].

4.3. Effect of Different Additives on the VFA Contents and Aerobic Stability of Whole-Plant Millet Silage

During the 7-day (168 h) monitoring of aerobic stability, it was observed that the temperature of all groups did not exceed room temperature by more than 2 °C. However, the contents of LA and AA in all treatment groups (except for group S) decreased after the aerobic stability test and the pH levels increased, though they remained significantly lower than those in the CK group. Upon aerobic exposure, air penetrates the silage, promoting the rapid growth and reproduction of yeasts and general bacteria, which utilize the remaining sugars and LA as fermentation substrates [45,46]. The AA produced by heterofermentative LAB metabolism in the silage has antifungal activity, which can enhance the aerobic stability of the silage [47,48]. Additionally, the experiment was conducted in winter, when the indoor temperature was relatively low (around 20 °C), which may have influenced the aerobic stability results.

Studies have found that the LA content in Experiment 2 is lower than in Experiment 1. This may be due to Experiment 2 being conducted in the second year based on the results of Experiment 1. Owing to differences in the climate and other environmental factors, the millet grown in different years may have varying growth conditions and durations, leading to differences in the composition and quality of silage materials. In Experiment 1, the WSC content of the dough stage raw material was 92.7 ± 10.4 g/kg DM, while in Experiment 2, it was 21.3 ± 3.96 g/kg DM. This difference in WSC content is likely the main reason for the variation in lactic acid content between the two experiments. The higher acetic acid content in the additive group is related to the significantly higher presence of *Lentilactobacillus* in the microbial composition compared to the control group. Studies have shown that *Lentilactobacillus*, such as *Lentilactobacillus buchneri*, can produce more AA [45]. Therefore, even though we added *Lactiplantibacillus plantarum*, the AA content in the additive group was significantly higher than in the control group.

Yeasts can convert organic matter into ethanol, which is then converted to acetic acid by acetic acid bacteria [49]. Under the stimulation of acetic acid, yeast cells can harden their cell walls to limit the entry of acetates, thereby increasing their tolerance to acetic acid [50]. This might explain why both the yeast content and AA (acetic acid) levels are relatively high in our experiment. Additionally, Cai et al. found that adding Lactobacillus plantarum to silage improved fermentation quality but did not inhibit the growth of yeasts in the silage [51]. This is consistent with the findings of our study.

Lactic acid bacteria are capable of generating amylase, which hydrolyzes starch into dextrin and ultimately into glucose. They utilize glucose as a carbon source to generate pyruvate through glycolysis and subsequently produce lactic acid under the action of lactic dehydrogenase [52]. Menglei Xia et al. demonstrated that acetic acid bacteria and lactic acid bacteria were negatively correlated during the fermentation process and that the metabolism of acetic acid bacteria inhibited the growth and metabolism of lactic acid bacteria [53]. Based on this determination, the relatively high concentration of acetic acid in Experiment 2 of this study might inhibit the activity of lactic acid bacteria and decrease the efficiency of lactic acid conversion, thereby retaining more WSCs.

4.4. Effect of Different Additives on the Bacterial Communities of Whole-Plant Millet Silage

The Goods_coverage index reached 1.00 in all groups, suggesting that the sequencing depth was sufficient to reliably analyze the bacterial communities. Post-ensiling, a decrease in the Chao1, Shannon, and Simpson indices across all treatment groups indicated a reduction in microbial diversity. This trend aligns with findings by Xin et al. [54], which attribute the reduction in diversity mainly to the acidic anaerobic environment of the silage inhibiting the growth of some microbes. Similarly, Wayne Polley et al. [55] reported that the microbial diversity tends to decrease when beneficial bacteria are abundant. In contrast, studies by Wang [56] and Ni [57] on soy and moringa leaf silage observed an increase in bacterial diversity post-ensiling, potentially influenced by the specific plant materials used. PCA analysis showed that the bacterial communities in the raw material were distinctly separated from those in the silage across all treatment groups, highlighting differences in microbial composition due to the ensiling process.

Microbial abundance analysis showed that Proteobacteria and Firmicutes dominated both the raw material and all treatment groups. Specifically, Proteobacteria accounted for 88.6% and 91.6% in the CK and raw material groups, respectively, while Firmicutes had relative abundances of 5.16% and 11.3%. These phyla likely flourished due to the low pH and anaerobic conditions favorable for their growth in silage. The use of additives significantly increased the relative abundance of Firmicutes, suggesting that additives can effectively alter the silage bacterial community. This observation aligns with the findings of Yuan [58] and Wang [59] and is possibly due to the increase in the *Lentilactobacillus* and *Lactiplantibacillus* genera in the additive groups, both of which belong to the Firmicutes phylum. During the silage fermentation process, *Lentilactobacillus* and *Lactiplantibacillus* were the dominant genera in the additive groups, differing from previous research results [60]. Both genera belong to the LAB group and played a dominant role during the silage fermentation process, initiating LA fermentation and improving fermentation quality. *Klebsiella*, a common pathogen in silage, can cause DM loss and produce carbon dioxide, also leading to mastitis and reduced milk production [61]; *Enterobacter* competes with LAB for sugars, degrades proteins, ferments AA, lowers the quality of silage, and produces toxic compounds [57,62]. Compared to the CK group, the relative abundances of *Klebsiella* and *Enterobacter* genera decreased in all three additive groups, with the most significant reductions observed in the LP and MIX groups. This decrease supports the findings of Zhang [63] and Jiang [64], highlighting the effectiveness of these additives in suppressing undesirable bacteria in silage.

Additives can inhibit the growth of harmful microbes in silage, with LP showing a superior effect compared to adding sucrose alone. The reduction in *Enterobacter* in silage reflects a comprehensive change across various factors, including the availability of nutrients and water, the efficiency of converting nutrients to fermentation products, the activity of LAB, pH levels, and temperature—all contributing to more favorable conditions within the silage environment [65]. Post-ensiling, the relative abundance of the *Pantoea* genus in all treatment groups decreased compared to the raw material. This is consistent with the findings of Cheng [66]. The role of the *Pantoea* genus in silage remains ambiguous. Research, such as that by Ogunade [67] and Zhang [68], suggests that the *Pantoea* genus is negatively correlated with ammonia nitrogen content, reducing ammonia nitrogen content and helping to preserve proteins. Conversely, Li [69] and You [39] consider the *Pantoea* genus as an undesirable microbe in silage as it competes with LAB for nutrients. Given these conflicting roles, further research is necessary to clarify the function of microbes like *Pantoea* in silage, potentially paving the way for their development as microbial inoculants.

5. Conclusions

Harvesting whole-plant millet at the dough stage and applying a combined additive of *Lactiplantibacillus plantarum* and sucrose can improve fermentation quality and aerobic stability by optimizing microbial composition.

Author Contributions: N.Z. and X.H.: Both authors contributed equally to the original drafting of the manuscript, methodology formulation, conducting investigations, performing formal analysis, and curating data. M.Y.: Contributed to reviewing and curating data and conducting investigations. C.L.: Responsible for experimental design and manuscript review. C.W.: Acquired funding, curated data, and contributed to editing. H.H.: Involved in editing, supervision, funding acquisition, data curation, and conceptualization of the research. All authors have read and agreed to the published version of the manuscript.

Funding: This research was funded by "the earmarked fund for China Agriculture Research System, CARS-38".

Institutional Review Board Statement: Not applicable.

Data Availability Statement: Data are contained within the article.

Conflicts of Interest: The authors declare no conflicts of interest.

References

1. Lu, H.; Zhang, J.; Liu, K.B.; Wu, N.; Li, Y.; Zhou, K.; Ye, M.; Zhang, T.; Zhang, H.; Yang, X.; et al. Earliest domestication of common millet (*Panicum miliaceum*) in East Asia extended to 10,000 years ago. *Proc. Natl. Acad. Sci. USA* **2009**, *106*, 7367–7372. [CrossRef] [PubMed]
2. Hassan, Z.M.; Sebola, N.A.; Mabelebele, M. The nutritional use of millet grain for food and feed: A review. *Agric. Food Secur.* **2021**, *10*, 16. [CrossRef] [PubMed]
3. Guo, Q.; Hao, X.; Li, Y.; Zhang, Q.; Wang, C.; Han, H. Microbial Communities and Metabolites of Whole Crop Corn Silage Inoculated with *Lentilactobacillus plantarum* and *Lentilactobacillus buchneri*. *Processes* **2022**, *10*, 2369. [CrossRef]
4. Costa, R.; Costa, K.; Souza, W.; Epifânio, P.; Santos, C.; Silva, J.; Oliveira, S. Production and quality of silages pearl millet and paiaguas palisadegrass in monocropping and intercropping in different forage systems. *Biosci. J.* **2018**, *34*, 957–967. [CrossRef]
5. Amer, S.; Mustafa, A.F. Short communication: Effects of feeding pearl millet silage on milk production of lactating dairy cows. *J. Dairy Sci.* **2010**, *93*, 5921–5925. [CrossRef] [PubMed]
6. Santos, R.D.; Neves, A.L.A.; Pereira, L.G.R.; Sollenberger, L.E.; Muniz, E.N.; Souza, E.Y.B.; Sobral, A.J.S.; Costa, N.V.; Gonçalves, L.C. Performance, agronomic traits, ensilability and nutritive value of pearl millet cultivar harvested at different growth stages. *J. Agric. Sci.* **2020**, *158*, 225–232. [CrossRef]
7. Hill, G.M.; Utley, P.R.; Gates, R.N.; Hanna, W.W.; Johnson, J.C., Jr. Pearl Millet Silage for Growing Beef Heifers and Steers. *J. Prod. Agric.* **1999**, *12*, 653–658. [CrossRef]
8. Nussio, L.G.; Park, R.S.; Stronge, M.D. *Silage Production from Tropical Forages*; Wageningen Academic: Wageningen, The Netherlands, 2005.
9. Hassanat, F.; Mustafa, A.F.; Seguin, P. Effects of inoculation on ensiling characteristics, chemical composition and aerobic stability of regular and brown midrib millet silages. *Anim. Feed. Sci. Technol.* **2007**, *139*, 125–140. [CrossRef]
10. Muck, R.E.; Nadeau, E.M.G.; McAllister, T.A.; Contreras-Govea, F.E.; Santos, M.C.; Kung, L., Jr. Silage review: Recent advances and future uses of silage additives. *J. Dairy Sci.* **2018**, *101*, 3980–4000. [CrossRef]
11. Rinne, M.; Franco, M.; Tapio, I.; Stefanski, T.; Bayat, A.; Mäntysaari, P. Effects of Grass Silage Additive Type and Barley Grain Preservation Method on Rumen Fermentation, Microbial Community and Milk Production of Dairy Cows. *Agriculture* **2022**, *12*, 266. [CrossRef]
12. Wang, X.; Liu, H.; Xie, Y.; Zhang, Y.; Lin, Y.; Zheng, Y.; Yang, X.; Wang, N.; Ni, K.; Yang, F. Effect of Sucrose and Lactic Acid Bacteria Additives on Fermentation Quality, Chemical Composition and Protein Fractions of Two Typical Woody Forage Silages. *Agriculture* **2021**, *11*, 256. [CrossRef]
13. Kasmaei, K.M.; Dicksved, J.; Spörndly, R.; Udén, P. Separating the effects of forage source and field microbiota on silage fermentation quality and aerobic stability. *Grass Forage Sci.* **2017**, *72*, 281–289. [CrossRef]
14. Mu, L.; Xie, Z.; Hu, L.; Chen, G.; Zhang, Z. Cellulase interacts with Lactobacillus plantarum to affect chemical composition, bacterial communities, and aerobic stability in mixed silage of high-moisture amaranth and rice straw. *Bioresour. Technol.* **2020**, *315*, 123772. [CrossRef]
15. Fang, D.; Dong, Z.; Wang, D.; Li, B.; Shi, P.; Yan, J.; Zhuang, D.; Shao, T.; Wang, W.; Gu, M. Evaluating the fermentation quality and bacterial community of high-moisture whole-plant quinoa silage ensiled with different additives. *J. Appl. Microbiol.* **2022**, *132*, 3578–3589. [CrossRef] [PubMed]
16. Buxton, D.R.; O'Kiely, P. Preharvest Plant Factors Affecting Ensiling. In *Silage Science and Technology*; Agronomy Monographs; Springer: Cham, Switzerland, 2003; pp. 199–250.
17. Khan, S.H.; Azim, A.; Sarwar, M.; Khan, A.G. Effect of maturity on comparative nutritive value and fermentation characteristics of maize, sorghum and millet silage. *Pak. J. Bot.* **2011**, *43*, 2967–2970.
18. Allen, M.; Coors, J.G.; Roth, G. Silage Science and Technology. Agronomy Monograph 42. *Corn Silage* **2003**, *42*, 547–608.
19. Oskey, M.; Velasquez, C.; Peña, O.M.; Andrae, J.; Bridges, W.; Ferreira, G.; Aguerre, M.J. Yield, nutritional composition, and digestibility of conventional and brown midrib (BMR) pearl millet as affected by planting and harvesting dates and interseeded cowpea. *Animals* **2023**, *13*, 260. [CrossRef]
20. Kiliçalp, N. Cultivar and Harvest Stage Effects on Nutritive Value of Whole Crop Oat (*Avena sativa* L.) Silages. *J. Agric. Nat.* **2022**, *26*, 437–449. [CrossRef]
21. Hassanat, F.; Mustafa, A.F.; Seguin, P. Effect of the brown midrib trait and stage of development at harvest on cell wall composition and degradability of forage pearl millet leaves and stems. *Can. J. Anim. Sci.* **2007**, *87*, 421–429. [CrossRef]
22. Morales, J.U.; Alatorre, J.A.H.; Nieto, C.A.R.; José, F.; Becerra, C. Forage production and nutritional content of silage from three varieties of pearl millet (*Pennisetum glaucum*) harvested at two maturity stages. *J. Anim. Plant Sci.* **2015**, *27*, 4161–4169.
23. Kurniawan, W.; Napirah, A.; Isnaeni, P.; Bain, A. Nutrient qualities hydroponic corn green fodder (HCGF) of yellow-maize on different harvest time. *IOP Conf. Ser. Earth Environ. Sci.* **2024**, *1341*, 012062. [CrossRef]
24. Ferraretto, L.; Shaver, R.; Luck, B. Silage review: Recent advances and future technologies for whole-plant and fractionated corn silage harvesting. *J. Dairy Sci.* **2018**, *101*, 3937–3951. [CrossRef]
25. Chu, P.F.; Yu, Z.W.; Wang, D.; Zhang, Y.L.; Xu, Z.Z. Effects of irrigation stage and amount on winter wheat fructan accumulation and translocation after anthesis and water use efficiency. *Ying Yong Sheng Tai Xue Bao* **2009**, *20*, 2691–2698.
26. Abid, M.; Ali, S.; Qi, L.K.; Zahoor, R.; Tian, Z.; Jiang, D.; Snider, J.L.; Dai, T. Physiological and biochemical changes during drought and recovery periods at tillering and jointing stages in wheat (*Triticum aestivum* L.). *Sci. Rep.* **2018**, *8*, 4615. [CrossRef] [PubMed]

27. Rosa, M.; Prado, C.; Podazza, G.; Interdonato, R.; González, J.A.; Hilal, M.; Prado, F.E. Soluble sugars. *Plant Signal. Behav.* **2009**, *4*, 388–393. [CrossRef]
28. Afzal, S.; Chaudhary, N.; Singh, N.K. Role of Soluble Sugars in Metabolism and Sensing Under Abiotic Stress. *Plant Growth Regul. Signal. Under Stress Cond.* **2021**, *14*, 305–334.
29. Webster, J. *The Biochemistry of Silage*; McDonald, P., Henderson, A.R., Heron, S.J.E., Marlow, B., Eds.; Chalcombe Publications: Southampton, UK, 1991; p. 340. ISBN 0-948617-225.
30. Kung, L. Silage fermentation and additives. *Arch. Latinoam. Prod. Anim.* **2018**, *26*, 61–66.
31. Gerlach, K.; Daniel, J.L.P.; Jobim, C.C.; Nussio, L.G. A data analysis on the effect of acetic acid on dry matter intake in dairy cattle. *Anim. Feed. Sci. Technol.* **2021**, *272*, 114782. [CrossRef]
32. Мамаев, A.; Осипян, Б.; Козлова, B. Role of acetic acid in the sustainability of silage from cereal grasses to adverse micloflore at air access. *Adapt. Fodd. Prod.* **2019**, *2019*, 48–57. [CrossRef]
33. Daniel, J.L.P.; Amaral, R.C.; Sá Neto, A.; Cabezas-Garcia, E.H.; Bispo, A.W.; Zopollatto, M.; Cardoso, T.L.; Spoto, M.H.F.; Santos, F.A.P.; Nussio, L.G. Performance of dairy cows fed high levels of acetic acid or ethanol. *J. Dairy Sci.* **2013**, *96*, 398–406. [CrossRef]
34. Danner, H.; Holzer, M.; Mayrhuber, E.; Braun, R. Acetic acid increases stability of silage under aerobic conditions. *Appl. Environ. Microbiol.* **2003**, *69*, 562–567. [CrossRef] [PubMed]
35. Zi, X.; Liu, Y.; Chen, T.; Li, M.; Zhou, H.; Tang, J. Effects of Sucrose, Glucose and Molasses on Fermentation Quality and Bacterial Community of Stylo Silage. *Fermentation* **2022**, *8*, 191. [CrossRef]
36. Kang, J.; Tang, S.; Zhong, R.; Tan, Z.; Wu, D. Alfalfa Silage Treated with Sucrose Has an Improved Feed Quality and More Beneficial Bacterial Communities. *Front. Microbiol.* **2021**, *12*, 670165. [CrossRef] [PubMed]
37. Shao, T.; Zhang, Z.X.; Shimojo, M.; Wang, T.; Masuda, Y. Comparison of Fermentation Characteristics of Italian Ryegrass (*Lolium multiflorum Lam.*) and Guineagrass (*Panicum maximum Jacq.*) during the Early Stage of Ensiling. *Asian-Australas. J. Anim. Sci.* **2005**, *18*, 1727–1734. [CrossRef]
38. Rezaei, J.; Rouzbehan, Y.; Fazaeli, H. Nutritive value of fresh and ensiled amaranth (*Amaranthus hypochondriacus*) treated with different levels of molasses. *Anim. Feed. Sci. Technol.* **2009**, *151*, 153–160. [CrossRef]
39. You, J.; Zhang, H.; Zhu, H.; Xue, Y.; Cai, Y.; Zhang, G. Microbial Community, Fermentation Quality, and In Vitro Degradability of Ensiling *Caragana* with Lactic Acid Bacteria and Rice Bran. *Front. Microbiol.* **2022**, *13*, 804429. [CrossRef]
40. He, L.; Chen, N.; Lv, H.; Wang, C.; Zhou, W.; Chen, X.; Zhang, Q. Gallic acid influencing fermentation quality, nitrogen distribution and bacterial community of high-moisture mulberry leaves and stylo silage. *Bioresour. Technol.* **2020**, *295*, 122255. [CrossRef]
41. Nsereko, V.L.; Smiley, B.K.; Rutherford, W.M.; Spielbauer, A.; Forrester, K.J.; Hettinger, G.H.; Harman, E.K.; Harman, B.R. Influence of inoculating forage with lactic acid bacterial strains that produce ferulate esterase on ensilage and ruminal degradation of fiber. *Anim. Feed. Sci. Technol.* **2008**, *145*, 122–135. [CrossRef]
42. Bao, J.; Wang, L.; Yu, Z. Effects of Different Moisture Levels and Additives on the Ensiling Characteristics and In Vitro Digestibility of Stylosanthes Silage. *Anim. Open Access J.* **2022**, *12*, 1555. [CrossRef]
43. Larsen, S.U.; Hjort-Gregersen, K.; Vazifehkhoran, A.H.; Triolo, J.M. Co-ensiling of straw with sugar beet leaves increases the methane yield from straw. *Bioresour. Technol.* **2017**, *245 Pt A*, 106–115. [CrossRef]
44. Zhang, H.; Wen, B.; Liu, Y.; Du, G.; Wei, X.; Imam, K.M.S.U.; Zhou, H.; Fan, S.; Wang, F.; Wang, Y.; et al. A reverse catalytic triad Asp containing loop shaping a wide substrate binding pocket of a feruloyl esterase from Lactobacillus plantarum. *Int. J. Biol. Macromol.* **2021**, *184*, 92–100. [CrossRef] [PubMed]
45. Mugabe, W.; Shao, T.; Li, J.; Dong, Z.; Yuan, X. Effect of hexanoic acid, Lactobacillus plantarum and their combination on the aerobic stability of napier grass silage. *J. Appl. Microbiol.* **2020**, *129*, 823–831. [CrossRef] [PubMed]
46. Yuan, X.J.; Wang, J.; Guo, G.; Wen, A.Y.; Desta, S.T.; Shao, T. Effects of ethanol, molasses and Lactobacillus plantarum on fermentation characteristics and aerobic stability of total mixed ration silages. *Grass Forage Sci.* **2016**, *71*, 328–338. [CrossRef]
47. Dong, Z.; Li, J.; Wang, S.; Dong, D.; Shao, T. Time of Day for Harvest Affects the Fermentation Parameters, Bacterial Community, and Metabolic Characteristics of Sorghum-Sudangrass Hybrid Silage. *Msphere* **2022**, *7*, e00168-22. [CrossRef] [PubMed]
48. Ben-Dov, E.; Shapiro, O.H.; Siboni, N.; Kushmaro, A. Advantage of Using Inosine at the 3′ Termini of 16S rRNA Gene Universal Primers for the Study of Microbial Diversity. *Appl. Environ. Microbiol.* **2006**, *72*, 6902–6906. [CrossRef] [PubMed]
49. Li, Y.; He, D.; Niu, D.; Zhao, Y. Acetic acid production from food wastes using yeast and acetic acid bacteria micro-aerobic fermentation. *Bioprocess Biosyst. Eng.* **2015**, *38*, 863–869. [CrossRef] [PubMed]
50. Ribeiro, R.A.; Vitorino, M.V.; Godinho, C.P.; Bourbon-Melo, N.; Robalo, T.T.; Fernandes, F.; Rodrigues, M.S.; Sá-Correia, I. Yeast adaptive response to acetic acid stress involves structural alterations and increased stiffness of the cell wall. *Sci. Rep.* **2021**, *11*, 12652. [CrossRef]
51. Cai, Y.; Benno, Y.; Ogawa, M.; Kumai, S. Effect of applying lactic acid bacteria isolated from forage crops on fermentation characteristics and aerobic deterioration of silage. *J. Dairy Sci.* **1999**, *82*, 520–526. [CrossRef]
52. Wang, Y.; Wu, J.; Lv, M.; Shao, Z.; Hungwe, M.; Wang, J.; Bai, X.; Xie, J.; Wang, Y.; Geng, W. Metabolism Characteristics of Lactic Acid Bacteria and the Expanding Applications in Food Industry. *Front. Bioeng. Biotechnol.* **2021**, *9*, 612285. [CrossRef]
53. Xia, M.; Zhang, X.; Xiao, Y.; Sheng, Q.; Tu, L.; Chen, F.; Yan, Y.; Zheng, Y.; Wang, M. Interaction of acetic acid bacteria and lactic acid bacteria in multispecies solid-state fermentation of traditional Chinese cereal vinegar. *Front. Microbiol.* **2022**, *13*, 964855. [CrossRef]

54. Xin, Y.; Chen, C.; Zhong, Y.; Bu, X.; Huang, S.; Tahir, M.; Du, Z.; Liu, W.; Yang, W.; Li, J.; et al. Effect of storage time on the silage quality and microbial community of mixed maize and faba bean in the Qinghai-Tibet Plateau. *Front. Microbiol.* **2022**, *13*, 1090401. [CrossRef]
55. Wayne Polley, H.; Wilsey, B.J.; Derner, J.D. Dominant species constrain effects of species diversity on temporal variability in biomass production of tallgrass prairie. *Oikos* **2007**, *116*, 2044–2052. [CrossRef]
56. Wang, Y.; Wang, C.; Zhou, W.; Yang, F.-Y.; Chen, X.-Y.; Zhang, Q. Effects of Wilting and Lactobacillus plantarum Addition on the Fermentation Quality and Microbial Community of Moringa oleifera Leaf Silage. *Front. Microbiol.* **2018**, *9*, 1817. [CrossRef]
57. Ni, K.; Wang, F.; Zhu, B.; Yang, J.; Zhou, G.-A.; Pan, Y.; Tao, Y.; Zhong, J. Effects of lactic acid bacteria and molasses additives on the microbial community and fermentation quality of soybean silage. *Bioresour. Technol.* **2017**, *238*, 706–715. [CrossRef]
58. Yuan, X.; Li, J.; Dong, Z.; Shao, T. The reconstitution mechanism of napier grass microiota during the ensiling of alfalfa and their contributions to fermentation quality of silage. *Bioresour. Technol.* **2020**, *297*, 122391. [CrossRef]
59. Wang, C.; Zheng, M.; Wu, S.; Zou, X.; Chen, X.; Ge, L.; Zhang, Q. Effects of Gallic Acid on Fermentation Parameters, Protein Fraction, and Bacterial Community of Whole Plant Soybean Silage. *Front. Microbiol.* **2021**, *12*, 662966. [CrossRef] [PubMed]
60. Zhang, Q.; Yang, H.-J.; Yu, Z. Effects of sucrose, formic acid and lactic acid bacteria inoculant on quality, in vitro rumen digestibility and fermentability of drooping wild ryegrass (*Elymus nutans* Griseb.) silage. *J. Anim. Feed. Sci.* **2017**, *26*, 26–32. [CrossRef]
61. Lv, H.; Pian, R.; Xing, Y.; Zhou, W.; Yang, F.; Chen, X.; Zhang, Q. Effects of citric acid on fermentation characteristics and bacterial diversity of Amomum villosum silage. *Bioresour. Technol.* **2020**, *307*, 123290. [CrossRef]
62. Muck, R.E. Silage microbiology and its control through additives. *Rev. Bras. Zootec.* **2010**, *39*, 183–191. [CrossRef]
63. Zhang, T.; Li, L.; Wang, X.; Zeng, Z.; Hu, Y.; Cui, Z.-J. Effects of Lactobacillus buchneri and Lactobacillus plantarum on fermentation, aerobic stability, bacteria diversity and ruminal degradability of alfalfa silage. *World J. Microbiol. Biotechnol.* **2009**, *25*, 965–971. [CrossRef]
64. Jiang, F.-G.; Cheng, H.-J.; Liu, D.; Wei, C.; An, W.-J.; Wang, Y.-F.; Sun, H.-T.; Song, E.-L. Treatment of Whole-Plant Corn Silage with Lactic Acid Bacteria and Organic Acid Enhances Quality by Elevating Acid Content, Reducing pH, and Inhibiting Undesirable Microorganisms. *Front. Microbiol.* **2020**, *11*, 593088. [CrossRef] [PubMed]
65. Pahlow, G.; Muck, R.E.; Driehuis, F.; Elferink, S.J.W.H.O.; Spoelstra, S.F. Microbiology of Ensiling. In *Silage Science and Technology*; American Society of Agronomy: Madison, WI, USA, 2003; pp. 31–93.
66. Cheng, Q.; Chen, L.; Chen, Y.; Li, P.; Chen, C. Effects of LAB Inoculants on the Fermentation Quality, Chemical Composition, and Bacterial Community of Oat Silage on the Qinghai-Tibetan Plateau. *Microorganisms* **2022**, *10*, 787. [CrossRef] [PubMed]
67. Ogunade, I.M.; Jiang, Y.; Pech Cervantes, A.A.; Kim, D.H.; Oliveira, A.S.; Vyas, D.; Weinberg, Z.G.; Jeong, K.C.; Adesogan, A.T. Bacterial diversity and composition of alfalfa silage as analyzed by Illumina MiSeq sequencing: Effects of Escherichia coli O157:H7 and silage additives. *J. Dairy Sci.* **2018**, *101*, 2048–2059. [CrossRef] [PubMed]
68. Zhang, J.; Liu, Y.; Wang, Z.; Bao, J.; Zhao, M.; Si, Q.; Sun, P.; Ge, G.; Jia, Y. Effects of Different Types of LAB on Dynamic Fermentation Quality and Microbial Community of Native Grass Silage during Anaerobic Fermentation and Aerobic Exposure. *Microorganisms* **2023**, *11*, 513. [CrossRef] [PubMed]
69. Li, L.; Yuan, Z.; Sun, Y.; Kong, X.; Dong, P.; Zhang, J. A reused method for molasses-processed wastewater: Effect on silage quality and anaerobic digestion performance of Pennisetum purpureum. *Bioresour. Technol.* **2017**, *241*, 1003–1011. [CrossRef]

agriculture

MDPI

Article

Fibrolytic Enzymes and Lactic Acid Bacteria Improve the Ensiling Characteristics of Ramie and Elephant Grass Mixed Silage

Mengwei Li [1,†], Faiz-ul Hassan [2,†], Muhammad Uzair Akhtar [2], Lijuan Peng [1], Fang Xie [1], Qian Deng [1], Huapei Zhong [1], Kelong Wei [1] and Chengjian Yang [1,*]

[1] Guangxi Key Laboratory of Buffalo Genetics, Reproduction and Breeding, Guangxi Buffalo Research Institute, Nanning 530001, China; lmw1607@163.com (M.L.); lijuanpeng2000@163.com (L.P.); xf_0730@sina.com (F.X.); 22017054@zju.edu.cn (Q.D.); 15878178301@163.com (H.Z.); wei151281938@163.com (K.W.)

[2] Faculty of Animal Production and Technology, Cholistan University of Veterinary and Animal Sciences, Bahawalpur 63100, Pakistan; f.hassan@cuvas.edu.pk (F.-u.H.); uzairakhtar@cuvas.edu.pk (M.U.A.)

* Correspondence: ycj0746@sina.com; Tel.: +86-15777162502

† These authors contributed equally to this work.

Abstract: Understanding the effects of bacteria and enzyme addition on mixed crop silage is imperative for dairy producers to make informed decisions. The current study evaluated the chemical changes in silage prepared from different ramie and elephant grass ratios (30:70, 50:50, 70:30, and 100:0) in response to bacteria (0, 100, 200, and 300 mg/kg) and enzyme addition (0, 10, 20, and 30 mg/kg) in a complete randomized design. The results indicated that the proportion of ramie in silage ($p < 0.01$), level of bacteria ($p < 0.05$), and level of enzyme added ($p = 0.05$) affected the CP, fiber, volatile fatty acids, and lactic acid contents and pH of silage. By comprehensive analysis, low lignin content and pH of silage with high lactic acid content was observed with a 30% ramie proportion. High CP and lactic acid contents with low ADF, lignin, and pH values were observed with the addition of bacteria ($p < 0.05$). High lactic acid with low ADF content was observed with the addition of enzyme in silage ($p \leq 0.05$). The optimum quality of silage was observed when the ramie, bacteria, and enzymes were added at the levels of 30%, 200 mg/kg, and 20 mg/kg, respectively, in ramie and elephant grass mixed silage.

Keywords: forage ramie; silage additive; nutritional evaluation; feed quality; elephant grass

Citation: Li, M.; Hassan, F.-u.; Akhtar, M.U.; Peng, L.; Xie, F.; Deng, Q.; Zhong, H.; Wei, K.; Yang, C. Fibrolytic Enzymes and Lactic Acid Bacteria Improve the Ensiling Characteristics of Ramie and Elephant Grass Mixed Silage. *Agriculture* **2024**, *14*, 1746. https://doi.org/10.3390/agriculture14101746

Academic Editor: Zoltán Győri

Received: 6 August 2024
Revised: 26 September 2024
Accepted: 1 October 2024
Published: 3 October 2024

1. Introduction

Ruminant nutritionists are in a continuous search for quality forages for sustainable livestock production. The efficient utilization of forages not only enhances animal productivity but also reduces production costs. In this context, the use of mixed crop silage has been observed to improve silage quality and the stability of fermentation conditions [1–3], exhibiting a positive impact on the quantity and quality of milk [4].

Ramie (*Boehmeria nivea* L. Gaud) is a non-conventional fiber crop that is used for livestock feeding due to its important nutritional profile. It is produced in various countries including China, India, Indonesia, Brazil, Japan, Korea, and Vietnam [5]. The crude protein (CP), lysine, and calcium contents of ramie are 22, 1.02, and 4.07%, respectively, on a dry matter (DM) basis, which indicates its potential as a suitable alternative to alfalfa [6]. It was reported that replacing alfalfa hay with ramie up to a certain limit in the diet of black goats did not compromise their growth performance [7]. A study conducted by Chen et al. [8] indicated that the quality of mixed silage was improved when a proportion of 20:80 rice straw and fresh ramie was used. Moreover, mixed silage prepared by the addition of ramie beneficially modulated the ruminal pH and increased DM digestibility in Chinese Holstein cows [4]. Elephant grass or Napier (*Pennisetum purpureum* Schum.) is a perennial, tall erect grass. This grass is extensively cultivated in different regions and is primarily grown in cut and carry systems in various dairy and feedlot operations. Napier grass holds many advantages such as high DM content (25–40%), excellent yield per cultivated area, water use efficiency, and high resistance to drought, in addition to the presence of reasonable soluble carbohydrates, making it convenient to preserve as silage for successful ruminant production systems [9,10].

Inoculation with bacterial strains and the addition of fibrolytic enzymes has been introduced as another strategy to improve ensiling characteristics, ultimately leading to the improved digestibility of silage [11–13]. Various inoculants and enzymes are being used for this purpose to improve silage quality by stimulating the fermentation process and improving preservation. A combination of organic acids, lactic acid bacteria, and fibrolytic enzymes may promote the silage quality of different grasses [14,15]. It has been reported that the lignin content of silage is reduced by a combination of lactic acid bacteria and cellulase [16]. The oxidation of lignin-related compounds is catalyzed by laccase [17], especially when combined with lactic acid bacteria [18]. Bacterial inoculants may ensure a decrease in pH and reduce proteolysis [19], while fibrolytic enzymes may increase carbohydrate availability separately and synergistically with bacterial combinations [20,21].

Most previous studies investigated the effects of bacteria and enzymes on silage prepared from different but single crops [22,23]. The literature is lacking on the response of mixed crop silage to the inclusion of bacterial inoculants and enzymes and their optimum dose levels, especially for ramie- and elephant grass-based silage. We hypothesized that the addition of bacteria and enzymes in mixed crop silage would improve the ensiling characteristics of the silage. The present study aimed to evaluate the effects of lactic acid-producing bacteria and fibrolytic enzymes and find the optimum combinations of ramie, enzyme, and bacteria levels in terms of mixed crop silage quality prepared from ramie and elephant grass.

2. Materials and Methods

2.1. Experimental Materials

Three cultures of bacterial species (viz: *Lactobacillus plantarum*, *Lactobacillus brucelli*, and *Pediococcus pentosaceus*, known as *Lactiplantibacillus plantarum*, *Apilactobacillus*, and *Lactobacillus pentosaceus* according to Zheng et al. [24]) were used (ratio of 1:1:1) as fermentation strains in this study. The total number of viable bacteria was 1.35×10^{10}, 1.39×10^{10}, and 1.38×10^{10} cfu/g for *Lactobacillus plantarum*, *Lactobacillus brucelli*, and *Pediococcus pentosaceus*, respectively. These strains were sourced in freeze-dried form and purchased from Shandong Zhongke Jiayi Bio-engineering Co. Ltd., Weifang, China, for the execution of the current study. Three fibrolytic enzymes, i.e., cellulase, xylanase, and laccase (ratio of 7:7:1) with enzymatic activities of 50,000, 50,000, and 1000 units/g, respectively, were taken from the Solarbio Science and Technology Co., Beijing, China.

2.2. Experimental Design

The experiment was designed in a $4 \times 4 \times 4$ factorial arrangement in a completely randomized design with five replicates per treatment group. Details of the experimental arrangement are presented in Table 1. Fermentation conditions varied, four ratios of ramie to elephant grass in silage (30:70, 50:50, 70:30, and 100:0), four levels of bacterial addition (0, 100, 200, and 300 mg/kg), and four levels of compound enzyme addition (0, 10, 20, and 30 mg/kg). Chopped elephant grass was thoroughly mixed with ramie according to the treatment proportions before packing it in plastic bags. The composite bacteria (LAB) and fibrolytic enzymes used in the experiment were dissolved in ultrapure water and uniformly sprayed on the silage. The spraying volume for each group was balanced by water simultaneously. An equivalent quantity of ultrapure water was sprayed on the samples in the control group. After uniform mixing, the plastic bags were vacuum-packed and stored at a temperature of around 20–22 °C. The samples were then ensiled for 45 days.

Table 1. Experimental grouping of the treatments.

Ramie/Elephant Grass Ratio	Bacteria Added (mg/kg)	Enzyme Added (mg/kg)	Ramie/Elephant Grass Ratio	Bacteria Added (mg/kg)	Enzyme Added (mg/kg)
30:70	0	0	70:30	0	0
30:70	0	10	70:30	0	10
30:70	0	20	70:30	0	20
30:70	0	30	70:30	0	30
30:70	100	0	70:30	100	0
30:70	100	10	70:30	100	10
30:70	100	20	70:30	100	20
30:70	100	30	70:30	100	30
30:70	200	0	70:30	200	0
30:70	200	10	70:30	200	10
30:70	200	20	70:30	200	20
30:70	200	30	70:30	200	30
30:70	300	0	70:30	300	0
30:70	300	10	70:30	300	10
30:70	300	20	70:30	300	20
30:70	300	30	70:30	300	30
50:50	0	0	100:0	0	0
50:50	0	10	100:0	0	10
50:50	0	20	100:0	0	20
50:50	0	30	100:0	0	30
50:50	100	0	100:0	100	0
50:50	100	10	100:0	100	10
50:50	100	20	100:0	100	20
50:50	100	30	100:0	100	30
50:50	200	0	100:0	200	0
50:50	200	10	100:0	200	10
50:50	200	20	100:0	200	20
50:50	200	30	100:0	200	30
50:50	300	0	100:0	300	0
50:50	300	10	100:0	300	10
50:50	300	20	100:0	300	20
50:50	300	30	100:0	300	30

2.3. Sampling and Chemical Analysis

The fresh samples of ramie and elephant grass were obtained by harvesting from 5 cm above ground level and chopped to the length of 1–2 cm at the Guangxi Buffalo Research Institute pasture (Nanning, China). A cutter mill (Puverisette 15, Fritsch GmbH, Idar-Oberstein, Germany) was used for this purpose. The plantation was sown in the month of April and harvesting was performed in September. The average temperature during the growing phase was 28.7 °C and relative humidity was 79.5%. The planting land was hilly terrain with an average rainfall of 1304.2 mm. These samples (500 g each) were stored in silage bags (150 mm × 250 mm) and sealed using a vacuum sealer (DZ500, Gzrifu Co. Ltd., Guangzhou, China) after the respective treatments. Raw material and silage samples were analyzed in duplicate for DM, CP, neutral detergent fiber (NDF), acid detergent fiber (ADF), crude ash, and lignin contents following AOAC [25]. Samples were analyzed for DM at 65 °C in a forced oven (LABO-250, STIK Co. Ltd., Shanghai, China) until a constant weight was achieved [26]. The dried samples were ground to pass a 1 mm screen size (FE220, Beijing Zhongxingweiye Instrument Co. Ltd., Beijing, China) for the analysis of DM (method 934.01) and ash (method 942.05) contents. The CP content was analyzed through the total nitrogen (method 954.01) using an automatic Kjeldahl apparatus (Kjeltec-8400, FOSS, Hillerød, Denmark). The NDF, ADF, and lignin (method 973.18) contents were analyzed using the fiber analyzer (ANKOM Technology Corp., Macedon, NY, USA) without ash correction or treatment with amylase and sodium sulfite, following the method described by Van Soest et al. [27]. To determine the pH value after fermentation, each sample (20 g) was stirred with distilled water (180 mL) in a blender for one minute and filtered with four layers of cheesecloth [28]. The filtrates were used to measure the pH of samples using a pH meter (Hanna Instruments Italia Sril, Padova, Italy). The gross energy of the samples was determined using a bomb calorimeter (PARR-6400, automatic bomb calorimeter, Moline, Illinois, USA), as mentioned previously [29]. Metabolizable energy was calculated using gross energy and NDF content following Sung and Kim [30]. The lactic acid content was estimated by the enzymatic method using commercially available kits (A019-2, Nanjing Jiancheng Bioengineering Institute, Nanjing, China) according to the manufacturer's instructions. Briefly, 0.02 mL H_2O, 1 mL enzyme working solution, and 0.2 mL chromogen were added to a blank tube, while 0.02 mL standard, 1 mL enzyme working solution, and 0.2 mL chromogen were added to the test tube. After 10 min of incubation in the water bath at 37 °C, 2 mL terminal solution was added to each tube and the absorbance was noted at 530 nm [31]. The following formula was used for the calculations:

$$\text{Lactic acid content (mmol/L)} = (\text{Test tube OD} - \text{blank tube OD})/(\text{standard tube OD} - \text{blank tube OD}) \times 3\ \text{mmol/L}$$

2.4. Statistical Analysis

The obtained data were subjected to the generalized linear model using SAS 9.4 software using a completely randomized design with a 4 × 4 × 4 factorial arrangement of the experimental treatments. Means were compared using Tukey's multiple comparison test. Range analysis and the comprehensive balance method were used to analyze the total score of silage fermentation quality. Ki was the factor I (I = 1, 2, 3, 4); the greater the Ki, the better the level was. The range was the difference between the maximum and minimum of Ki. The greater the range, the greater the influence of this factor on silage quality. Data were expressed as means and results were declared significant at $p < 0.05$.

3. Results

For the nutrient composition of feed samples on an air-dried basis, the DM, CP, NDF, ADF, and ash contents were 21.9, 15.6, 42.4, 39.6, and 12.5% in ramie and 17.2, 6.89, 75.7, 48.5, and 8.89% in elephant grass, respectively.

3.1. Changes in Chemical Composition of Silage under Different Fermentation Conditions

The effects of different fermentation conditions on the chemical composition of silage are presented in Table 2. The proportion of ramie in the silage ($p = 0.001$) and the level of enzyme addition ($p = 0.001$) not only had independent effects but also interacted with each other and the bacteria level to affect the DM content. Interactions of R × B, E × B, and R × B × E were observed for the DM content of the silage ($p < 0.05$). Although silage CP content responded to the proportion of ramie in the silage ($p = 0.046$) and the level of bacterial inoculation ($p < 0.001$), with no direct effect of the level of enzyme addition ($p = 0.28$), interactions were observed for R × B, R × E, E × B, and R × B × E ($p < 0.05$). Silage energy content changed with the proportion of ramie ($p = 0.011$), level of bacteria added ($p = 0.012$), and interactions of R × E and E × B ($p < 0.05$). Ash content did not change by the enzyme and bacteria levels; however, effects of ramie proportion and E × B were observed ($p < 0.05$). The NDF content responded to bacteria level and R × B, R × E, and R × B × E interactions ($p < 0.05$). Silage ADF and lignin contents changed with ramie proportion, bacteria and enzyme levels, and their interactions ($p \leq 0.05$).

Table 2. Effects of different fermentation conditions on silage characteristics.

Item [1]	Ramie-to-Elephant-Grass Ratio	Bacteria, mg/kg				Enzyme, mg/kg				SEM	*p*-Value [2]						
		0	100	200	300	0	10	20	30		R	B	E	R × B	R × E	B × E	R × B × E
DM, %	30:70	20.3	22.5	21.6	18.6	20.3	18.6	22.5	21.6	1.15	0.001	0.76	0.001	0.001	0.051	0.001	0.017
	50:50	17.3	15.8	19.8	20.0	20.0	17.3	19.8	15.8								
	70:30	23.9	20.5	18.7	19.0	20.5	18.7	23.9	19.0								
	100:0	21.0	23.5	19.3	23.9	19.3	23.5	23.9	21.0								
CP, %	30:70	9.55	12.6	12.9	11.1	9.54	11.1	12.6	12.9	0.44	0.046	<0.001	0.28	0.01	<0.001	0.008	0.008
	50:50	10.3	10.3	11.7	11.1	11.1	10.3	11.7	10.3								
	70:30	11.3	11.7	11.7	11.8	11.7	11.7	11.3	11.8								
	100:0	10.7	13.1	12.0	11.7	12.0	13.1	11.7	10.8								
Metabolizable energy, cal/g	30:70	2145	2130	2245	2152	2145	2152	2130	2245	28.4	0.011	0.012	0.51	0.12	0.008	0.009	0.078
	50:50	2238	2322	2134	2205	2205	2238	2134	2320								
	70:30	2399	2321	2198	2200	2321	2198	2399	2200								
	100:0	2293	2295	2207	2149	2207	2295	2149	2293								
Ash, %	30:70	15.3	13.0	14.3	15.9	15.3	15.9	13.0	14.3	0.252	<0.001	0.15	0.72	0.25	0.09	<0.001	0.13
	50:50	15.4	15.9	15.2	15.5	15.5	15.4	15.2	15.8								
	70:30	16.0	15.8	15.6	16.3	15.8	15.6	16.0	16.3								
	100:0	16.2	16.8	17.2	17.8	17.2	16.8	17.8	16.2								
NDF, %	30:70	57.2	54.4	51.1	56.3	57.2	56.3	54.4	51.1	0.99	0.78	0.03	0.11	0.02	0.008	0.08	0.03
	50:50	54.6	54.0	54.8	55.8	55.8	54.6	54.8	54.1								
	70:30	52.9	54.5	54.7	55.1	54.5	54.7	52.9	55.1								
	100:0	53.6	52.1	54.4	56.4	54.4	52.1	56.4	53.6								
ADF, %	30:70	39.1	34.6	33.9	38.1	39.1	38.1	34.6	33.9	0.78	<0.001	0.01	0.05	0.017	0.005	<0.001	0.048
	50:50	39.8	39.5	38.3	39.7	39.7	39.8	38.3	39.5								
	70:30	39.4	40.7	39.4	39.3	40.7	39.4	39.4	39.2								
	100:0	41.6	40.0	40.6	41.8	40.0	39.9	41.8	41.7								
Lignin, %	30:70	14.4	12.2	11.0	12.4	14.4	12.4	12.2	11.0	0.74	0.001	0.001	0.002	<0.001	<0.001	<0.001	<0.001
	50:50	15.9	15.0	12.4	15.8	15.8	15.9	12.4	15.0								
	70:30	15.9	15.0	12.7	12.4	15.0	12.7	15.9	12.4								
	100:0	13.6	20.2	15.0	12.7	15.0	20.2	12.7	13.7								

[1] DM, dry matter; CP, crude protein; NDF, neutral detergent fiber; ADF, acid detergent fiber; SEM, standard error of the mean. [2] R = proportion of ramie in silage, B = level of bacterial inoculation, E = level of enzyme added.

3.2. Best Combinations of Fermentation Conditions for Silage Nutrient Composition

Ideal combinations of ramie proportion with bacteria and enzyme levels for different silage composition parameters are presented in Table 3. The factor ranking regarding the effect on DM content was ramie > enzyme > bacteria, where a ramie proportion of 100 with a bacteria level of 0 and a 20 mg/kg enzyme level was the best combination. The ranking of factors affecting the CP content was bacteria > ramie > enzyme, where a bacteria level of 200 mg/kg with a ramie proportion of 100 and a 20 mg/kg enzyme level was the ideal combination for CP content. The factor ranking regarding the effect on energy content was ramie > bacteria > enzyme, where a ramie proportion of 70 with 0 mg/kg bacteria and enzyme levels was the ideal combination for energy content. The factor ranking regarding the effect on ash content was ramie > bacteria > enzyme, where a ramie proportion of 0 with a 100 mg/kg bacteria level and a 20 mg/kg enzyme level was the best combination. The ranking of factors affecting NDF content was bacteria > enzyme > ramie, where a bacteria level of 300 mg/kg with a 0 mg/kg enzyme level and a ramie proportion of 50 was the best combination. The factor ranking regarding the effect on ADF content was ramie > enzyme > bacteria, where a ramie proportion of 50 with a 20 mg/kg enzyme level and a 200 mg/kg bacteria level was the ideal combination. The factor ranking regarding the effect on ash content was ramie > bacteria > enzyme, where a ramie proportion of 0 with a 100 mg/kg bacteria level and a 20 mg/kg enzyme level was the best combination. The factor ranking regarding the effect on lignin content was ramie > bacteria > enzyme, where a ramie proportion of 0 with 200 mg/kg bacteria and 30 mg/kg enzyme levels was the ideal combination for lignin content.

Table 3. The optimal combination of ramie-to-elephant-grass ratio, enzyme level, and bacteria inoculation level for various characteristics of mixed silage.

Item [1]		Fermentation Conditions [2]			Optimum Combination [3]
		A	B	C	
DM, %	K1 [4]	20.72	20.62	20.03	Factor ranking A > C > B A4C3B1
	K2	18.24	20.58	19.51	
	K3	20.52	19.84	22.53	
	K4	21.95	20.40	19.37	
	Range	3.71	0.78	3.16	
CP, %	K1	11.54	10.45	11.08	Factor ranking B > A > C B3A4C3
	K2	10.83	11.90	11.53	
	K3	11.59	12.07	11.81	
	K4	11.88	11.42	11.42	
	Range	1.05	1.62	0.73	
Metabolizable energy, cal/g	K1	2168	2268	2219	Factor ranking A > B > C A3B1C1
	K2	2224	2266	2220	
	K3	2278	2196	2203	
	K4	2235	2176	2264	
	Range	78.51	70.25	30.74	
Ash, %	K1	14.60	15.71	15.94	Factor ranking A > B > C A1B2C3
	K2	15.50	15.37	15.93	
	K3	15.93	15.57	15.51	
	K4	16.99	16.36	15.65	
	Range	2.39	0.99	0.43	
NDF, %	K1	54.78	54.61	55.51	Factor ranking B > C > A B4C1A2
	K2	54.84	53.80	54.46	
	K3	54.37	53.78	54.65	
	K4	54.15	55.93	53.50	
	Range	0.69	2.15	2.01	
ADF, %	K1	36.46	40.01	40.08	Factor ranking A > C > B A2C3B3
	K2	39.36	38.73	38.73	
	K3	39.72	38.07	38.07	
	K4	41.02	39.75	39.75	
	Range	4.56	1.94	2.01	

Table 3. *Cont.*

Item [1]		Fermentation Conditions [2]			Optimum Combination [3]
		A	B	C	
Lignin, %	K1	12.52	14.99	15.09	Factor ranking A > B > C A1B3C4
	K2	14.83	15.62	15.33	
	K3	14.02	12.83	13.33	
	K4	15.44	13.37	13.05	
	Range	2.92	2.79	2.28	

[1] DM, dry matter; CP, crude protein; NDF, neutral detergent fiber; ADF, acid detergent fiber. [2] Fermentation conditions represented by the ratio of ramie to elephant grass in silage (30:70, 50:50, 70:30, and 100:0; Factor A), level of bacteria addition (0, 100, 200, and 300 mg/kg; Factor B), and level of enzyme addition (0, 10, 20, and 30 mg/kg; Factor C). [3] A1, A2, A3, and A4 represent the proportion of ramie in the silage (30, 50, 70, and 100, respectively); B1, B2, B3, and B4 represent the level of added bacteria (0, 100, 200, and 300 mg/kg, respectively); and C1, C2, C3, and C4 represent the level of added enzyme (0, 10, 20, and 30 mg/kg, respectively). [4] Ki represents the average value at that specific level in an increasing order from K1 to K4.

3.3. Changes in the pH, Lactic Acid, Ammonia-N, and VFAs of Silage under Different Fermentation Conditions

The effects of different fermentation conditions on the pH, lactic acid, ammonia-N, and VFAs of silage are presented in Table 4. Silage pH responded to ramie proportion, bacteria level, and R × E and B × E interactions ($p < 0.05$). The proportion of ramie, bacteria level, enzyme level, and their interactions affected the lactic acid and ammonia-N contents of the silage ($p < 0.05$). Acetic acid and butyric acid changed with ramie proportion, enzyme level, their interactions with each other, and bacteria level ($p < 0.05$). Propionic acid content was affected by ramie proportion and the B × E interaction ($p < 0.05$).

3.4. Best Combinations of Fermentation Conditions for Silage Quality Parameters

Ideal combinations of ramie proportion with bacteria and enzyme levels for the pH, lactic acid, ammonia-N, and VFAs of mixed silage are presented in Table 5. The factor ranking regarding the effect on pH and lactic acid was ramie > bacteria > enzyme, where a ramie proportion of 0 with 200 mg/kg bacteria and 30 mg/kg enzyme levels was the ideal combination for silage pH and lactic acid content. The factor ranking regarding the effect on ammonia-N content was bacteria > ramie > enzyme, where a ramie proportion of 100 with a 200 mg/kg bacteria level and a 20 mg/kg enzyme level was the best combination. The factor ranking regarding the effect on acetic acid content was ramie > enzyme > bacteria, where a ramie proportion of 0 with a 30 mg/kg enzyme level and a bacteria level of 0 was the best combination. The factor ranking regarding the effect on propionic acid was ramie > enzyme > bacteria, where a ramie proportion of 100 with 0 mg/kg bacteria and 0 mg/kg enzyme levels was the ideal combination for propionic acid content. The factor ranking regarding the effect on butyric acid content was ramie > enzyme > bacteria, where a ramie proportion of 0 with a 20 mg/kg enzyme level and a 100 mg/kg bacteria level was the ideal combination for butyric acid.

3.5. Comprehensive Score Analysis of Silage Quality

The results of the comprehensive analysis are presented in Table 6. Silage pH, lactic acid, ash, lignin, acetic acid, and butyric acid were primarily affected by the ramie proportion in the silage, and a proportion of 30 was best for these indices. The DM, CP, and propionic acid contents were high with a ramie proportion of 100. A bacterial dose of 200 mg/kg was best for silage CP, ADF, lignin, pH, lactic acid, and ammonia-N. A bacterial dose of 100 mg/kg was ideal for ash and butyric acid contents. An enzyme level of 20 mg/kg was ideal for silage DM, CP, ash, ADF, and butyric acid, while 30 mg/kg enzyme was ideal for lignin, pH, lactic acid, and acetic acid. The best silage combination was achieved with a ramie proportion of 30%, compound bacteria of 200 mg/kg, and compound enzyme of 20 mg/kg.

Table 4. Analysis of pH, lactic acid, ammonia-N, and volatile fatty acid of mixed silage under different fermentation conditions [1].

Item	Ramie-to-Elephant-Grass Ratio	Bacteria, mg/kg				Enzyme, mg/kg				SEM	p-Value [2]						
		0	100	200	300	0	10	20	30		R	B	E	R × B	R × E	B × E	R × B × E
pH	30:70	4.95	3.97	3.66	4.44	4.95	4.44	3.97	3.66	0.217	0.001	0.017	0.449	0.192	0.024	0.004	0.136
	50:50	4.88	4.87	4.54	4.60	4.60	4.88	4.54	4.87								
	70:30	4.82	4.48	4.56	4.60	4.48	4.56	4.82	4.60								
	100:0	4.98	4.84	4.92	4.86	4.92	4.84	4.86	4.98								
Lactic acid, g/Kg DM	30:70	2.37	12.5	19.0	2.20	2.37	2.20	12.5	19.0	0.451	<0.001	<0.001	<0.001	<0.001	<0.001	<0.001	<0.001
	50:50	3.74	3.19	3.29	2.70	2.70	3.74	3.29	3.19								
	70:30	2.61	2.58	2.83	2.45	2.58	2.83	2.61	2.45								
	100:0	5.02	3.61	4.07	3.69	4.07	3.61	3.69	5.02								
NH_3-N, g/Kg total-N	30:70	4.37	2.33	3.04	3.69	4.37	3.69	2.33	3.04	0.212	<0.001	<0.001	<0.001	<0.001	<0.001	<0.001	<0.001
	50:50	3.46	3.31	1.42	1.45	1.45	3.46	1.42	3.31								
	70:30	3.08	2.45	1.14	2.42	2.45	1.14	3.08	2.42								
	100:0	2.54	1.96	0.90	2.17	0.90	1.96	2.17	2.54								
Acetic acid, g/Kg DM	30:70	10.9	4.95	4.45	9.48	10.9	9.48	4.95	4.45	1.515	<0.001	0.098	0.019	0.006	0.017	<0.001	0.026
	50:50	12.4	15.7	17.8	18.3	18.3	12.4	17.8	15.7								
	70:30	16.6	17.8	19.5	18.8	17.8	19.5	16.6	18.8								
	100:0	15.7	20.8	21.3	19.4	21.3	20.8	19.4	15.7								
Propionic acid, g/Kg DM	30:70	2.96	1.67	2.32	2.44	2.96	2.44	1.67	2.32	0.372	<0.001	0.788	0.322	0.201	0.353	<0.001	0.189
	50:50	4.17	4.48	3.69	4.02	4.02	4.17	3.69	4.48								
	70:30	4.89	4.56	4.35	4.19	4.56	4.35	4.89	4.19								
	100:0	6.66	7.25	7.59	6.98	7.59	7.25	6.98	6.66								
Butyric acid, g/Kg DM	30:70	3.31	1.77	5.29	3.45	3.31	3.45	1.77	5.29	0.441	<0.001	0.107	0.026	<0.001	<0.001	<0.001	<0.001
	50:50	6.88	5.71	5.28	4.65	4.65	6.88	5.28	5.71								
	70:30	5.36	5.31	4.41	5.24	5.31	4.41	5.36	5.24								
	100:0	4.90	5.17	5.32	4.97	5.32	5.17	4.97	4.90								

[1] DM, dry matter; SEM, standard error of the mean. [2] R = proportion of ramie in silage, B = level of bacterial inoculation, E = level of enzyme added.

Table 5. The optimal combination of ramie-to-elephant-grass ratio, enzyme level, and bacteria inoculation level for the pH, lactic acid, ammonia-N, and volatile fatty acid of mixed silage.

Item [1]		Fermentation Conditions [2]			Optimum Combination [3]
		A	B	C	
pH	K1 [4]	4.25	4.91	4.74	Factor ranking A > B > C A1B3C4
	K2	4.72	4.52	4.67	
	K3	4.61	4.42	4.54	
	K4	4.90	4.61	4.53	
	Range	0.65	0.49	0.18	
Lactic acid, g/Kg DM	K1	9.03	3.44	2.93	Factor ranking A > B > C A1B3C4
	K2	3.23	5.56	3.07	
	K3	2.62	7.31	5.62	
	K4	4.15	2.71	7.43	
	Range	6.41	4.60	4.50	
NH_3-N, g/Kg total-N	K1	3.46	3.36	2.29	Factor ranking B > A > C B3A4C3
	K2	2.41	2.54	2.59	
	K3	2.27	1.63	2.26	
	K4	1.87	2.45	2.83	
	Range	1.59	1.73	0.57	
Acetic acid, g/Kg DM	K1	7.44	13.89	17.06	Factor ranking A > C > B A1C4B1
	K2	16.06	14.48	15.25	
	K3	18.15	15.75	14.43	
	K4	19.20	16.35	13.67	
	Range	11.76	2.46	3.39	
Propionic acid, g/Kg DM	K1	2.35	4.67	4.78	Factor ranking A > C > B A4C1B1
	K2	4.09	4.34	4.41	
	K3	4.50	4.49	4.17	
	K4	7.12	4.27	4.41	
	Range	4.77	0.40	0.61	
Butyric acid, g/Kg DM	K1	3.46	5.11	4.65	Factor ranking A > C > B A1C3B2
	K2	5.63	4.45	4.97	
	K3	5.08	5.07	4.31	
	K4	5.09	4.56	5.28	
	Range	2.17	0.66	0.97	

[1] DM, dry matter. [2] Fermentation conditions represented by the ratio of ramie to elephant grass in silage (30:70, 50:50, 70:30, and 100:0; Factor A), level of bacteria addition (0, 100, 200, and 300 mg/kg; Factor B), and level of enzyme addition (0, 10, 20, and 30 mg/kg; Factor C). [3] A1, A2, A3, and A4 represent the proportion of ramie in the silage (30, 50, 70, and 100, respectively); B1, B2, B3, and B4 represent the level of added bacteria (0, 100, 200, and 300 mg/kg, respectively); and C1, C2, C3, and C4 represent the level of added enzyme (0, 10, 20, and 30 mg/kg, respectively). [4] Ki represents the average value at that specific level in an increasing order from K1 to K4.

Table 6. Levels of ramie, bacteria, and enzyme added in mixed silage corresponding to the highest values for silage quality parameters, [1] estimated through the comprehensive balance method.

Items	DM	CP	Metabolizable Energy	Ash	NDF	ADF	Lignin	pH	Lactic Acid	NH_3-N	Acetic Acid	Propionic Acid	Butyric Acid
Ramie proportion (%)	100	100	70	30	50	50	30	30	30	100	30	100	30
Compound bacteria dose (mg/kg)	0	200	0	100	300	200	200	200	200	200	0	0	100
Compound enzyme dose (mg/kg)	20	20	0	20	0	20	30	30	30	20	30	0	20

[1] DM, dry matter; CP, crude protein; NDF, neutral detergent fiber; ADF, acid detergent fiber.

4. Discussion

The nutrient profiles of ramie and elephant grass were close to those of previous reports [32]. However, the DM content of ramie was relatively lower than previously reported values, which might be attributable to the maturity level and cutting stage of the plant. Ensiling is a practical approach to preserve forages with high moisture content using lactic acid bacteria, which anaerobically convert carbohydrates into organic acids to inhibit the activity of spoilage organisms by decreasing the pH [22]. It is reported that the water-soluble carbohydrate content of ramie is low, which causes low-quality silage after fermentation [33]. This results from the lower multiplication and dominance of lactic acid bacteria [34]. Therefore, a combination of different forages could provide sufficient fermentation conditions and promote lactic acid bacteria. Several studies have reported that bacterial inoculation and enzyme addition affect carbohydrate content and decrease the fiber content of silage [35–37]. Therefore, it might be inferred that enzymes degrade the fiber content to improve the sugar content, promoting lactic acid bacteria, which ultimately decreases the pH, leading to the inhibition of non-lactic acid bacterial activity and plant enzymes for proteolysis [20]. Numerous previous studies have reported that enzymes and/or bacteria could improve fermentation by decreasing the pH and improving the aerobic stability of silage [38–40].

The proportion of ramie in the silage and the level of bacterial inoculation affected the DM content of the silage in the present study. Although no independent effect of the enzyme was detected, two-way and three-way interactions indicated that the enzyme levels interacted with both ramie proportion and bacteria levels to affect the DM content. The optimal combination for DM was observed when the ramie proportion was 100%, bacterial quantity was 0 mg/kg, and enzyme addition was 20 mg/kg. However, the DM content of silage was greater with 100% ramie due to the relatively higher initial DM of ramie compared with elephant grass, and no additional beneficial effect of bacterial quantity and enzyme addition was observed on DM content. Previously, improved ruminal degradability of silage or no change in the ruminal degradability of silage with bacteria and enzyme addition was reported. These differences in silage quality were attributed to the roughages used, as bacteria and enzymes affect the degradability of different roughages differently [20,21]. In the present study, the response of silage characteristics was also changed with the proportion of ramie in the mixed silage. Interestingly, different interactions observed in this study indicated that the introduction of enzymes and bacteria at the same time could be more beneficial through their synergistic effects than individual treatments with bacteria and enzymes [20]. For CP content, the optimum bacterial quantity was 200 mg/kg when the proportion of ramie silage was 100% and enzyme addition was 20 mg/kg. During the ensiling, changes in protein fractions may occur due to extensive proteolysis by plant and microbial proteases [8]. Similar to previous reports, the addition of bacteria might have reduced the proteolysis of high-moisture silage through decreased pH [8]. Better protein preservation in different silages is reported with bacterial addition compared with controls in the literature [41]. Another interesting finding of the current study observed through the three-way interaction of R × B × E revealed that the ability of bacteria to preserve the protein content of silage is influenced by silage enzyme addition and forage type. According to the best of our knowledge, this is the first study reporting the effects of forage type × bacteria × enzyme interaction on mixed crop silage quality, and further studies using different ratios of various forage sources are required for further understand the interaction of bacteria and enzymes with forage type and level. For energy, the optimum proportion of ramie silage was 70% when no bacteria and enzymes were added. Moreover, for crude ash, the optimum proportion of ramie silage was 30% when the quantity of bacterial culture and enzyme was 100 and 20 mg/kg, respectively. Crude ash and ADF contents were significantly decreased with enzyme addition, which is in agreement with the previous studies [42,43]. However, it is noteworthy that the role of forage type and bacteria level also needs to be considered for optimum ADF content when adding enzymes to mixed crop silage, as indicated by their interaction in the present study. For ADF, optimum conditions were achieved by 50% ramie silage with 200 mg/kg bacterial quantity and 20 mg/kg enzyme addition, whereas, for NDF, optimum conditions were achieved by 50% ramie silage with 300 mg/kg bacterial quantity, without any enzyme addition. For lignin content, the optimum proportion of ramie silage was 30% when the quantity of bacteria and enzyme was 200 mg/kg and 30 mg/kg, respectively. The enzymatic hydrolysis of plants is primarily controlled by the diffusion movements of enzyme solutions to the specific target spots of plant cells. Differences in moisture content may result in interference with the transportation system and promote the enzymatic hydrolysis of silages with optimum moisture contents, leading to a greater release of soluble carbohydrates during the fermentation process [43]. Furthermore, pH is also a crucial factor that affects enzymatic activity.

The pH of silage is an important indicator to evaluate the quality of ensiling [44]. The addition of bacteria accelerates the decrease in pH and effectively inhibits the activity of undesirable organisms [45]. No interaction of ramie with bacteria but a two-way interaction of ramie with enzyme level and an interaction between bacteria and enzyme level indicated that a lower ramie proportion provided better conditions for enzyme and bacteria to reduce the pH of silage by increasing the production of lactic acid. Consequently, the same optimal combination was observed for the lactic acid and pH of silage. Moreover, propionic acid and butyric acid contents increased with increasing ramie proportion in the silage, which is considered undesirable for silage quality [46] as their production results in extensive energy losses [43]. Butyric acid is generally produced by certain anaerobic bacteria, particularly clostridium [47], and increased butyric acid with a high ramie proportion was indicative of increased clostridium levels in this study. Although no direct effect of bacteria on butyric acid or enzyme on propionic acid was observed, both bacteria and enzyme interacted to affect the production of butyric acid and propionic acid in the present study. The lowest ammonia-N content as an indicator of the lowest protein degradation [26] was observed with a high ramie proportion combined with a 200 mg/kg bacteria level. This indicated that this bacterial combination was strong enough to keep working, even with the highest ramie level in the silage, and was not dependent upon the silage CP content. For pH and lactic acid content, the optimum proportion of ramie in the silage was 30% when the bacterial quantity was 200 mg/kg and the enzyme addition was 30 mg/kg in this study. Bacterial addition ensured a sufficient number of lactic acid-producing bacteria to decrease the pH, undesirable fermentation, and proteolysis [48], thereby improving forage preservation [49].

5. Conclusions

In the current study, the proportion of ramie, bacterial inoculation, and enzyme addition not only affected the silage quality independently but also interacted to alter the silage fermentation characteristics. Silage dry matter and protein contents increased with an increase in the ramie proportion and bacterial inoculation level in the silage. Acid detergent fiber and lignin contents decreased when bacteria and enzymes were added to the silage. Overall, the optimum quality of silage was observed when the ratio of ramie to elephant grass was 30:70 and the amounts of compound bacteria and enzyme were 200 mg/kg and 20 mg/kg, respectively. These findings infer that ensiling characteristics are affected by forage type/ratio, while bacteria and enzymes interact not only with each other but also with forage type to affect the composition of mixed silage prepared from ramie and elephant grass.

Author Contributions: Conceptualization, M.L., F.-u.H. and C.Y.; Data curation: M.L., F.-u.H., L.P. and F.X.; Validation, C.Y. and K.W.; Formal analysis, M.L., F.-u.H., L.P., H.Z. and F.X.; Investigation, M.L., F.-u.H., Q.D. and C.Y.; Methodology, M.L., F.-u.H., H.Z. and M.U.A.; Resources, C.Y., Q.D. and K.W.; Writing—original draft preparation, M.L., F.-u.H., M.U.A. and L.P.; writing—review and editing, F.-u.H., M.U.A. and C.Y.; Visualization, H.Z., Q.D. and K.W.; Supervision, C.Y.; Funding acquisition, M.L., F.-u.H. and C.Y. All authors have read and agreed to the published version of the manuscript.

Funding: This research was supported by the Guangxi Science and Technology Major Project under grant number GuiKeAA22068099, the Guangxi Natural Science Foundation Youth Science Fund under grant number 2021GXNSFBA196060, and the Guangxi Milk Buffalo Innovation Team at the National Modern Agricultural Industry Technology System under grant number nycytxgxcxtd-2021-21-03.

Institutional Review Board Statement: Not applicable.

Data Availability Statement: The data presented in this study are available from the corresponding author upon reasonable request.

Conflicts of Interest: The authors declare no conflicts of interest.

References

1. Larsen, S.U.; Hjort-Gregersen, K.; Vazifehkhoran, A.H.; Triolo, J.M. Co-ensiling of straw with sugar beet leaves increases the methane yield from straw. *Bioresour. Technol.* **2017**, *245*, 106–115. [CrossRef] [PubMed]
2. Jiang, Y.; Dennehy, C.; Lawlor, P.G.; Hu, Z.; Zhan, X.; Gardiner, G.E. Inactivation of enteric indicator bacteria and system stability during dry co-digestion of food waste and pig manure. *Sci. Total Environ.* **2018**, *612*, 293–302. [CrossRef] [PubMed]
3. Zeng, T.; Li, X.; Guan, H.; Yang, W.; Liu, W.; Liu, J.; Du, Z.; Li, X.; Xiao, Q.; Wang, X. Dynamic microbial diversity and fermentation quality of the mixed silage of corn and soybean grown in strip intercropping system. *Bioresour. Technol.* **2020**, *313*, 123655. [CrossRef]
4. Gao, S.; Liao, Y.; Li, Z.; Hou, Z.; Zhong, R.; Wu, D. Including ramie (*Boehmeria nivea* L. Gaud) in the diet of dairy cows: Effects on production performance, milk composition, rumen fermentation, and nutrient digestion. *Ital. J. Anim. Sci.* **2020**, *19*, 240–244. [CrossRef]
5. Mitra, S.; Saha, S.; Guha, B.; Chakrabarti, K.; Satya, P.; Sharma, A.; Gawande, S.; Kumar, M.; Saha, M. Ramie: The strongest bast fibre of nature. In *Technical Bulletin No. 8*; Central Research Institute for Jute and Allied Fibres, ICAR: Barrackpore, Kolkata, India, 2013; pp. 1–38.
6. Kipriotis, E.; Heping, X.; Vafeiadakis, T.; Kiprioti, M.; Alexopoulou, E. Ramie and kenaf as feed crops. *Ind. Crops Prod.* **2015**, *68*, 126–130. [CrossRef]
7. Tang, S.; He, Y.; Zhang, P.; Jiao, J.; Han, X.; Yan, Q.; Tan, Z.; Wang, H.; Wu, D.; Yu, L. Nutrient digestion, rumen fermentation and performance as ramie (*Boehmeria nivea*) is increased in the diets of goats. *Anim. Feed Sci. Technol.* **2019**, *247*, 15–22. [CrossRef]
8. Chen, G.X.; Xiang, H.; Zheng, X.; Hou, Z.P.; Wu, D.Q. Evaluation of mix ratio on quality and feeding value of silage ramie and rice straw silage. *China Feed* **2019**, *5*, 82–85.
9. Kabirizi, J.; Muyekho, F.; Mulaa, M.; Msangi, R.; Pallangyo, B.; Kawube, G.; Zziwa, E.; Mugerwa, S.; Ajanga, S.; Lukwago, G.; et al. Napier Grass Feed Resource: Production, Constraints and Implications for Smallholder Farmers in East and Central Africa. The Eastern Africa Agricultural Productivity Project (EAAPP). 2015. Available online: http://www.researchgate.net/publication/281556114 (accessed on 17 September 2024).
10. Rambau, M.D.; Fushai, F.; Callaway, T.R.; Baloyi, J.J. Dry matter and crude protein degradability of Napier grass (*Pennisetum purpureum*) silage is affected by fertilization with cow-dung bio-digester slurry and fermentable carbohydrate additives at ensiling. *Transl. Anim. Sci.* **2022**, *6*, txac075. [CrossRef] [PubMed]
11. Li, F.; Ding, Z.; Ke, W.; Xu, D.; Zhang, P.; Bai, J.; Mudassar, S.; Muhammad, I.; Guo, X. Ferulic acid esterase-producing lactic acid bacteria and cellulase pretreatments of corn stalk silage at two different temperatures: Ensiling characteristics, carbohydrates composition and enzymatic saccharification. *Bioresour. Technol.* **2019**, *282*, 211–221. [CrossRef]
12. Wang, Y.-L.; Wang, W.-K.; Wu, Q.-C.; Zhang, F.; Li, W.-J.; Yang, Z.-M.; Bo, Y.-K.; Yang, H.-J. The effect of different lactic acid bacteria inoculants on silage quality, phenolic acid profiles, bacterial community and in vitro rumen fermentation characteristic of whole corn silage. *Fermentation* **2022**, *8*, 285. [CrossRef]
13. Bao, J.; Wang, L.; Yu, Z. Effects of different moisture levels and additives on the ensiling characteristics and in vitro digestibility of stylosanthes silage. *Animals* **2022**, *12*, 1555. [CrossRef] [PubMed]
14. Michelena, J.; Senra, A.; Fraga, C. Effect of formic acid, propionic acid and pre-drying on the nutritive value of king grass (*Pennisetum purpureum*) silage. *Cuba. J. Agric. Sci.* **2002**, *36*, 231–236.
15. Zhao, M.-m.; Yu, Z. Effects of lactic acid bacteria and cellulase on napier grass silages. *Acta Agrestia Sin.* **2015**, *23*, 205.
16. Li, F.; Ke, W.; Ding, Z.; Bai, J.; Zhang, Y.; Xu, D.; Li, Z.; Guo, X. Pretreatment of *Pennisetum sinese* silages with ferulic acid esterase-producing lactic acid bacteria and cellulase at two dry matter contents: Fermentation characteristics, carbohydrates composition and enzymatic saccharification. *Bioresour. Technol.* **2020**, *295*, 122261. [CrossRef]
17. Nazar, M.; Xu, L.; Ullah, M.W.; Moradian, J.M.; Wang, Y.; Sethupathy, S.; Iqbal, B.; Nawaz, M.Z.; Zhu, D. Biological delignification of rice straw using laccase from *Bacillus ligniniphilus* L1 for bioethanol production: A clean approach for agro-biomass utilization. *J. Clean. Prod.* **2022**, *360*, 132171. [CrossRef]
18. Bao, X.; Feng, H.; Guo, G.; Huo, W.; Li, Q.; Xu, Q.; Liu, Q.; Wang, C.; Chen, L. Effects of laccase and lactic acid bacteria on the fermentation quality, nutrient composition, enzymatic hydrolysis, and bacterial community of alfalfa silage. *Front. Microbiol.* **2022**, *13*, 1035942. [CrossRef]
19. Tian, J.; Yu, Y.; Yu, Z.; Shao, T.; Na, R.; Zhao, M. Effects of lactic acid bacteria inoculants and cellulase on fermentation quality and in vitro digestibility of *Leymus chinensis* silage. *Grassl. Sci.* **2014**, *60*, 199–205. [CrossRef]
20. Li, M.; Zhou, H.; Zi, X.; Cai, Y. Silage fermentation and ruminal degradation of stylo prepared with lactic acid bacteria and cellulase. *Anim. Sci. J.* **2017**, *88*, 1531–1537. [CrossRef]
21. Zhang, Q.; Yu, Z.; Yang, H.; Na, R. The effects of stage of growth and additives with or without cellulase on fermentation and in vitro degradation characteristics of *Leymus chinensis* silage. *Grass Forage Sci.* **2016**, *71*, 595–606. [CrossRef]

22. He, L.; Zhou, W.; Wang, Y.; Wang, C.; Chen, X.; Zhang, Q. Effect of applying lactic acid bacteria and cellulase on the fermentation quality, nutritive value, tannins profile and in vitro digestibility of *Neolamarckia cadamba* leaves silage. *J. Anim. Physiol. Anim. Nutr.* **2018**, *102*, 1429–1436. [CrossRef]
23. Kaewpila, C.; Thip-Uten, S.; Cherdthong, A.; Khota, W. Impact of cellulase and lactic acid bacteria inoculant to modify ensiling characteristics and in vitro digestibility of sweet corn stover and cassava pulp silage. *Agriculture* **2021**, *11*, 66. [CrossRef]
24. Zheng, J.; Wittouck, S.; Salvetti, E.; Franz, C.M.A.P.; Harris, H.M.B.; Mattarelli, P.; O'Toole, P.W.; Pot, B.; Vandamme, P.; Walter, J.; et al. A taxonomic note on the genus *Lactobacillus*: Description of 23 novel genera, emended description of the genus *Lactobacillus Beijerinck* 1901, and union of *Lactobacillaceae* and *Leuconostocaceae*. *Int. J. Syst. Evol. Microbiol.* **2020**, *70*, 2782–2858. [CrossRef] [PubMed]
25. AOAC. *Official Methods of Analysis*, 18th ed.; Association of Official Analytical Chemists: Gaithersburg, MD, USA, 2005.
26. Li, D.; Xie, H.; Zeng, F.; Luo, X.; Peng, L.; Sun, X.; Wang, X.; Yang, C. An assessment on the fermentation quality and bacterial community of corn straw silage with pineapple residue. *Fermentation* **2024**, *10*, 242. [CrossRef]
27. Van Soest, P.v.; Robertson, J.B.; Lewis, B.A. Methods for dietary fiber, neutral detergent fiber, and nonstarch polysaccharides in relation to animal nutrition. *J. Dairy Sci.* **1991**, *74*, 3583–3597. [CrossRef]
28. Zhang, Q.; Li, X.; Zhao, M.; Yu, Z. Lactic acid bacteria strains for enhancing the fermentation quality and aerobic stability of *Leymus chinensis* silage. *Grass Forage Sci.* **2016**, *71*, 472–481. [CrossRef]
29. Xie, H.; Peng, L.; Li, M.; Guo, Y.; Liang, X.; Peng, K.; Yang, C. Effects of mixed sugarcane tops and napiergrass silages on fermentative quality, nutritional value, and milk yield in water buffaloes. *Anim. Sci. J.* **2023**, *94*, e13824. [CrossRef] [PubMed]
30. Sung, J.Y.; Kim, B.G. Prediction equations for digestible and metabolizable energy concentrations in feed ingredients and diets for pigs based on chemical composition. *Anim. Biosci.* **2021**, *34*, 306. [CrossRef] [PubMed]
31. Pan, X.; Yang, L.; Xue, F.; Xin, H.; Jiang, L.; Xiong, B.; Beckers, Y. Relationship between thiamine and subacute ruminal acidosis induced by a high-grain diet in dairy cows. *J. Dairy Sci.* **2016**, *99*, 8790–8801. [CrossRef]
32. Huifen, D.; Hong, L.; Renke, H.; Baizhong, Z.; Kangle, Y.; Ao, S. Study on mixed silage of forage ramie and *Pennisetum purpureum* (Schum.) in different proportions. *Pak. J. Agric. Sci.* **2023**, *60*, 593–597.
33. Despal, D.; Permana, I.; Safarina, S.; Tatra, A. Addition of water soluble carbohydrate sources prior to ensilage for ramie leaves silage qualities improvement. *Media Peternak.* **2011**, *34*, 69–76. [CrossRef]
34. Chen, X.-Z.; Liu, Y.; Gao, C.-F.; Zhang, X.-P.; Li, W.-Y.; Dong, X.-N.; Weng, B.-q. Effect of mixing ratio of ramie and hybrid pennisetumonon quality of fermented silage. *Fujian J. Agric. Sci.* **2015**, *30*, 836–840.
35. Zhu, Y.; Nishino, N.; Xusheng, G. Chemical changes during ensilage and in sacco degradation of two tropical grasses: Rhodesgrass and guineagrass treated with cell wall-degrading enzymes. *Asian-Australas. J. Anim. Sci.* **2011**, *24*, 214–221. [CrossRef]
36. Guo, G.; Yuan, X.; Li, L.; Wen, A.; Shao, T. Effects of fibrolytic enzymes, molasses and lactic acid bacteria on fermentation quality of mixed silage of corn and hulless–barely straw in the Tibetan Plateau. *Grassl. Sci.* **2014**, *60*, 240–246. [CrossRef]
37. Li, M.; Zi, X.; Zhou, H.; Hou, G.; Cai, Y. Effects of sucrose, glucose, molasses and cellulase on fermentation quality and in vitro gas production of king grass silage. *Anim. Feed Sci. Technol.* **2014**, *197*, 206–212. [CrossRef]
38. Moselhy, M.A.; Borba, J.P.; Borba, A.E. Improving the nutritive value, in vitro digestibility and aerobic stability of *Hedychium gardnerianum* silage through application of additives at ensiling time. *Anim. Feed Sci. Technol.* **2015**, *206*, 8–18. [CrossRef]
39. Napasirth, V.; Napasirth, P.; Sulinthone, T.; Phommachanh, K.; Cai, Y. Microbial population, chemical composition and silage fermentation of cassava residues. *Anim. Sci. J.* **2015**, *86*, 842–848. [CrossRef]
40. Chen, L.; Guo, G.; Yuan, X.; Zhang, J.; Li, J.; Shao, T. Effects of applying molasses, lactic acid bacteria and propionic acid on fermentation quality, aerobic stability and in vitro gas production of total mixed ration silage prepared with oat–common vetch intercrop on the Tibetan Plateau. *J. Sci. Food Agric.* **2016**, *96*, 1678–1685. [CrossRef]
41. Contreras-Govea, F.E.; Muck, R.E.; Mertens, D.R.; Weimer, P.J. Microbial inoculant effects on silage and in vitro ruminal fermentation, and microbial biomass estimation for alfalfa, bmr corn, and corn silages. *Anim. Feed Sci. Technol.* **2011**, *163*, 2–10. [CrossRef]
42. Su, R.; Ni, K.; Wang, T.; Yang, X.; Zhang, J.; Liu, Y.; Shi, W.; Yan, L.; Jie, C.; Zhong, J. Effects of ferulic acid esterase-producing *Lactobacillus fermentum* and cellulase additives on the fermentation quality and microbial community of alfalfa silage. *PeerJ* **2019**, *7*, e7712. [CrossRef]
43. Xu, J.; Zhang, K.; Lin, Y.; Li, M.; Wang, X.; Yu, Q.; Sun, H.; Cheng, Q.; Xie, Y.; Wang, C. Effect of cellulase and lactic acid bacteria on the fermentation quality, carbohydrate conversion, and microbial community of ensiling oat with different moisture contents. *Front. Microbiol.* **2022**, *13*, 1013258. [CrossRef]
44. Ren, X.; Tian, H.; Zhao, K.; Li, D.; Xiao, Z.; Yu, Y.; Liu, F. Research on pH value detection method during maize silage secondary fermentation based on computer vision. *Agriculture* **2022**, *12*, 1623. [CrossRef]
45. Kung, L., Jr.; Shaver, R.; Grant, R.; Schmidt, R. Silage review: Interpretation of chemical, microbial, and organoleptic components of silages. *J. Dairy Sci.* **2018**, *101*, 4020–4033. [CrossRef] [PubMed]
46. Dong, L.; Zhang, H.; Gao, Y.; Diao, Q. Dynamic profiles of fermentation characteristics and bacterial community composition of *Broussonetia papyrifera* ensiled with perennial ryegrass. *Bioresour. Technol.* **2020**, *310*, 123396. [CrossRef] [PubMed]
47. Wang, C.; He, L.; Xing, Y.; Zhou, W.; Yang, F.; Chen, X.; Zhang, Q. Fermentation quality and microbial community of alfalfa and stylo silage mixed with Moringa oleifera leaves. *Bioresour. Technol.* **2019**, *284*, 240–247. [CrossRef]

48. Kung, L., Jr.; Taylor, C.; Lynch, M.; Neylon, J. The effect of treating alfalfa with *Lactobacillus buchneri* 40788 on silage fermentation, aerobic stability, and nutritive value for lactating dairy cows. *J. Dairy Sci.* **2003**, *86*, 336–343. [CrossRef]
49. Pahlow, G. Microbiology of ensiling. In *Silage Science and Technology*; American Society of Agronomy, Crop Science Society of America, Soil Science Society of America: Madison, WI, USA, 2003; pp. 31–93.

agriculture

Article

Screening Lactic Acid Bacteria Strains for Their Tolerance to Increased Osmotic Pressure and Their Suitability to Ensile High Dry Matter Forages

Siriwan D. Martens [1,*], Wolfgang Wagner [2], Mariana Schneider [3], Klaus Hünting [4], Susanne Ohl [5] and Christof Löffler [6]

1 Department of Animal Husbandry, Saxon State Office for Environment, Agriculture and Geology (LfULG), Am Park 3, 04886 Köllitsch, Germany
2 Agricultural Technology Center (LTZ) Augustenberg, Nesslerstr. 25, 76227 Karlsruhe, Germany
3 Bavarian State Research Center for Agriculture (LfL), Institute for Animal Nutrition and Feed Management, Prof.-Dürrwaechter-Platz 3, 85586 Poing, Germany
4 Experimental and Training Centre Agriculture Haus Riswick, Chamber of Agriculture North Rhine-Westphalia, Elsenpass 5, 47533 Kleve, Germany
5 Research and Training Centre Futterkamp, Chamber of Agriculture Schleswig-Holstein, Gutshof, 24327 Blekendorf, Germany
6 Agricultural Centre for Cattle Production, Grassland Management, Dairy Food, Wildlife and Fisheries Baden-Württemberg (LAZBW), Atzenberger Weg 99, 88326 Aulendorf, Germany
* Correspondence: siriwan.martens@smekul.sachsen.de

Abstract: Lactic acid bacteria (LAB) should not only survive, but also perform under increased osmotic pressure in the process of ensiling, which results from the best practice of wilting forage. Simple laboratory protocols are needed to select suitable LAB strains as inoculants for high dry matter (DM) conditions. The aim of this study was to simulate conditions of high osmolality without inducing salt stress and to select a suitable indicator of LAB performance. For that, an MRS medium was enriched with increasing concentrations of glucose and fructose plus a maximum of 28 g KCl/L until achieving an osmolality of 2.4 osmol/kg. Both, growth in the inoculated medium and pH decline, were then compared to the LAB performance in the basic medium. The latter was clearly delayed in the new medium. Finally, the method was validated by comparing the pH of small-scale grass silages of 30–35 and 45–49% target DM after 3–5 days of ensiling to the pH values of the microbiological growth medium. The pH levels of treatments with the homofermentative LAB were clearly attributable to the dry matter or the sugar concentration, respectively. The developed liquid growth medium sufficiently approximates high DM conditions to select for the osmotolerant homofermentative LAB.

Keywords: biological silage additives; ensiling; growth medium; osmolality; protocol; strain selection; wilting; validation

Citation: Martens, S.D.; Wagner, W.; Schneider, M.; Hünting, K.; Ohl, S.; Löffler, C. Screening Lactic Acid Bacteria Strains for Their Tolerance to Increased Osmotic Pressure and Their Suitability to Ensile High Dry Matter Forages. *Agriculture* **2024**, *14*, 825. https://doi.org/10.3390/agriculture14060825

Academic Editor: Siran Wang

Received: 30 April 2024
Revised: 22 May 2024
Accepted: 23 May 2024
Published: 25 May 2024

1. Introduction

The best ensiling practice includes the rapid wilting of forages to above 300 g dry matter (DM)/kg. This reduces silage effluent, decreases respiration and proteolysis, increases the concentration of water-soluble carbohydrates as fermentation substrate, and gives preference to lactic acid bacteria (LAB) compared to enterobacteria and clostridia in order to reduce the risk of butyric acid fermentation under anaerobic conditions [1–4]. Facultatively heterofermentative (colloquially referred to as "homofermentative") LAB are often added to promote lactic acid fermentation during ensiling. *Lactiplantibacillus plantarum* as one representative of this group tends to have a higher resistance to osmotic stress than *Clostridium butyricum*, especially at lower pH [5,6]. Osmotic stress is increased by wilting [7]. The suppliers of biological silage additives face the challenge of offering LAB strains, which are particularly suitable for high dry matter conditions.

There have been attempts to simulate high osmotic pressure in microbiological culture media by increasing the concentration of osmotically active particles in the solution, which is the osmolality. Hoedtke [7] investigated the osmolality of various temperate forages and their silages. For *Lolium perenne* wilted to 46% DM, she determined an osmolality of 2.4 osmol/kg plant extract. Whereas some components of forages such as mono- and disaccharides and salts have an osmotic effect, others such as proteins and starch have very little effect [8,9]. To simulate the increased osmotic pressure and decreased water activity, salts such as NaCl have been added to a solution [10]. Moreover, in a test to approve biological silage additives for osmotolerance, KCl has been added to MRS agar (97 g KCl/L) as an elective agent and to simulate the reduced water activity at around 45% dry matter in forages [11]. Similarly, this is done in the Rostock fermentation test, a rapid in vitro test to evaluate the effectiveness of silage inoculants using minced forage material in an aqueous solution [12]. However, the observation of the strong inhibition of either the growth or activity of various LAB strains when adding salt was the starting point of the presented study to look for reasons and alternatives. Glasker et al. [13] found that both NaCl and KCl induce salt stress in *Lactiplantibacillus plantarum*, a species which is currently represented in many silage inoculants. Thus, these salts are less suitable to simulate hyperosmotic conditions in a biological system. In order to cultivate osmophilic microorganisms, mostly the use of glucose or sometimes saccharose is recommended [14]. Plate count is favored as a relatively simple and cost-effective standardized method. So, it was used in the above-mentioned approval scheme for ensiling agents [11]. However, in contrast to the objective of pure growth, for ensiling, the fermentation effectiveness is of eminent importance. As metabolic activity is sometimes less inhibited than cell propagation at increased osmolality [15,16], it is advised to rather measure an indicator for activity than to estimate cell counts. The Rostock fermentation test is one example of activity measurement in a simulated ensiling environment [12,17–19]. In the initial phase of its development, it was proven that the main agent that caused a pH decrease was lactic acid produced by the inoculated LAB. Consequently, pH was employed as an indicator of LAB activity in the present study.

The aim of this study was to develop a laboratory method to screen lactic acid bacteria for their lactic acid fermentation capacity in highly wilted forage. This approach included three steps:

(1) Developing a standardized medium (liquid as broth and solid with agar) for LAB with an osmolality of 2.4 osmol/kg (simulating about 45% DM in a temperate grass) without provoking salt stress.
(2) Testing the growth of different LAB strains on the solid medium developed in step (1) compared to an unmodified MRS (de Man, Rogosa, Sharpe [20]) medium, and measuring the acidification in the liquid medium (from step (1)) compared to the unmodified MRS broth. This two-track approach was followed to indicate metabolic activity in possible contrast to cell propagation.
(3) Validating the ensiling capacity of the selected LAB strains in low and high DM forage compared to the laboratory results, i.e., whether there were statistical effects of the DM level or medium on the final pH across all strains.

Hypothesis: *The modified medium provides a sufficiently accurate approach to simulate high osmolality conditions, which prevail in high DM forages, to reflect the metabolic activity of LAB strains in the initial anaerobic ensiling phase, indicated by pH development.*

2. Materials and Methods

2.1. Step 1 Development of a Medium for LAB with Increased Osmolality and Limited Salt Concentration

As a starting point, the minimum, maximum, and mean contents of sugars and mineral salts (Na, K, and Cl) as osmotically active substances in fresh grass of 70 samples from Saxony for the years 2018–2019 were calculated (Table 1).

Table 1. The content of sugars and mineral salts in fresh grass from pastures in Saxony (2018–2019), n = 70, calculated for 45% dry matter (DM).

Item	Mean	SEM	Median	Minimum	Maximum
Sugar (g/kg)	83.5	3.09	82.9	34.9	153
Sodium (g/kg)	0.34	0.039	0.20	0.05	1.80
Potassium (g/kg)	11.7	0.36	11.8	5.51	18.0
Chloride (g/kg)	2.56	0.178	2.06	0.53	6.09
Salts (sum) (g/kg)	14.6		14.1	6.09	25.8

To cover the nutrient requirements of LAB, an MRS medium was chosen as the basic medium (MRS Broth pH 5.7, Art. No. HP64.1, Carl Roth GmbH + Co. KG, Karlsruhe, Germany).

In a step-by-step approach, the osmolality of the medium was adapted to 2.4 osmol/kg using the increasing concentrations of glucose (D(+)-glucose-monohydrate, Art. No. 108342, Merck KGaA, Darmstadt, Germany) and fructose (D(−)-fructose, Art. No. 4981.4, Carl Roth GmbH + Co. KG, Karlsruhe, Germany) plus a maximum of 28 g KCl (Potassium chloride for analysis, Art. No. 1.04936, Merck KGaA, Darmstadt, Germany) per liter distilled water (Table 2).

Table 2. Increasing concentrations of glucose, fructose, and KCl (g) added to the MRS broth (containing 20 g glucose/L) and 1.0 L of distilled water to obtain ~2.4 osmol/kg in the medium.

Level	+Glucose	+Fructose	+KCl
0	0	0	0
1	10	0	0
2	10	10	0
3	10	20	0
4	10	30	0
5	10	30	14
6	10	30	28
7	30	30	0
8	30	60	0
9	30	60	14
10	30	60	28
11	60	30	0
12	60	60	0
13	60	60	14
14	60	60	28
15	80	80	28
16	100	100	28
17	110	110	28
18	120	120	28
19	130	130	28

The osmolality was determined using Gonotec® Osmomat® (Model 010, ELITech-Group Inc., Logan, UT, USA).

2.2. Step 2 Evaluation of Growth of and Acidification by Different LAB Strains in the Modified Media

The final medium was prepared both as a solid medium using bacteriological agar (GranaCult^{TM} MRS-Agar, Art. No. 1.10660, Merck KGaA, Darmstadt, Germany) and as a broth (see step 1). In total, LAB from 31 biological silage additives were grown. Of those, 10 products contained obligate heterofermentative LAB exclusively or predominantly (no. 10) (Table 3), in the following referred to as "heterofermentative". The other 21 products containing facultatively heterofermentative and obligately homofermentative strains are referred to as "homofermentative" in the following. The products were diluted in serial dilutions and plated on Petri dishes by the pour plate method. They were incubated aerobically at 30 °C for 3 d in the case of the standard MRS and 3 to 10 d in the high-sugar (HS) medium, depending on the growth of colonies, before counting. For pH measurement, the amount of product was inoculated in 10 mL broth as recommended by the manufacturer (g/g fresh matter = g/mL broth), and the tubes were incubated for a maximum of 72 h at 30 °C [20]. The pH was determined at 0, 24, 48, and 72 h using a pH meter (inoLab pH 720, electrode SenTix 81, WTW GmbH, Weilheim, Germany).

Table 3. List of biological additives tested *in vitro* and partly *in situ*, and their declared active lactic acid bacteria (LAB) species.

Additive No.	[1] *L. buchneri*	*L. plantarum*	*E. faecium*	*P. acidilactici*	*L. paracasei*	*L. lactis*	*L. rhamnosus*	*P. pentosaceus*	[1] *L. brevis*	[1] *L. diolivorans*	*L. kefir.*
1	x										
2	x										
3	x										
4	x										
5 *	x										
6	x										
7 #	x										
8	x										
9	x										
10	x						x			x	
11		x									
12 +#		x	x					x			
13		x						x			
14 *		x									
15	x	x					x				
16 +#				x	x	x					
17	x	x						x			
18	x	x					x	x	x		
19		x	x					x			
20 +#		x	x					x			
21		x						x			
22 +#		x	x					x			
23 *	x	x						x			x
24 +#		x	x								
25		x	x					x			
26	x	x					x				
27 +#		x									
28	x	x									
29		x									
30		x									
31		x									

[1] obligately heterofermentative species, x species contained in the additive, * additives were tested in situ at 4 institutes in 2022, + additives were tested in situ at 1 institute in 2022, and # additives were tested in situ at 3–5 institutes in 2023.

2.3. Step 3 Comparison of In Vitro pH Decreases in Growth Medium and In Situ in Temperate Grass at Two Wilting Levels

Year 1: From the 31 products tested in step 2, three additives were selected (nos. 5, 14, and 23) and tested on the grass at four different locations (institutes, see authors affiliations) in the south, east, north, and west of Germany (48°10′–54°17′ N; 6°10′–13°7′ E; 15–516 m ASL) in 2022. The forage was cut from different types of grassland. All four institutions located in Bavaria, North Rhine-Westphalia, Schleswig-Holstein, and Saxony used first-cut grass from from the vegetative to the generative phase. One of the tested additives was a heterofermentative LAB (no. 5, Table 3). Additionally, another two products were tested in one of the three institutes each (nos. 22, 27; 20, 24; 12, 16). Thus, a total of 9 additives were tested in situ in laboratory-scale silages. A DM content of around 30–35% was aimed at for a low osmolality, while a DM content of 45–47% was set as the target for high osmolality. It was then wilted to achieve the target DM (field-dried up to 48 h) and chopped to about 2–4 cm in length. The additives were diluted with water and the different treatments were applied according to the dosage recommended by the manufacturer using a pump sprayer. The treatments comprised a negative control without inoculants (CON) to check the potential of the forage with the native microflora. After thorough mixing, the ensiling material was either manually compacted in WECK® jars (0.5–1.0 L volume) and then closed using a rubber seal and clips or filled in small vacuum bags (13 × 22 cm) and vacuum-sealed in triplicates. After three days at around 22–25 °C the samples were removed from the storage room. The silos were opened and the pH was measured in an aliquot of the forage (10 g) soaked in distilled water (100 mL).

Year 2: In 2023, seven products (nos. 7, 12, 16, 20, 22, 24, and 27; Table 3) containing one purely obligately heterofermentative (no. 7) were grown in vitro in the sterilized standard MRS broth and the HS medium at 30 °C for 48 h (n = 3). The change in pH was measured at 0, 24, and 48 h. Grass silages were prepared at five different locations, see year 1, plus Baden-Wuerttemberg (47°57′ N; 9°38′ O; 590 m ASL), including the first to the fourth cut from May to September. For low osmolality, a DM of about 30–35% was aimed at, and for high osmolality, 47–49% in order to obtain a clear differentiation between the levels. The heterofermentative product (no. 7, Table 3) was inoculated at all five trial sites. All other products (nos. 12, 16, 20, 24, and 27) were applied at three sites each, and product no. 22 at four sites. The ensiling procedure was the same as described for year 1. The only difference was that there was a second opening date on day 5 for the heterofermentative treatments (another three replicates). The theoretical ensilability of the fresh forage material was characterized by analyzing the sugar content (WSC) and buffering capacity (BC) [21] and calculating the fermentability coefficient (FC) [22].

2.4. Statistical Analysis

For step 3, a variance analysis considering the osmolality, adjusted by either wilting or sugar level of the medium, respectively, as effect on final pH was performed using the procedure GLM, followed by a Tukey test post hoc (SAS Studio 3.82, SAS Institute, Cary, NC, USA).

$$Y_i = \mu + OSMOL_i + \varepsilon_i$$

where μ = general mean, i = 1, 2, 3, 4 (osmolality adjusted by low/high DM or sugar level respectively), and ε_i = residual error.

3. Results

3.1. Step 1 A Medium for LAB with Increased Osmolality

The final medium with an increased osmolality (2.41 osmol/kg) consisted of MRS as the dehydrated culture media (54 g) plus 100 g glucose and fructose each, along with 28 g KCl dissolved in 1.0 L distilled water (Figure 1, medium 16).

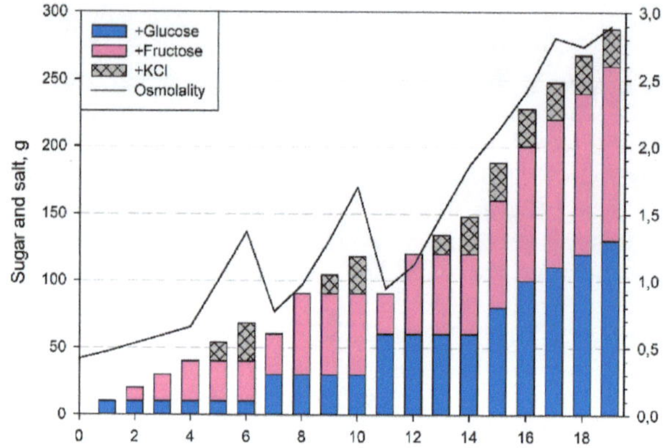

Figure 1. Amount of sugars and KCl added to 1.0 L distilled water and MRS broth and the resulting osmolality of the medium (composition see Table 2).

3.2. Step 2 Growth of and Acidification by Different LAB Strains in the Modified Media

On the high-sugar MRS agar, all of the 10 heterofermentative products exhibited similar growth compared to the standard MRS agar (Figure 2). The colony counts were at least 55% of those on the standard agar. In contrast, only two-thirds of the homofermentative products met this criterion, while two products (nos. 26 and 28) had less than 15% of the colony counts on the high-sugar agar compared to the standard agar (Figure 2).

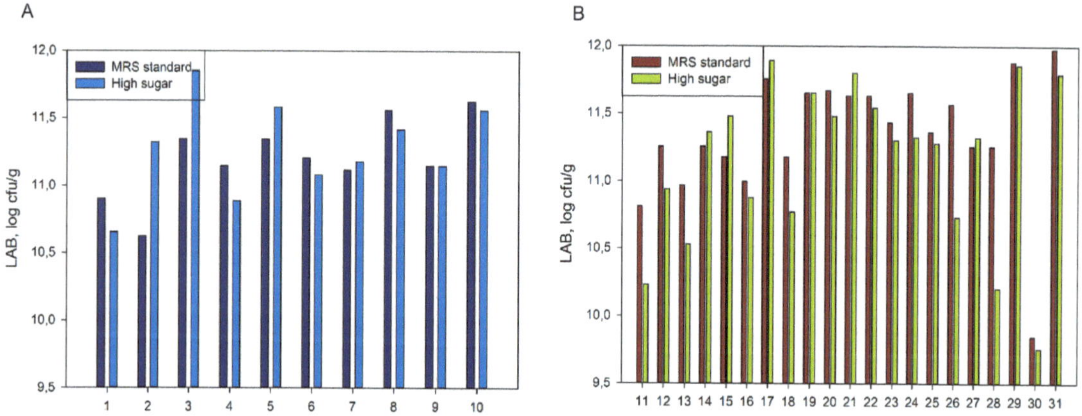

Figure 2. Numbers of lactic acid bacteria (LAB) colonies (cfu/g product, see Table 3) counted on standard and high-sugar MRS agar, (**A**) heterofermentative products, (**B**) homofermentative products.

The initial pH of the standard medium was 5.7 while it was 5.2 in the high-sugar medium. The pH decline was generally more rapid with the homofermentative products in contrast to the heterofermentative ones and obviously delayed in the high-sugar medium (Figure 3). For the homofermentative products, the impeding effect had almost vanished by the end of the second day while for most of the heterofermentative products, there were still differences at that time.

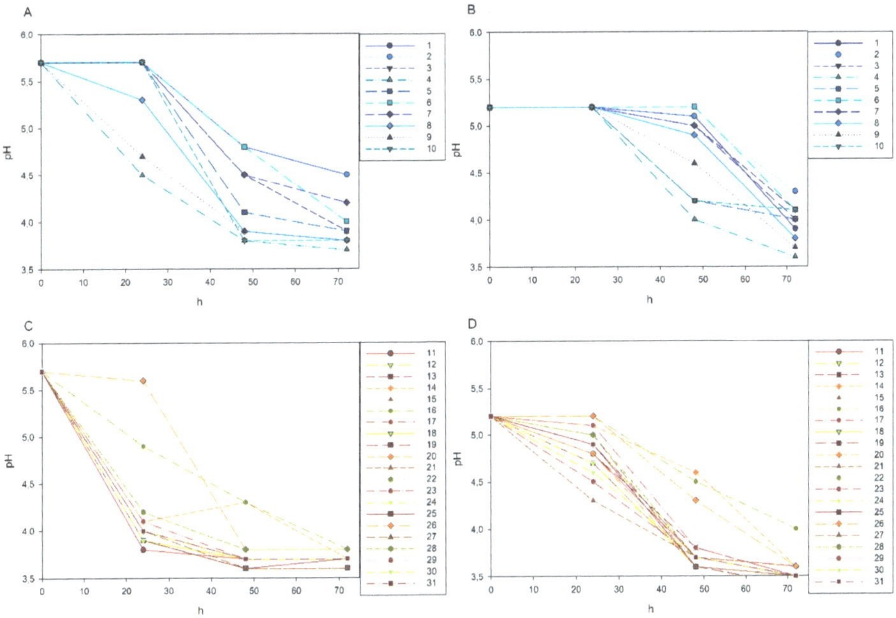

Figure 3. pH development in MRS broth of heterofermentative LAB (**A**,**B**) and homofermentative LAB (**C**,**D**), and in standard MRS (**A**,**C**) and high-sugar MRS (**B**,**D**). LAB species see Table 3.

3.3. *Step 3 In Vitro pH Decrease in Growth Medium and In Situ*

In the first experimental year, the target DM levels both in the low (30–35%) and in the high level (45–47 or 47–49%) were met more exactly with a mean of 336 and 475 g DM/kg than in the second year with 347 and 532 g DM/kg (Supplementary file S1: Table S1). Here, the limit of 500 g DM/kg was mostly exceeded at the high wilting level. The fermentability coefficient, which considers sugar content, buffering capacity, and DM content as a measure for a potentially butyric acid-free fermentation, varied widely even within the same DM level (35–74 for low DM and 56–82 for high DM) (Supplementary file S2: Table S2).

3.3.1. Products Containing Homofermentative LAB Species

With respect to the achieved forage DM, the correlation to the pH of the inoculated silages after three days of ensiling is presented in Figure 4.

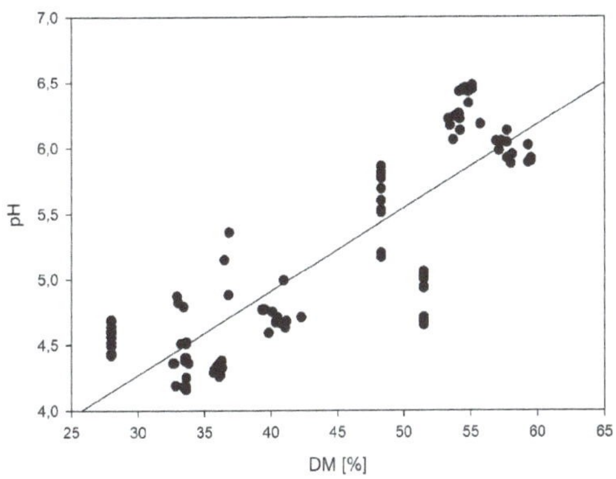

Figure 4. pH value of inoculated grass silages (homofermentative LAB) after 3 days of ensiling as a function of DM concentration (years 2022–2023, n = 117): $f(x) = 0.064\,x + 2.362$, $r^2 = 0.70$, $p < 0.001$.

Table 4 summarizes the comparison of the pH after 3 d of all inoculated silages in the two different DM levels to the pH values after 24 h of all inoculated media (standard and HS MRS). In the first year, the standard medium and the low DM were similar with a pH of around 4.1. On the other hand, the HS medium and the high DM ranged around pH 4.8. In the second year, there was the following ranking between the four groups: high DM (5.8) > HS medium (5.0) > low DM (4.5) > standard medium (4.2).

Table 4. Dry matter (DM) and pH (3 d) of grass silages and pH of growth medium (24 h) inoculated with homofermentative LAB in years 2022 and 2023.

	2022		2023		2022–2023	
Medium/DM Level	**DM (g/kg)**	**pH**	**DM (g/kg)**	**pH**	**DM (g/kg)**	**pH**
Standard medium		4.08 ^{b}		4.19 ^{d}		4.13 ^{c}
Low DM	336 ^{b}	4.12 ^{b}	343 ^{b}	4.51 ^{c}	339 ^{b}	4.32 ^{c}
High-sugar medium		4.72 ^{a}		4.95 ^{b}		4.83 ^{b}
High DM	472 ^{a}	4.83 ^{a}	533 ^{a}	5.79 ^{a}	503 ^{a}	5.31 ^{a}
SEM	0.377	0.039	0.308	0.033	0.235	0.028
p-value						
Medium (DM level)	<0.001	<0.001	<0.001	<0.001	<0.001	<0.001
Year					<0.001	<0.001
Medium × Year					<0.001	<0.001

$^{a–d}$ different superscript letters within a column mean significant differences ($p < 0.05$).

3.3.2. Products Containing Predominantly Heterofermentative LAB Species

After 24 h of the in vitro incubation of product no. 5 and 7, the mean pH in the HS medium (5.14 ± 0.05) was equal or very close to the original at 0 h, which was pH 5.2. Overall, this was similar for the in situ high DM treatments after 3 d of ensiling (mean of 5.58 ± 0.72), having only marginally decreased the pH of the fresh forage of >6.0. Thus, in the second year, 48 h of incubation and 5 d of ensiling were additionally compared using product no. 7. While in the overall evaluation there was a clear difference in pH between the high DM level (mean of 532 g DM/kg) versus the low DM level (mean of 347 g DM/kg) and the standard medium, the pH of the HS medium was lower than in the high DM level (Table 5).

Table 5. Descriptive statistics comparing the pH values of in vitro (after 48 h incubation in growth medium at two osmolality levels) and in situ (after 120 h of anaerobic storage at two DM levels) when inoculated with an obligately heterofermentative silage additive (no. 7, Table 3) in year 2.

Product No. 7	Mean	SD
Standard medium	4.60 ^{b}	0.232
Low DM	4.64 ^{b}	0.100
High-sugar medium	4.90 ^{b}	0.232
High DM	5.67 ^{a}	0.100
p	<0.001	

^{a,b} different superscript letters within a column mean significant differences ($p < 0.05$). DM dry matter.

4. Discussion

The concentration of sugars and salts chosen initially (levels 1–6) for the development of the modified medium were derived from the concentrations prevailing in fresh grass wilted to 450 g DM/kg. However, the results showed that the increased osmolality measured in wilted grass [7] is obviously not only a result of the inherent sugar and salt contents. Hoedtke [7] introduced the unit osmol/kg DM as a measurand to compare different plant species independent from their DM content with regard to their osmolality. She found out that even varieties and cultivars of the same species could differ as a result of their inherent metabolic peculiarities, while location, fertilization, and weather conditions could play another role. In the study presented here, sugars had to be increased by a multiple to achieve the targeted osmolality while KCl was restricted to 28 g to avoid salt stress. The final sugar concentration of the HS medium is similar to MY20 agar or glucose agar with 25% glucose [14], which were developed for the cultivation of osmophilic bacteria.

The HS medium obviously delayed the decrease in pH. This shows that the increased osmolality had the expected effect on the lactic acid production of the LAB. On the other hand, the effect was not a total inhibition, which can be a problem for microorganisms if electrolytes such as NaCl or KCl are used as osmolytes [23–25].

The connection between a reduced percentage of growth and the rate of pH reduction in the HS medium in general was not clear. The product no. 26, containing both homo- and heterofermentative strains, was the only one with both poor colony growth on HS medium (15%) compared to the standard agar, and a slow pH decline. Two other homofermentative products were rather slow in pH decrease (no. 14 and 16). However, there was no noticeable growth depression observed in the solid high-sugar medium. All other homofermentative products had a comparable pH development on the HS medium, irrespective of colony numbers (Figures 2 and 3). Thus, there seems to be no stringent connection between the growth and fermentation activity of LAB, which again confirms the observations of Marauska et al. and Beker et al. [15,16]. Their investigation showed that the lactic acid production of *L. plantarum* ceased at a significantly higher osmolality than its cell growth. This shows the necessity to work with a meaningful and robust indicator for the metabolic activity of LAB, which is the focus of interest when developing silage starter cultures, such as the pH.

The aim of ensiling grass at two different wilting levels with the selected products was to validate whether the developed medium could sufficiently simulate high dry matter conditions to evaluate in vitro the potential osmotolerance of LAB products in the future. The MRS standard medium can be considered as a medium, which provides optimal growth conditions for the majority of LAB species in silage additives. Thus, the fermentation capacity should be unlimited until the nutrients are depleted or until a too-low pH (autoacidification) has become the limiting factor [26]. In the high-sugar medium, the osmotic pressure should be the only initial obstacle to overcome. In contrast, a heterogeneous material such as grass consisting of different species at varying growth stages from diverse locations offers the LAB variable growth and fermentation conditions. Moreover, inoculants also have to compete with the native microflora. The forage utilized at the different institutions showed a high variability in its potential ensilability characterized by the sugar content and buffering capacity. Furthermore, it was not very easy to achieve the desired target DM contents due to the variations in solar radiation and wind strength. This was especially true for the second experimental year, highlighting the need for standardized conditions when selecting LAB for high osmotolerance. In the first year, there was no significant pH difference between the low DM and standard medium, and between the high DM and HS medium, indicating a high degree of comparability. In the second year, the statistical grouping resulted in four groups. However, as in the first year, the pH values in the low DM/standard medium groups were lower than in the high DM/HS medium groups. It has to be taken into account that the realized DM was above the targeted one, resulting in a higher osmolality than in the HS medium. For example, *Lolium perenne* with 600 g DM/kg had an osmolality of around 3.7 osmol/kg DM [7]. Consequently, this resulted in a higher pH in the high DM silage. Considering the high variation in the forage material, still specific pH ranges could be attributed to the different DM levels and compared to the growth media with different osmolality levels.

Obligately heterofermentative LAB exhibit a slower decrease in pH due to their metabolic pathway, which is split into the production of both lactic and acetic acid, the latter representing a weaker acid, or ethanol, depending on the substrate [27]. Heterofermentative LAB are used as silage additives with the aim to enhance aerobic stability, mainly because of their acetic acid production in the long run, which decelerates yeast activity [28]. When evaluating their osmotolerance based on the pH decrease within the first days of fermentation, it can be questioned whether this is the appropriate indicator to measure their metabolic activity under the given conditions as they exhibit their desired properties only after 7–8 weeks of ensiling [29]. In the presented *in vitro* trial, the lag phase was clearly pronounced. In situ, other products than no. 7 could be evaluated in the future after 5 d of ensiling with the mentioned reservation. An alternative here could be to continue the plate counting method on the two types of solid medium to evaluate colony growth at low and high osmolality.

5. Conclusions

For homofermentative LAB, evaluating the pH decrease after 24 h of incubation in the newly developed liquid HS medium is a suitable method to select strains for fermentation activity at high osmotic pressure compared to a standard MRS medium, which confirms the initial hypotheses. The HS medium provides a sufficiently accurate approximation. It simulates high DM conditions in forages as a first step to select for osmotolerance in the laboratory prior to field evaluations. For obligately heterofermentative LAB, this method needs further evaluation.

Supplementary Materials: The following supporting information can be downloaded at: https://www.mdpi.com/article/10.3390/agriculture14060825/s1, Table S1: DM contents of the silages of the low and high DM level in 2022 and 2023 at the participating institutes; Table S2: Ensilability parameters of forages in 2023.

Author Contributions: Conceptualization, S.D.M. and W.W.; methodology, S.D.M. and W.W.; validation, S.D.M., W.W., M.S., K.H., S.O. and C.L.; formal analysis, S.D.M. and M.S.; investigation, S.D.M., W.W., M.S., K.H., S.O. and C.L.; writing—original draft preparation, S.D.M.; writing—review and editing, S.D.M., W.W., M.S., K.H., S.O. and C.L. All authors have read and agreed to the published version of the manuscript.

Funding: The DLG Service GmbH supported parts of the trials financially.

Institutional Review Board Statement: Not applicable.

Data Availability Statement: The original data can be provided by the corresponding author upon reasonable request.

Acknowledgments: We kindly thank the DLG (German Agricultural Society) Technical Committee for Silage Additives and Preservatives for providing the platform for the scientific exchange to develop the protocol. Thanks go to Barbara Misthilger (LfL) who organized the osmolality measurement of the media.

Conflicts of Interest: The authors declare no conflicts of interest.

Abbreviations

ASL, above sea level, BC, buffering capacity [g lactic acid/kg DM]; DM, dry matter; *E. faecium*, *Enterococcus faecium*; FC, Fermentability coefficient [FC = 8 * S/BC + DM, %]; HS, high sugar; LAB, lactic acid bacteria; *L. brevis*, *Levilactobacillus brevis*; *L. buchneri*, *Lentilactobacillus buchneri*; *L. diolivorans*, *Lentilactobacillus diolivorans*; *L. kefir.*, *Lactobacillus kefiranofaciens*; *L. paracasei*, *Lacticaseibacillus paracasei*; *L. plantarum*, *Lactiplantibacillus plantarum*; *L. rhamnosus*, *Lacticaseibacillus rhamnosus*; MRS, growth medium according to de Man, Rogosa, Sharpe [20]; *P. acidilactici*, *Pediococcus acidilactici*; *P. pentosaceus*, *Pediococcus pentosaceus*; S/BC, Sugar: buffering capacity quotient

References

1. Buxton, D.R.; O'Kiely, P. Preharvest Plant Factors Affecting Ensiling. In *Silage Science and Technology*; Buxton, D.R., Muck, R.E., Harrison, J.H., Eds.; American Society of Agronomy, Crop Science Society of America, Soil Science Society of America: Madison, WI, USA, 2003; pp. 199–250.
2. Muck, R.E.; Moser, L.E.; Pitt, R.E. Postharvest Factors Affecting Ensiling. In *Silage Science and Technology*; Buxton, D.R., Muck, R.E., Harrison, J.H., Eds.; American Society of Agronomy, Crop Science Society of America, Soil Science Society of America: Madison, WI, USA, 2003; pp. 251–304.

3. DLG e.V. *Praxishandbuch Futter—Und Substratkonservierung*; DLG-Verlag: Frankfurt, Germany, 2012; p. 416.
4. Mitrik, T. *Ensilage*; Wageningen Academic Publishers: Wageningen, The Netherlands, 2021; p. 280.
5. Wieringa, G.W. The Effect of Wilting on Butyric Acid Fermentation in Silage. *Neth. J. Agric. Sci.* **1958**, *6*, 204–210. [CrossRef]
6. Lanigan, G.W. Silage Bacteriology I. Water Activity and Temperature Relationships of Silage Strains of *Lactobacillus plantarum*, *Lactobacillus brevis*, and *Pediococcus cerevisiae*. *Aust. J. Biol. Sci.* **1963**, *16*, 606–615. [CrossRef]
7. Hoedtke, S. Die Quantifizierung der Osmolalität in Futterpflanzen und ihre Veränderung in verschiedenen Stadien der Silierung. Ph.D. Thesis, University of Rostock, Rostock, Germany, 2008.
8. Atkins, P.W. *Einführung in Die Physikalische Chemie. Ein Lehrbuch Für Alle Naturwissenschaftler*; VCH Verlagsgesellschaft: Weinheim, Germany, 1994; p. 472.
9. Levine, I.N. *Physical Chemistry*; McGraw-Hill: New York, NY, USA, 2022; p. 989.
10. Weissbach, F. Beziehungen zwischen Ausgangsmaterial und Gärungsverlauf bei der Grünfuttersilierung. Post-Doctoral Thesis, University of Rostock, Rostock, Germany, 1968.
11. DLG TestService GmbH. *DLG Testing Guidelines for the Award and Use of the DLG Quality Mark for Ensiling Agents*; DLG TestService GmbH: Frankfurt, Germany, 2018.
12. Pieper, B.; Hoedtke, S.; Wensch-Dorendorf, M.; Korn, U.; Wolf, P.; Zeyner, A. Validation of the Rostock Fermentation Test As an *in-vitro* method to estimate ensilability of forages using glass jar model silages as a basis for comparison. *Grass Forage Sci.* **2016**, *72*, 568–580. [CrossRef]
13. Glaasker, E.; Tjan, F.S.B.; Ter Steeg, P.F.; Konings, W.N.; Poolman, B. Physiological response of *Lactobacillus plantarum* to salt and nonelectrolyte stress. *J. Bacter.* **1998**, *180*, 4718–4723. [CrossRef] [PubMed]
14. Atlas, R.M. *Handbook of Microbiological Media*; CRC Press: Boca Raton, FL, USA, 2010; p. 2040.
15. Beker, M.E.; Vigant, A.K.; Marauska, M.K.; Klintsare, A.A. Osmotic sensitivity of the bacterium *Lactobacillus casei var. alactosus*. *Prikl. Biokhimiia Mikrobiol.* **1998**, *34*, 161–163.
16. Marauska, M.; Vigants, A.; Klincare, D.; Upite, D.; Kaminska, E.; Bekers, M. Influence of Water Activity and Medium Osmolality on the Growth and Acid Production of *Lactobacillus casei* var. *alactosus*. *Proc. Latv. Acad. Sci. Sect. B* **1996**, *50*, 144–146.
17. Schuster, M.; Kolln, K.; Baranowski, A.; Richter, W.I.F.; Spiekers, H. Method and range of applications of the "Rostocker Fermentation Test". *Ubers. Zur Tierernährung* **2007**, *35*, 103–116.
18. Zierenberg, B. *In Vitro Method to Determine the Fermentation Performance of Lactic Acid Bacteria and Their Influence on the Metabolic Activity of Other Microorganisms Relevant for Ensiling Under Varying Fermentation Conditions*; University of Rostock: Rostock, Germany, 2000.
19. Pieper, B.; Müller, T.; Robowsky, K.-D.; Seyfarth, W. Rapid Fermentation Test as a Method for Assessing the Ensiling Potential of Herbage. In Proceedings of the XIth International Silage Conference, Wales, UK, 8–11 September 1996; pp. 120–121.
20. de Man, J.C.; Rogosa, M.; Sharpe, M.E. A medium for the cultivation of *Lactobacilli*. *J. Appl. Bacteriol.* **1960**, *23*, 130. [CrossRef]
21. Weissbach, F. Die Bestimmung der Pufferkapazität in Futterpflanzen und ihre Bedeutung für die Beurteilung der Vergärbarkeit. *Tagungsbericht* **1967**, *92*, 211–220.
22. Pahlow, G. Application of a New Concept for the Estimation of the Ensiling Potential of Forages for a Range of Crops. In Proceedings of the XIIIth International Silage Conference, Auchincruive, Scotland, UK, 11–13 September 2002; pp. 361–371.
23. Baird-Parker, A.C.; Freame, B. Combined effect of water activity, pH and temperature on the growth of *Clostridium botulinum* from spore and vegetative cell inocula. *J. Appl. Bacteriol.* **1967**, *30*, 420–429. [CrossRef] [PubMed]
24. Rehacek, J.; Sozzi, T.; Studer, P. Effect of water activity on the development of lactic acid bacteria and yeast utilized in the food industry. *Milchwissenschaft* **1982**, *37*, 151–154.
25. Strong, D.H.; Foster, E.F.; Duncan, C.L. Influence of water activity on the growth of *Clostridium perfringens*. *Appl. Microbiol.* **1970**, *19*, 980–987. [CrossRef] [PubMed]
26. Vinderola, G.; Champagne, C.P.; Desfosses-Foucault, E. The Production of Lactic Acid Bacteria Starters and Probiotic Cultures: An Industrial Perspective. In *Lactic Acid Bacteria*; CRC Press: Boca Raton, FL, USA, 2019; pp. 317–336.
27. von Wright, A.; Axelsson, L. Lactic Acid Bacteria. An Introduction. In *Lactic Acid Bacteria: Microbiological and Functional Aspects*, 5th ed.; Vinderola, G., Ouwehand, A., Salminen, S., von Wright, A., Eds.; CRC Press: Boca Raton, FL, USA, 2019; Chapter 1, pp. 1–16.
28. Elferink, S.J.W.H.O.; Driehuis, F.; Krooneman, J.; Gottschal, J.C.; Spoelstra, S.F. *Lactobacillus buchneri* can improve the aerobic stability of silage via a novel fermentation pathway: The anaerobic degradation of lactic acid to acetic acid and 1,2-propanediol. In Proceedings of the XII International Silage Conference, Uppsala, Sweden, 5–7 July 1999; pp. 266–267.
29. Kleinschmit, D.H.; Kung, L. The effects of *Lactobacillus buchneri* 40788 and *Pediococcus pentosaceus* R1094 on the fermentation of corn silage. *J. Dairy Sci.* **2006**, *89*, 3999–4004. [CrossRef] [PubMed]

agriculture

Article

Impact of Monensin Sodium and Essential Limonene Oil on the Fermentation and Chemical Composition of Total Mixed Ration Silages with Moisture Variations

Ronnie Coêlho de Andrade [1], Marco Antonio Previdelli Orrico Junior [1,*], Yara América da Silva [1], Marciana Retore [2], Tatiane Fernandes [3], Ana Carolina Amorim Orrico [1], Fernando Miranda de Vargas Junior [1] and Isabele Paola de Oliveira Amaral [1]

[1] College of Agricultural Science, Federal University of Grande Dourados, Dourados 79804-970, MS, Brazil; ronnier2.coelho@hotmail.com (R.C.d.A.); yaraamerica603@gmail.com (Y.A.d.S.); anaorrico@ufgd.edu.br (A.C.A.O.); fernandojunior@ufgd.edu.br (F.M.d.V.J.); isabelepoliveira@gmail.com (I.P.d.O.A.)
[2] EMBRAPA Agropecuária Oeste, Dourados 79804-970, MS, Brazil; marciana.retore@embrapa.br
[3] School of Animal Science, Virginia Polytechnic Institute, State University, Blacksburg, VA 3290, USA; fernandest@vt.edu
* Correspondence: marcojunior@ufgd.edu.br

Abstract: Monensin and essential oils have antimicrobial properties that may impact silage fermentation. The present study was divided into two trials to evaluate the effects of monensin (MON) and essential limonene oils (ELO) as additives in the ensiling of total mixed ration (TMR). In the first assay, TMR was tested with sheep in growth (65% dry matter—DM) using the following treatments: control (no additive), MON35 (35 mg of monensina per kg of DM), MON45 (45 mg of monensina per kg of DM), ELO300 (300 mg of essential limonene oil per kg of DM), and ELO600 (600 mg of essential limonene oil per kg of DM). In the second assay, the same treatments were used in TMR for lactating cows under two moisture conditions (30% and 40% DM). The parameters assessed included fermentative losses, short-chain fatty acid profiles, aerobic stability (hours needed for silage to reach 2 °C above ambient), chemical composition, and in vitro DM digestibility of the silages. Treatment averages were compared using the Scott–Knott test at 5% significance. In the first assay, the treatments with ELO had the lowest ($p < 0.05$) pH values and the highest ($p < 0.05$) lactic acid concentrations, with treatment ELO600 leading to the highest ($p < 0.05$) aerobic stability (297.88 h). Only the starch contents of the ELO treatments were lower ($p < 0.05$) than the others. In the second assay, the silages with the highest moisture contents and ELO600 exhibited the lowest ($p < 0.05$) values of DM recovery, lactic acid, and pH. The highest ($p < 0.05$) lactic acid:acetic acid ratios were observed in the silages with the most moisture added with MON35 and MON45. The use of MON and ELO increased aerobic stability, with the highest ($p < 0.05$) values observed for ELO600 and MON35. The treatments with MON and ELO resulted in silages with the lowest ($p < 0.05$) fiber contents and highest ether extract and starch contents when compared with control. Thus, MON and essential oils improve fermentative quality but ELO should be used in lower doses in humid silages to avoid negative fermentation impacts.

Keywords: dairy cows; ionophore; limonene; moisture; total diets

Citation: de Andrade, R.C.; Orrico Junior, M.A.P.; da Silva, Y.A.; Retore, M.; Fernandes, T.; Orrico, A.C.A.; Vargas Junior, F.M.d.; Amaral, I.P.d.O. Impact of Monensin Sodium and Essential Limonene Oil on the Fermentation and Chemical Composition of Total Mixed Ration Silages with Moisture Variations. *Agriculture* **2024**, *14*, 1319. https://doi.org/10.3390/agriculture14081319

Academic Editor: Siran Wang

Received: 8 July 2024
Revised: 30 July 2024
Accepted: 31 July 2024
Published: 9 August 2024

1. Introduction

The use of total mixed rations (TMR) in animal feed is widely adopted, which allows a nutritionally balanced diet to be provided and reduces feed selection by the animals [1]. However, the need to mix the ingredients daily can add cost to the productive process [2]. Moreover, the use of humid agro-industry coproducts is limited due to their rapid spoilage and unviable extended storage.

TMR silages offer various economic benefits, including the use of labor and machinery only during production, eliminating the need for daily mixing of ingredients [2]. They also have the potential to incorporate less palatable and more stable wet by-products from food industries [3]. TMR silages maintain high aerobic stability after silo opening and present opportunities for commercialization, particularly when stored in smaller structures such as bales, sacks, and bags [2]. Several studies have explored the effects of different ingredients and microbial additives on the fermentative and nutritional quality of TMR silage [2,4,5]. Nevertheless, few have assessed the influence of ruminal fermentation modulators such as ionophores on the fermentation process of ensiled TMR.

Monensin is one of the most studied and utilized ionophores in ruminant nutrition, acting by inhibiting Gram-positive bacteria in the rumen, which leads to lower methane production and increased animal efficiency [6]. However, its use has been banned in some European countries due to potential risks to public health and the environment [7]. Therefore, there is a search for new ruminal fermentation modulators that exert selective action on rumen microorganisms without leaving the residual effects of ionophores. Among these new modulators, essential oils (EOs) from plants have shown positive results [8].

Essential oils are concentrated plant extracts obtained from various parts of plants through methods like steam distillation or cold pressing [9]. Essential oils can help control microorganisms such as bacteria, fungi, and viruses due to their complex chemical composition, which includes compounds like terpenes, phenolics, and aldehydes [10].

Satisfactory fermentation patterns have been observed in TMR silages containing 33 mg/kg DM monensin (MON) [11]. However, it should be pointed out that all treatments tested by [11] had the same MON dose, and the lack of a control treatment prevented concluding whether MON impacts the fermentation process. Another noteworthy aspect of the research by [11] was the high dry matter (DM) contents in the TMRs at an average of 60%, since it is known that water activity is directly linked to the intensity of fermentation [12]. Since MON [13] and essential oils [10] often act by unbalancing ion concentrations in microbial cells, it is supposed that the higher the water activity in the silage (more humid silages), the greater will be the action of those additives on the microorganisms of the medium.

Moreover, under higher moisture conditions, essential oils can penetrate the forage more uniformly, which leads to a more consistent antimicrobial effect [9]. Appropriate moisture levels also ensure that the essential oils are adequately mixed throughout the silage, thus preventing areas where spoilage microorganisms can thrive [14].

In this context, the present research was based on the following hypotheses: (1) The use of MON and essential limonene oil (ELO) in TMR formulation inhibits microbial growth, thus reducing lactic acid production and improving aerobic stability without significantly altering the nutritional value of the silage. (2) The effects of MON and ELO on fermentation will be more pronounced in silages with higher moisture content. This study aimed to assess the influence of adding MON and ELO on the quality of the TMR ensilage process, and to identify the main changes in TMR fermentation at different moisture contents.

2. Materials and Methods

2.1. Experimental Assays

The research was split into two assays, with the first focusing only on identifying the possible changes in the fermentation parameters of TMR silages with the addition of MON and ELO, and the second testing the effect of adding MON and ELO under two moisture conditions.

The first assay was carried out in April 2022 at Embrapa (*Agropecuária Oeste*) (22°16′44″ S, 54°49′10″ W), located in the municipality of Dourados, MS, Brazil. Since that research institution focuses on lamb rearing, the TMR was formulated to meet the nutritional requirements of lambs (20 kg mean body weight) in the growth phase (mean weight gain of 300 g/day) with the intake of 0.7 kg DM/day according to the recommendations by Ref. [15] (Table 1).

Table 1. Proportions of ingredients and chemical compositions of the TMRs formulated in each of the experimental assays.

Ingredients	First Assay	Second Assay
	% of DM	% of DM
Sorghum	35.00	46.56
Ground corn kernel	41.74	28.67
Soybean meal	21.76	22.46
Calcitic lime	0.93	1.18
Dicalcium phosphate	0.56	1.11
Salt	0.01	0.02
Total	100.00	100.00
DM, %	62.45	41.00
CP, % DM	18.10	16.42
NDF, % DM	30.54	39.53
ADF, % DM	21.21	25.14
Starch, % DM	27.05	16.71
EE, % DM	2.21	2.38
Crude Ash, % DM	8.18	6.33
Lignin, % DM	2.80	4.04
NFC, % DM	42.28	36.08
BC, meqg NaOH/100 g DM	24.23	29.32

DM = dry matter; CP = crude protein; NDF = neutral detergent fiber; ADF = acid detergent fiber; EE = ether extract; NFC = non-fiber carbohydrate; BC = buffer capacity.

The first assay was conducted using a completely randomized design with four repetitions per treatment for a total of 20 experimental units. The treatments tested in that first assay were control (no additive), MON35 (35 mg of monensina per kg of DM), MON45 (45 mg of monensina per kg of DM), ELO300 (300 mg of essential limonene oil per kg of DM), and ELO600 (600 mg of essential limonene oil per kg of DM). The dose of 35 mg of MON per kg of DM was based on the manufacturer's recommendation. The dose of 45 mg of MON was selected to assess the impact of an overdose on the fermentative quality of TMR silage. The doses of 300 and 600 mg of ELO per kg of DM were determined based on previous experiments [16–18].

The second assay was developed in March 2023 at the Experimental Farm of the Federal University of Grande Dourados (UFGD) (22°13′ 52.44′95″ S, 54°59′ 10.53′72″ W) in the municipality of Dourados, MS, Brazil. In the second assay, the TMR was formulated to meet the nutritional requirements of dairy cows (500 kg body weight) in lactation (15 kg milk/day) and mean intake of 14 kg DM/day according to the recommendations by [19] (Table 1).

The second assay followed a completely randomized design in a 5 × 2 factorial arrangement with four repetitions per treatment combination for a total of 40 experimental units. The main factors corresponded to the same treatments of the first assay at two TMR moisture levels (30% and 40% DM). In order to provide a higher moisture level and maintain the proportion of ingredients in TMR formulation, distilled water had to be added when mixing the ingredients in part of the treatments of the second assay.

Both assays used sorghum (*Sorghum bicolor* L. Moench) as a source of roughage and concentrates based on ground corn, soybean meal, dicalcium phosphate, calcitic lime, and table salt. Sorghum was used in the TMR formulation because it is a crop widely used for silage production in the central region of Brazil. The proportion of ingredients used and the chemical composition of the diet in each of the experimental assays are presented in Table 1.

The additives and their respective doses were previously added to the concentrate to facilitate mixing and ensure homogenous distribution in the silage mass. In the first assay, the mix of roughage with concentrate resulted in TMR with mean contents of 62.45% DM due to the higher proportion of concentrate needed to meet the requirements of lambs. In the second assay, roughage had a higher participation and resulted in TMR with mean contents of 41% DM, very close to the desired DM content of 40%; thus, this TMR was readily used in the experimental silos. To obtain TMR with a content of 30% DM, 171 mL of distilled water was added to the material used in each experimental silo (average of 4.25 kg material per silo). To prevent effluent formation, the material was homogenized while adding the distilled water.

Both assays employed experimental silos built using PVC pipes (10 cm diameter and 50 cm height), with a useful volume of 3.8 L. The material was manually compacted using wood rods. At the bottom of each silo, a layer of approximately 4.5 cm of sand (300 g) was placed for effluent drainage. A fine cotton fabric mesh was used to keep the forage from touching the sand. After filling, the experimental silos were sealed with double-faced (black and white) plastic film and adhesive tape and stored in the laboratory at room temperature (average of 25.15 °C) for 90 days.

During silo filling, TMR samples were collected from each treatment. The first sample, of approximately 300 g, was used to determine the chemical composition, and the second, of approximately 70 g, was frozen for later processing and to determine pH values and buffer capacity.

2.2. Methodologies Employed in the Assays

To calculate fermentation loses, all components in the silos (silo, sand, and fabric) and the TMR mass ensiled were weighed both before and after ensiling. Dry matter recovery, gas losses, and effluent production were calculated according to the equations in [20]. Dry matter recovery was calculated using the following formula:

$$DMR = 100 - \left(\frac{DMI - DMF}{DMI} \times 100\right) \quad (1)$$

where DMR = dry matter recovery (% of the initial dry mass), DMI = initial dry mass (kg DM placed into the silos), and DMF = final dry mass (kg DM removed from the silos).

Gas losses were calculated using the following formula:

$$GL = \frac{WSI - WSC}{DMI} \times 100 \quad (2)$$

where GL = gas losses during storage (% of initial dry mass), WSI = initial weight of the closed silo (kg), WSC = weight of the closed silo when opened (kg), and DMI = initial dry mass (kg DM placed into the silos).

Effluent production was calculated using the following formula:

$$EP = \frac{Wf - Wi}{DMI} \times 1000 \quad (3)$$

where EP = effluent production (kg/t dry matter), Wf = final combined weight (silo + sand + fabric) in kg, Wi = initial combined weight (silo + sand + fabric) in kg, and DMI = initial dry mass (kg DM placed into the silos).

Then, the silos were opened and the material inside them was removed and homogenized for sample collection. Chemical analyses were performed using an NIRs^{TM} DS2500 F (Foss NIR Systems, Laurel, MD, USA) spectrophotometer, using a large ring cup and the factory calibrations provided with the instrument (FOSS global calibration). Before spectral measurements, a standard protocol to ensure accuracy was performed; the spectral device was preheated for 30 min, followed by calibration with a white spectral on the calibration panel, followed by measurement of known standards to validate the model by a separate set of samples. For the measurements, labeled samples were filled into the cup and flattened to avoid translucent cavities. All near infrared spectra were collected at wavelengths between 850 and 2500 nm, registering absorbance values log (1/R) (where r = reflectance) at 2 nm intervals for each sample. Concentrations of dry matter (DM), crude ash, crude protein (CP), soluble protein (SP), neutral detergent insoluble protein (NDIP), acid detergent insoluble protein (ADIP), neutral detergent fiber (NDF), acid detergent fiber (ADF), lignin, ether extract (EE), non-fiber carbohydrates (NFC), and starch were obtained via FossManager^{TM} (Hilleroed, Denmark).

A 25 g sample of TMR was diluted in 225 mL of distilled water and manually homogenized for approximately 20 min for the production of an aqueous extract. The extract was used to measure pH both before and after silage, buffer capacity before silage, and the profile of short-chain organic acids. pH was determined using a digital potentiometer (mPA210 MS Tecnopon, Piracicaba, SP, Brazil), and buffer capacity was determined according to [21].

A portion of that extract was filtered using a paper filter, centrifuged for 15 min at 10,000 rpm, and the supernatant was frozen at −20 °C for later analysis of volatile organic acids. The organic acids were determined using gas chromatography with a mass detector (GCMS QP 2010 Plus, Shimadzu, Kyoto, Japan) using a capillary column (Stabilwax, Restek, Bellefonte, PA, USA, 60 m, 0.25 mm Ø, 0.25 μm Crossbond Carbowax polyethylene glycol). The lactic acid concentration was determined using the colorimetric method proposed by [22]. The concentration of ammoniacal nitrogen (NH_3-N) was determined by the Kjeldahl method [23]. In vitro DM digestibility (IVDMD) was determined according to [24], with samples incubated for 48 h.

Aerobic stability (AS) was determined in all silos after they were opened. Samples (2 ± 0.005 kg) of each replicate of each treatment were freely placed in the clean experimental silos. Temperature sensors were placed in the geometric centers of the silages, and a double layer of gauze was placed on top of each experimental silo to prevent drying and contamination while allowing for air penetration. Ambient temperature, as well as the temperature of each silage, was recorded every minute, and the average was calculated every 20 min using a datalogger (RC-4, Elitech®, Canoas, RS, Brazil), which was calibrated before starting the measurements. Aerobic stability was defined as the number of hours required for the silage temperature to reach 2 °C above ambient temperature [25].

2.3. Statistical Data Analysis

The data of the first assay were analyzed using the statistical program Sisvar 5.8 (build 92, 2018). Before the actual analysis, data were explored to seek disparate information ("outliers") and for normality of residuals using the Shapiro–Wilk test. An individual observation was considered an outlier if it exceeded 3 standard deviations from the average. Data that did not attend to the normality premise were subjected to logarithmic [Log (X + 1)] or square root [RQ (X + 1/2)] transformation.

The data of the first assay were analyzed according to the following model:

$$Yik = \mu + Ai + \varepsilon ik \quad (4)$$

where Yik = dependent variable, μ = overall average, Ai = effect of different additives (fixed effect; i = Control, MON35, MON45, ELO300, and ELO600), and εijk = random error associated with each observation (k = 4). The statistical significance was set at a *p*-value lower than 0.05. Post hoc means separation was conducted if the main effect of treatment was significant ($p \leq 0.05$), using the clustering algorithm of the Scott–Knott test [26], and indicated by letters in the table.

The data of the second assay were analyzed using the statistical program Sisvar 5.8 (Build 92, 2018). Assay data were analyzed according to the following model:

$$Yijk = \mu + Ai + Mj + A \times Mij + \varepsilon ijk \quad (5)$$

where Yijk = dependent variable, μ = overall average, Ai = effect of different additives (fixed effect; i = Control, MON35, MON45, ELO300, and ELO600), Mj = moisture effect (fixed effect; j = higher and lower), A * Mij = effect of the interaction between additives and moisture, and εijk = random error associated with each observation (k = 4). The statistical significance was set at a *p*-value lower than 0.05.

Post hoc means separation was conducted if the main effect of treatment was significant ($p \leq 0.05$), using the Scott–Knott test, and indicated by letters in the table. When the interaction of factors was significant ($\alpha \leq 0.05$), the factor combinations were read as individual treatment, clustered by Scott–Knott test, and indicated by letters in the figures.

3. Results

3.1. First Assay

No difference ($p > 0.05$) was observed between the treatments tested for gas losses, effluent production, and DM recovery (Table 2). In the first assay, the gas losses varied from 2.03 to 2.88% ensiled DM while mean effluent productions were 5.1 kg/ton DM and DM recovery losses were above 97%.

Table 2. Fermentation losses, final pH, DM recovery, aerobic stability, and silage fermentation profiles of TMRs submitted to different doses of MON and ELO (first assay).

Parameters	Treatment					SEM	*p*
	Control	MON35	MON45	ELO300	ELO600		
GL, % DM	2.88	2.03	2.27	2.18	2.28	0.35	0.50
EL, kg/t DM	6.27	4.14	5.78	5.89	3.38	1.40	0.53
DMR, %	97.38	98.18	97.92	97.95	98.16	0.26	0.23
Final pH	4.75 a	4.75 a	4.76 a	4.74b	4.73 b	0.01	0.01
Lactic ac., % DM	3.61 b	3.79 b	4.19 b	5.06 a	4.84 a	0.26	<0.01
Ethanol, % DM	0.16 b	0.14 b	0.16 b	0.21 a	0.13 b	0.01	0.03
Acetic ac., % DM	0.76 b	0.97 a	1.05 a	1.12 a	1.08 a	0.05	<0.01
Propionic ac., mg/kg DM	26.33 e	45.09 d	61.52 c	74.84 b	88.78 a	3.19	<0.01
Butyric ac., mg/kg DM	7.45	9.12	7.90	8.01	8.83	1.14	0.83
Isobutyric ac., mg/kg DM	3.10	2.05	1.90	2.93	2.07	0.52	0.36
Isovaleric ac., mg/kg DM	3.37 c	4.72 b	4.56 b	5.22 a	6.20 a	0.39	<0.01
Valeric ac., mg/kg DM	1.06	1.35	1.37	1.22	1.17	0.14	0.55
AS, hours	213.65 c	259.28 b	242.92 b	245.63 b	297.88 a	18.2	<0.01
NH_3-N, % TN	10.48 a	8.85 b	8.12 b	8.29 b	7.75 c	0.19	<0.01

TMR with no additive (control); TMR with 35 mg MON/kg DM (MON35); TMR with 45 mg MON/kg DM (MON45); TMR with 300 mg ELO/kg DM (ELO300); TMR with 600 mg ELO/kg DM (ELO600); GL = gas losses; EL = effluent losses; DMR = dry matter recovery; AS = aerobic stability; NH_3-N = ammoniacal nitrogen; TN = total nitrogen; ac. = acid. Means followed by different letters differ according to Scott–Knott test at 5% probability. SEM = standard error of the mean; p = p-value.

The lowest ($p < 0.05$) final pH values were observed for treatments ELO600 and ELO300, with no differences ($p > 0.05$) in pH values between the control and the treatments added with monensin (Table 2). The silages that received additives ELO300 and ELO600 had the highest concentrations ($p < 0.05$) of lactic acid (average of 4.95% of DM). No significant differences were observed for lactic acid content in the silages from treatments MON35, MON45, or control (Table 2). Acetic acid concentrations were higher ($p < 0.05$) for the silages with additives when compared with the control silages (Table 2). On average, the silages added with MON or ELO had 38.8% higher acetic acid contents when compared with the control treatment. Propionic acid concentrations increased with the use of additives, with the lowest concentration observed for the control treatment followed by MON35, MON45, ELO300, and ELO600, respectively (Table 2).

During the first assay, no difference ($p > 0.05$) was found between the treatments regarding contents of butyric acid, isobutyric acid, and valeric acid, which were, on average, 8.2 mg/kg DM, 2.41 mg/kg DM, and 1.23 mg/kg DM, respectively. Isovaleric acid concentrations were higher ($p < 0.05$) for silages ELO300 and ELO600 at 5.71 mg/kg DM on average.

Aerobic stability was impacted by the additives tested (Table 2). The silage with treatment ELO600 had the highest ($p < 0.05$) AS value (average of 297.88 h), followed by treatments ELO300, MON35, and MON45. The control treatment had the lowest AS at 231.65 h.

The NH_3-N contents were the lowest for the treatments with some type of additive (Table 2). The control treatment had the highest NH_3-N value with an average of 10.48% of TN. Treatment ELO600 had the lowest NH_3-N value. The highest ethanol production was observed for treatment ELO300 (0.21% DM), whereas the other treatments had no significant differences between each other (Table 2).

During the first assay, differences ($p < 0.05$) were observed only for contents of DM, starch, and NFC between the treatments tested (Table 3).

Table 3. Chemical compositions of TMR silages submitted to different doses of MON and ELO (first assay).

Parameters	Treatment					SEM	*p*
	Control	MON35	MON45	ELO300	ELO600		
DM, %	63.78 a	61.94 b	62.51 b	62.14 b	61.90 b	0.22	<0.01
CP, % DM	16.58	18.00	17.60	19.16	19.18	0.74	0.11
SP, % CP	25.00	33.60	28.00	40.40	38.60	4.15	0.07
DP, % CP	57.20	61.00	56.60	64.40	65.00	2.65	0.09
NDIP, % DM	4.60	4.62	4.58	4.66	4.46	0.19	0.96
ADIP, % DM	1.04	1.02	1.12	0.98	1.10	0.09	0.84
NDF, % DM	28.20	30.86	29.56	31.66	32.42	1.21	0.14
ADF, % DM	20.09	21.08	19.85	21.25	22.25	1.14	0.21
Starch, % DM	30.12 a	27.18 a	28.26 a	25.01 b	24.76 b	1.31	0.04
EE, % DM	2.32	2.10	2.18	2.34	2.14	0.14	0.68
Crude Ash, % DM	8.00	8.24	8.04	8.30	8.36	0.14	0.31
Lignin, % DM	2.38	2.70	2.72	3.00	3.22	0.22	0.13
NFC, % DM	46.32 a	42.31 b	43.14 b	40.11 b	39.54 b	1.56	0.03
IVDMD % DM	85.47	84.26	83.23	83.96	85.05	3.21	0.98

TMR with no additive (control); TMR with 35 mg MON/kg DM (MON35); TMR with 45 mg MON/kg DM (MON45); TMR with 300 mg ELO/kg DM (ELO300); TMR with 600 mg ELO/kg DM (ELO600); DM = dry matter; CP = crude protein; SP = soluble protein; DP = degradable protein; NDIP = neutral detergent insoluble protein; ADIP = acid detergent insoluble protein; NDF = neutral detergent fiber; ADF = acid detergent fiber; EE = ether extract; NFC = non-fiber carbohydrates; IVDMD = in vitro dry matter digestibility. Means followed by different letters differ according to Scott–Knott test at 5% probability. SEM = standard error of the mean; *p* = *p*-value.

The use of additives (MON and ELO) led to silages with lower DM contents when compared with the control treatment. Starch contents were lower ($p < 0.05$) for treatments ELO300 and ELO600 (average of 24.88% of DM) when compared with MON35, MON45, and control (average of 28.52% of DM), which did not differ ($p > 0.05$) (Table 2). Contents of NFC were lower ($p < 0.01$) in the silages with some type of additive (MON or ELO), regardless of the dose employed.

3.2. Second Assay

In the second assay, differences ($p < 0.05$) were observed between the additives for gas losses (Table 4). The highest values were found for treatments MON35 and ELO600, which were 77.28% higher than in the other treatments tested. The more humid silages had gas loss values of 134.57% higher when compared with those with lower moisture. No differences ($p > 0.05$) were observed between the additives for effluent production, which were the highest in the silages with high moisture contents (average of 59.2 kg/t DM) (Table 4).

Table 4. Fermentation losses, pH, aerobic stability, and silage fermentation profiles of TMRs with different moisture contents and different doses of MON and ELO (second assay).

Parameters	Additives (A)					Moisture (M)		SEM	*P*		
	Control	MON35	MON45	ELO300	ELO600	High	Low		A	M	A * M
GL, % DM	1.37 b	2.13 a	1.32 b	1.39 b	2.72 a	2.51	1.07	0.22	0.03	<0.01	0.06
EL, kg/t DM	28.88	36.24	35.83	37.90	32.85	59.20	4.68	2.45	0.25	<0.01	0.41
DMR, % DM	94.58	95.29	95.95	94.06	92.66	91.45	97.56	0.38	<0.01	<0.01	<0.01
Final pH	4.01	4.07	4.05	3.96	4.17	4.06	4.04	0.03	<0.01	0.58	<0.01
Lactic ac., % DM	5.50	6.41	6.58	7.19	5.79	6.51	6.08	0.18	<0.01	0.01	<0.01
Acetic ac., % DM	0.88	0.77	0.76	0.97	1.54	1.05	0.91	0.04	<0.01	<0.01	<0.01
Lat/Ac ratio	6.50	8.4	8.94	7.49	4.77	7.64	6.8	0.4	<0.01	0.02	<0.01
Ethanol, % DM	0.23	0.28	0.31	0.26	0.25	0.33	0.20	0.02	0.37	0.01	0.42
Propionic ac., mg/kg DM	243.81 a	118.59 b	119.77 b	67.28 b	111.51 b	85.39	179.00	28.62	<0.01	<0.01	0.08
Butyric ac., mg/kg DM	76.00 a	122.89 a	44.64 b	17.42 b	18.33 b	75.66	36.07	19.27	<0.01	0.02	0.35
Isobutyric ac., mg/kg DM	38.89 c	59.11 b	75.54 a	49.90 c	38.43 c	44.27	60.49	3.04	<0.01	<0.01	0.06
Valeric ac., mg/kg DM	51.44 a	19.49 b	10.59 b	8.32 b	6.26 b	17.56	20.89	4.19	<0.01	0.37	0.58
Valeric ac., mg/kg DM	55.71 a	20.99 b	8.72 c	5.74 c	3.66 c	18.72	19.22	4.72	<0.01	0.90	0.57
AS, hours	83.94	91.56	89.13	88.94	92.35	88.55	89.81	0.607	<0.01	0.02	<0.01
NH_3-N, % of TN	8.46 b	8.56 b	8.58 b	8.58 b	9.34 a	9.09	8.48	0.21	<0.01	<0.01	0.06

TMR with no additive (control); TMR with 35 mg MON/kg DM (MON35); TMR with 45 mg MON/kg DM (MON45); TMR with 300 mg ELO/kg DM (ELO300); TMR with 600 mg ELO/kg DM (ELO600); GL = gas losses; EL = effluent losses; DMR = dry matter recovery; ac. = acid; Lat/Ac ratio = acetic/lactic acid ratio; AS = aerobic stability; NH_3-N = ammoniacal nitrogen; TN = total nitrogen; high moisture = 30% of DM; low moisture = 40% of DM. Means followed by different letters differ according to Scott–Knott test at 5% probability. SEM = standard error of the mean.

Significant interaction was found between TMR silage humidity and the additives tested for DMR (Figure 1). Overall, the silages with high moisture content had the lowest ($p < 0.05$) DM recovery values, especially ELO600 at only 88% DM recovery. Among the silages produced with high moisture content, the treatments using MON had the highest DM recovery values. No differences ($p > 0.05$) were observed in DM recovery for the silages with low moisture content, which were all above 97%.

An interaction ($p < 0.05$) was found between the additives tested and the moisture contents of the TMR silages, where ELO600 in the TMR with the highest moisture resulted in the highest pH values (average of 4.31) (Figure 1). The other treatments tested, irrespective of moisture (including ELO600 in the silages with lower moisture) did not differ ($p > 0.05$), with a mean pH of 4.02.

The silage with the highest ($p < 0.05$) lactic acid content (average of 8.1% of DM) was ELO300, with higher moisture (Figure 1). Only treatment ELO600 had higher ($p < 0.05$) lactic acid production in the drier silage when compared with the more humid one. For the other additives, the silages with higher moisture content always had higher lactic acid production.

Lactic acid concentration was higher ($p < 0.05$) for silage ELO600 with higher moisture (2.11% of DM) (Figure 1). The opposite behavior was observed for the control, MON35, and MON45 treatments, which had the lowest acetic acid values for the silages with higher moisture content. On average, treatment ELO600 in silages with higher moisture produced 2.45 times more acetic acid compared with the average of the other treatments.

The silages with the highest ($p < 0.05$) lactic acid/acetic acid ratios were the ones added with MON with higher moisture (Figure 1). Only ELO600 had a higher ($p < 0.05$) lactic acid:acetic acid ratio in the silage with lower moisture when compared with the silage with higher moisture. For the other additives tested, the lactic acid: acetic acid ratios were always higher or similar in the silages with higher moisture when compared with those with lower moisture. The silages with higher moisture added with ELO600 had the lowest lactic acid:acetic acid ratios (average of 2.3).

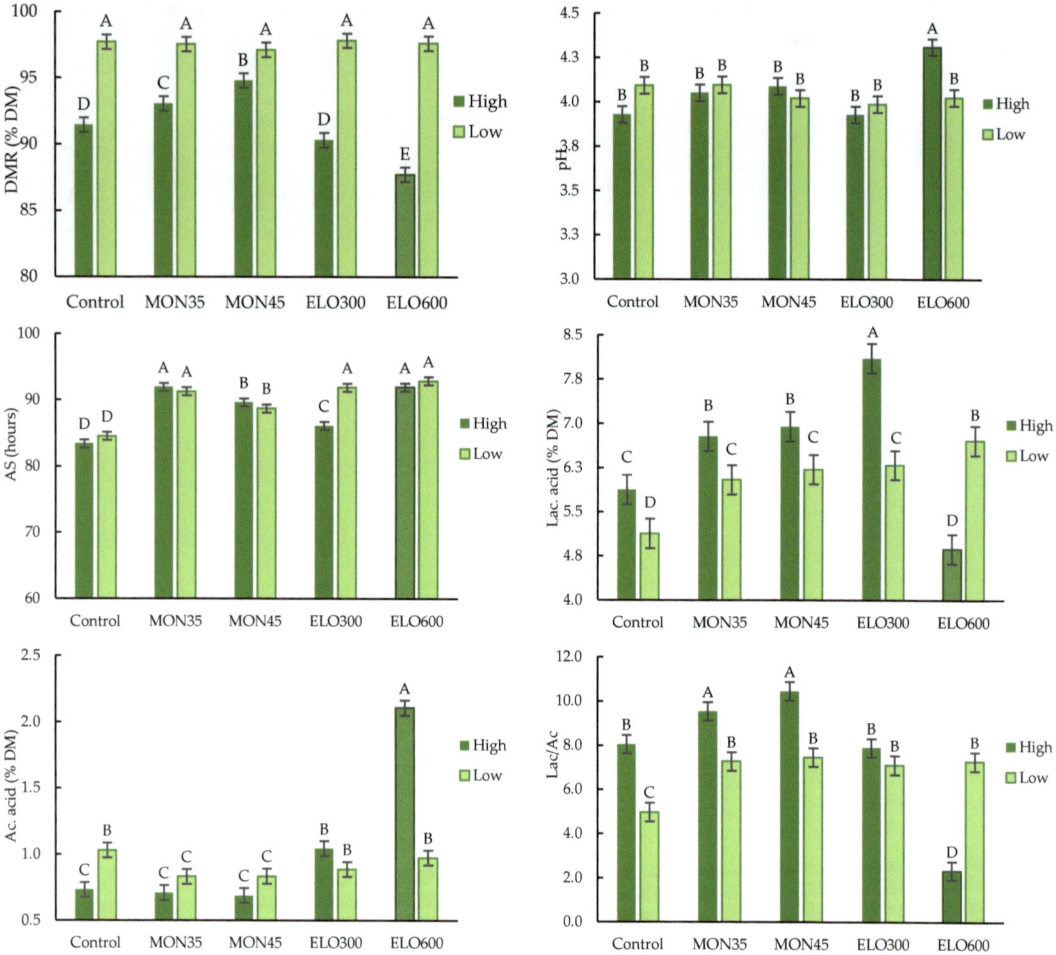

Figure 1. DM recovery (DMR), potential of hydrogen (pH), aerobic stability (AS), acetic acid content (Ac. acid), lactic acid content (Lac. acid), and acetic acid:lactic acid ratio (Lac/Ac) of the TMR silages with higher and lower moisture and MON and ELO doses (second assay). High moisture = 30% of DM; low moisture = 40% of DM. Means followed by different letters differ according to Scott–Knott test at 5% probability. I = standard error of the mean.

The additives tested did not impact ($p > 0.05$) ethanol production, while higher ethanol values were found for silages with higher moisture compared with those with lower moisture (Table 4). A difference ($p < 0.05$) was found between the silages with higher and lower moisture for propionic and butyric acid contents (Table 4). The highest butyric acid concentrations (75.66 mg/kg DM) were observed for the silages with higher moisture, whereas the opposite comparison was observed for the concentrations of propionic and isobutyric acids. Overall, the control treatment had the highest concentrations of propionic, butyric, isovaleric, and valeric acids, with a significant reduction in these acids observed when MON and ELO were added.

The NH_3-N contents in the silages tested were impacted ($p < 0.05$) by the additives and moisture contents of the TMRs. Overall, the silages with higher moisture had the highest NH_3-N values (9.09% of TN) compared with those with lower moisture (8.48% of TN). The control, MON35, MON45, and ELO300 treatments had the lowest NH_3-N contents, while ELO600 was the treatment with the highest NH_3-N value.

Aerobic stability was affected by the interaction ($p < 0.05$) between the additives and moisture contents of the TMRs tested (Figure 1). Regardless of the moisture tested, the lowest AS values were observed in the control treatment, with an average of 83.5 h for AS failure. Both ELO600 and MON35 had the best AS values (>92 h), irrespective of silage moisture. The silages with higher moisture added with ELO300 had lower AS (average of 85.33 h) compared with those with higher moisture (average of 91.84 h).

The silages with lower moisture had the highest ($p < 0.05$) DM, NDF, and ADF contents, whereas those with higher moisture had the highest ($p < 0.05$) EE values (Table 5). All additives tested resulted in lower fiber contents and higher starch contents. Ether extract contents were higher in the treatments added with MON (Table 5). The higher IVDMD coefficient (77.56% of DM) was observed in the silages added with MON35, followed by MON45, ELO300, and ELO600 with intermediate values and the control treatment with the lowest IVDMD value (71.09% of DM) (Table 5).

Table 5. Chemical compositions of TMR silages with different doses of MON and ELO submitted to higher and lower moisture (second assay).

Parameter	Additives (A)					Moisture (M)		SEM	*p*-Value		
	Control	MON35	MON45	ELO300	ELO600	High	Low		A	M	A * M
DM, %	36.28	37.05	36.35	35.51	35.55	30.49	41.81	0.54	0.27	<0.01	0.39
Crude Ash, % DM	6.43	6.69	6.95	6.73	6.88	7.13	6.34	0.21	0.46	0.06	0.12
CP, % DM	15.62	16.27	16.86	16.97	16.43	16.76	16.52	0.21	0.24	0.43	0.38
SP, % CP	47.38	47.13	49.25	47.25	48.25	50.85	44.85	1.26	0.73	0.62	0.87
NDIP, % DM	19.04	18.53	18.58	18.34	18.73	17.26	20.03	0.81	0.98	0.14	0.64
ADIP, % DM	5.41	6.24	6.05	5.81	5.96	5.12	6.67	0.43	0.73	0.24	0.70
NDF, % DM	41.21 a	38.25 b	38.01 b	37.67 b	37.07 b	37.75	39.54	0.91	0.02	0.03	0.90
ADF, % DM	26.78 a	24.40 b	24.80 b	24.75 b	23.99 b	24.62	25.67	0.52	0.01	0.03	0.89
Lignin, % DM	4.16	3.71	3.88	3.84	3.95	3.77	4.05	0.22	0.69	0.17	0.27
Starch, % DM	16.36 b	18.55 a	18.72 a	18.45 a	19.72 a	20.01	16.72	0.68	0.02	<0.01	0.15
NFC, % DM	34.50	36.06	36.16	36.71	37.23	36.18	36.09	0.72	0.11	0.88	0.61
EE, % DM	2.77 b	3.11 a	3.21 a	2.98 b	2.87 b	3.59	2.39	0.07	<0.01	<0.01	0.42
IVDMD, % DM	71.09 c	77.56 a	73.09 b	72.73 b	72.38 b	74.26	72.79	0.86	<0.01	0.06	0.07

TMR with no additive (control); TMR with 35 mg MON/kg DM (MON35); TMR with 45 mg MON/kg DM (MON45); TMR with 300 mg ELO/kg DM (ELO300); TMR with 600 mg ELO/kg DM (ELO600); DM = dry matter; CP = crude protein; SP = soluble protein; NDIP = neutral detergent insoluble protein; ADIP = acid detergent insoluble protein; NDF = neutral detergent fiber; ADF = acid detergent fiber; NFC = non-fiber carbohydrates; EE = ether extract; IVDMD = in vitro dry matter digestibility; high moisture = 30% of DM; low moisture = 40% of DM. Means followed by different letters differ according to Scott–Knott test at 5% probability. SEM = standard error of the mean.

4. Discussion

4.1. First Assay

The gas loss values observed for all treatments tested were less than 3% of the ensiled DM, which is characteristic of silages with good fermentative quality (prevalence of lactic fermentation), according to [12]. Effluent production in this assay was minimal, which is common in silages with high DM content [11,27]. According to [2], high DM contents (above 40%) contribute to the conservation of the ensiled material and increase DM recovery values. Similar data were reported by [11], who found DM recovery values above 96% when ensiling TMR with DM contents between 60 and 62%.

pH is an important parameter to assess the silage fermentation quality, as it is directly related with the total organic acids produced [28,29]. Silages with higher DM values, i.e., low water activity, result in higher final pH values (lower fermentation intensity) [30]. This explains the pH values observed in this assay (average of 4.74). According to [4,5,31], pH values between 4.0 and 5.0 are common in silages with DM contents above 40%, with such a pH range being sufficient to control the growth of undesirable microorganisms (particularly (mold, yeast, and *Clostridium* spp.).

When assessing several studies in the literature, [18] concluded in their meta-analysis that essential oils act by inhibiting the growth of undesirable microorganisms in the silages (particularly molds and yeasts), thus improving AS. This matches the results obtained in this research, especially with the highest ELO dose (600 mg/kg DM). To a lesser extent, MON also improved silage AS compared with the control treatment The authors of [11] observed that TMR silages with 33 mg MON/kg DM had AS values above 240 h, which corroborated the data obtained in this assay.

Another aspect that helps explain the higher AS of the treatments with MON and ELO was acetic acid production. According to [25], a major characteristic of acetic acid is to inhibit the growth of molds and yeasts and, at moderate amounts, it can contribute to improving silage stability. The mean normal concentration of acetic acid in the silage is 3 to 4% of DM, a value that is lower (0.5 to 2%) in silages with higher DM contents (45 to 55% DM) [25].

The treatments added with ELO, at either dose, had the highest lactic acid concentrations; this contradicts most data found in the literature on the use of essential oils in silages [18]. However, Ref. [32] observed that adding 300 mg/kg cumin EO had a selective antimicrobial effect, decreasing the number of harmful microorganisms (molds and yeasts) and increasing the number of beneficial microorganisms (lactic acid bacteria). Such a situation led to higher lactic acid productions and lower pH values, which corroborated the data obtained in the present research.

The control treatment had the highest butyric acid and NH_3-N contents, suggesting the additives tested helped inhibit undesirable fermentations. In the research by [32] with cumin EO in oat silage, those authors also observed a reduction in NH_3-N contents when adding essential oil. Essential oils help reduce proteolysis in silages primarily due to their antimicrobial and antioxidant properties. Compounds such as terpenes, phenols, and aldehydes in essential oils inhibit the growth of proteolytic microorganisms and the activity of proteolytic enzymes responsible for protein degradation during fermentation [10]. Additionally, by improving the aerobic stability of the silage, essential oils reduce the proliferation of microorganisms that cause deterioration when the silage is exposed to air, thus preserving the original quality of the proteins [18].

The antioxidant action of essential oils also plays a crucial role in protecting proteins from oxidative damage. Antioxidants present in essential oils neutralize free radicals and inhibit oxidation reactions that could degrade proteins and other nutrients in the silage [33]. Consequently, essential oils significantly contribute to preserving the nutritional quality of the silage, resulting in a final product with higher nutritional value for animals.

Similar to essential oils, monensin inhibits protein degradation by targeting Gram-positive proteolytic bacteria [34]. This antimicrobial effect fosters the growth of Gram-negative bacteria, which produce propionate rather than ammonia, thereby preserving more dietary protein for absorption [6]. While the NH_3-N level is an indicator of silage quality, it is noteworthy that the values obtained in this study for all treatments were close to the ideal range (10% of total nitrogen), as recommended in the literature [12,25,29].

According to [29], some heterofermentative bacteria can produce ethanol from the fermentation products or substrates present in the medium, which explains the higher ethanol contents in the treatments added with MON and ELO. Nonetheless, the ethanol values obtained in both assays can be considered low and did not negatively impact the quality of the fermentation process [35].

Using the ELO and MON additives led to a slight reduction ($p < 0.05$) in the DM contents of the silages. It is known that determining DM in a forced-air oven causes partial loss of the volatile fatty acids produced during ensiling; hence, lower DM values are commonly observed in silages with higher organic acid contents [36]. This may help explain why the silages added with ELO and MON (higher organic acid productions) had the lowest DM values.

In the literature, some studies also observed reductions in starch and NFC contents during TMR ensiling, similarly to the findings in the present study. Researchers in [3] found reductions by 5.5% in starch contents in TMR based on corn kernels. According to those authors, some types of microorganisms present in the silages are able to use the starch as substrate for the production of lactic acid, acetic acid, and ethanol, and the concentrations of those compounds commonly increase in the final product. The silages added with ELO300 and ELO600 had the highest lactic acid concentrations (37.11% higher than in the control treatment) and the highest starch reductions, which corroborates the statements by [3].

The small variations in nutrient levels observed in the first trial can be explained by the lower fermentation intensity in silages with low moisture content. Moisture limitation restricts the activity of microorganisms responsible for undesirable fermentations, leading to better preservation of nutrients in the silage. Additionally, in low moisture conditions, effluent production is reduced, which decreases the loss of soluble nutrients through leaching.

4.2. Second Assay

4.2.1. Effect on Fermentation Parameters

Although higher, the GL in treatments MON35 and ELO600 were below 3% of the ensiled DM, which is considered adequate according to [12]. Ref. [4] observed that TMR with higher moisture (23% of DM) associated with heterofermentative microbial additives resulted in slightly higher GL (3.7% of DM), while drier TMR silages with no inoculants (32% of DM) had lower GL values (3.1% of DM), which corroborated the data obtained in this research.

Since silage moisture is directly correlated with EL, higher EL values were expected for the more humid TMRs. According to [12], excess effluent is avoided in most cases when the DM content of the material is above 35%. The higher effluent production in the more humid silages likely contributed to their lower DM recovery values.

According to [12], lactic fermentation results in silages with minimum DM and energy losses, whereas acetic, alcoholic, and butyric fermentations result in silages with higher DM and energy losses, i.e., lower DM recovery values. Treatment ELO600, with higher moisture, experienced more intense heterofermentation and used the sugars present in the medium for higher acetic acid production, which explains the lower DM recovery values.

The more humid silages had higher lactic acid concentrations in the control, MON35, MON45, and ELO300 treatments. However, treatment ELO600, with higher moisture, resulted in a significant reduction in lactic acid production and increased pH, which was not found when the same dose was employed in drier TMR silages. The higher EO dose (600 mg/kg DM), in association with the higher medium moisture, likely potentialized antimicrobial action and favored the growth of heterofermentative bacteria (more resistant to ELO), in detriment of homofermentative bacteria (less resistant to ELO).

Limonene, an important compound of several citric oils, acts as a bactericidal and fungicidal agent by disrupting the cell membrane, increasing its permeability, and causing the loss of essential ions, leading to cell death [16]. Additionally, it inhibits ergosterol synthesis in fungal cells and can induce oxidative stress, generating reactive oxygen species that damage DNA, proteins, and lipids. These mechanisms make limonene an effective antimicrobial agent with applications in food preservation, agriculture, and medicine. Studies have shown that limonene has significant antibacterial activity against several bacterial species, including food-related microorganisms such as *Lactobacillus plantarum* and *Lactobacillus brevis*, common in silage [16].

The meta-analysis by [18] concluded that not only do essential oils inhibit the growth of bacteria harmful to the silage (especially cumin and oregano), but they also inhibit the growth of bacteria that produce the main organic acids. However, most studies used by [18] were conducted on silages with moisture contents between 22 and 35%, i.e., silages with higher moisture that likely had more effective antimicrobial action by the EOs, particularly at higher doses, which corroborates the data observed in the present research.

The higher lactic acid:acetic acid ratios observed when adding MON allow concluding that the additive modulated fermentation in the TMRs and prioritized homofermentation.

That is a positive result, since it allows using MON in the formulation of TMR to be ensiled aiming at improving the fermentation process. Moreover, [11] found that MON does not undergo microbial breakdown during the ensiling process and remains active in the diets for later action in the rumen.

A synergistic effect likely took place between the low moisture of the silages and the antimicrobial action of limonene, thus reducing the activity of undesirable bacteria, which explains the lower butyric acid values in the silages with ELO300 and ELO600. Although the control treatment had the highest butyric acid and NH_3-N concentrations, they were within the acceptable range (1 g butyric acid/kg DM and 10% TN) for silages with good nutritional and sanitary quality [12,28,29]. The divergent results regarding NH_3-N content can be explained by the difference in fermentation intensity of the two trials. The DM content of the first trial contributes to a lower population of *Clostridium* spp., an NH_3-N-forming bacteria [37]. In the second trial, it is possible that the additive interacted with the moisture content of the material, which could be favorable for the population of saccharolytic Clostridium (ferment sugars into butyric acid) in the cases of the control and MON treatments, or proteolytic Clostridium (NH_3-forming) in the case of ELO use [25]. This makes it clear that the choice of additive should be linked to the dry matter content of the material to be ensiled.

4.2.2. Aerobic Stability

As in the first assay, adding ELO and MON likely helped reduce the growth of undesirable microorganisms after the silages were opened (especially molds and yeasts), thus improving their AS values, especially in treatments ELO600 and MON35. It is noteworthy that, in the second assay, the silages had higher moisture and, therefore, the AS values were much lower than those obtained in the first assay. Therefore, it can be said that using MON and ELO is a more interesting strategy to preserve the quality of TMR silage, with higher moisture after opening.

Greater aerobic stability in silage offers several advantages, including preserving nutritional value by preventing the growth of undesirable microorganisms, extending the silage life after opening, reducing waste, improving animal health due to lower presence of mycotoxins, saving on feed costs, and ensuring consistency in animal diets, leading to better performance in milk and meat production [38]. These benefits make aerobic stability a crucial factor for producing high-quality silages efficiently.

4.2.3. Impact on Nutrient Composition

The reductions in NDF and ADF contents when adding MON and ELO were likely due to higher activity of fibrolytic enzymes during the fermentation process. In an experiment with alfalfa and cumin EO at 300 and 500 mg/kg DM, Ref. [32] observed reductions in NDF and ADF contents in relation to the control treatment. According to those authors, the doses employed stimulated enzyme activities that promoted cell-wall breakage and released saccharose into the medium. Such saccharose release was likely used as substrate by the LAB to produce lactic acid [29]. The same behavior was observed by [39] when adding cumin EO at 200, 300, and 500 mg/kg to oat silages. Those authors observed that cumin EO significantly reduced the cellulolytic fraction of the silages.

Ref. [2] reported that the partial break of the cell wall is favored in silages with higher moisture and that the cell wall components can be used as substrate for silage fermentation. This helps explain why, in the present experiment, NDF and ADF contents were lower in the silages with higher moisture when compared with those with lower moisture.

The increase in starch content in the TMR silages added with MON and ELO is likely related to the reduction in prolamin, the layer that protects the starch in the seed. When reviewing TMR ensiling for ruminants, Ref. [2] argued that TMR silage fermentation promotes a positive effect on protease activity, favoring prolamin reduction and, consequently, increasing starch availability in the silage. The lower prolamin content may increase the efficiency of starch detection, since the standard technique depends on the hydrolysis of starch granules [40]. Therefore, if prolamin impacts starch detection, that error will be intrinsic in the NIR calibration equations as they are derived from the extension of the conditions observed in the standard technique [41]. Another factor that may help explain the higher starch contents is the dilution effect, since the decrease in fiber contents increases the proportion of other non-fiber nutrients in the silage, such as starch.

The higher EE contents observed in the treatments with ELO were also reported in other research in the literature [42–44]. When studying the origin of lipolysis in alfalfa silages, Ref. [45] found that the decrease in total fatty acid contents during ensiling took place mainly due to the breakdown of fatty acids C18:2n-6 and C18:3n-3. According to those authors, plant enzymes play a major role in lipolysis during alfalfa ensiling; however, several epiphyte microorganisms in alfalfa contributed much more to lipolysis. In that same research, the authors showed that homofermentative LAB purely inoculated in alfalfa silages did not impact the lipolysis of the material, but rather only the epiphytic flora contributed to it. This finding by [45] may help explain why treatments MON35 and MON45 preserved EE contents more, since, during TMR ensiling with MON, homolactic fermentation (higher lactic acid:acetic acid ratios) prevailed, thus decreasing the lipolytic action of other bacterial groups.

The increase in IVDMD in the TMR silage added with MON35 may be related to the fiber contents in that silage. Since the activity of fibrolytic enzymes was higher in that treatment, the production of other more digestible compounds, among which is lactic acid, may also have been higher. Following that assumption, the silages added with MON45, ELO300, and ELO600 may also have had the same stimulus for greater breakdown of the fibrous fraction, albeit at lower intensity. Improving feed efficiency in ruminants leads to greater weight gain or milk production, thus reducing production costs. These benefits result in more efficient, economical, and sustainable animal production.

The data obtained in this research are important, as they show growth-promoting additives can be added to TMR formulations with no harm to the fermentation process or nutritional quality of the silage. However, the effect of ionophores and essential oils on the dynamics of microbial populations over the ensiling process must still be better studied. Moreover, further studies are needed with other types of ionophores (lasalocid) and essential oils (eugenol, carvacrol, thymol, and cumin) commonly used in animal diets.

5. Conclusions

Strategic use of MON and ELO may effectively improve TMR silage quality by increasing acids production and aerobic stability, and have potential benefits for ruminant nutrition and production profitability. We demonstrated that ELO interacts with the silage moisture content on its effect on fermentation quality. Therefore, when silages present DM content above 40%, it is recommended to add the higher dose (ELO600), and for TMR silages with DM content below 30%, it is recommended to add the lower dose (ELO300).

Author Contributions: M.A.P.O.J.: Conceptualization and methodology. R.C.d.A.: Data curation and formal analysis. T.F. and A.C.A.O.: Editorial reviewers. M.R.: Visualization. F.M.d.V.J.: Resources. Y.A.d.S. and I.P.d.O.A.: Formal analysis. All authors have read and agreed to the published version of the manuscript.

Funding: This research was funded by Coordination for the Improvement of Higher Education Personnel (CAPES; Brasília, DF, Brazil–nº0001), the Development of Education, Science and Technology (FUNDECT; Mato Grosso do Sul, MS, Brazil–TO 007/2023 SIAFEM:32817), and, PDPG–Estados FUNDECT/CAPES (TO: 133/2023/SIAFEM: 33108).

Institutional Review Board Statement: Not applicable.

Data Availability Statement: All data generated or analyzed during this study are included in this published article.

Acknowledgments: We would like to thank the Brazilian Agricultural Research Corporation (EMBRAPA, Mato Grosso do Sul, MS, Brazil) for providing support and assistance.

Conflicts of Interest: The authors declare there are no conflicts of interest.

Abbreviations

TMR with no additive (control); TMR with 35 mg MON/kg DM (MON35); TMR with 45 mg MON/kg DM (MON45); TMR with 300 mg ELO/kg DM (ELO300); TMR with 600 mg ELO/kg DM (ELO600); GL = gas losses; EL = effluent losses; DMR = dry matter recovery; Ac. = acid; Lat/Ac ratio = acetic/lactic acid ratio; AS = aerobic stability; NH_3-N = ammoniacal nitrogen; TN = total nitrogen; DM = dry matter; CP = crude protein; SP = soluble protein; NDIP = neutral detergent insoluble protein; ADIP = acid detergent insoluble protein; NDF = neutral detergent fiber; ADF = acid detergent fiber; EE = ether extract; NFC = non-fiber carbohydrates; IVDMD = in vitro dry matter digestibility.

References

1. Schingoethe, D.J. A 100-Year Review: Total Mixed Ration Feeding of Dairy Cows. *J. Dairy Sci.* **2017**, *100*, 10143–10150. [CrossRef] [PubMed]
2. Bueno, A.V.I.; Lazzari, G.; Jobim, C.C.; Daniel, J.L.P. Ensiling Total Mixed Ration for Ruminants: A Review. *Agronomy* **2020**, *10*, 879. [CrossRef]
3. Miyaji, M.; Matsuyama, H.; Nonaka, K. Effect of Ensiling Process of Total Mixed Ration on Fermentation Profile, Nutrient Loss and In Situ Ruminal Degradation Characteristics of Diet. *Anim. Sci. J.* **2017**, *88*, 134–139. [CrossRef] [PubMed]
4. Alves, J.P.; Galeano Jara, E.S.; Orrico Junior, M.A.P.; Fernandes, T.; Retore, M.; Silva, M.S.J.; Orrico, A.C.A.; Lopes, L.S. The Influence of Plant Age and Microbes-Enzymatic Additives on Fermentation of Total Mixed Ration Silages of Capiaçu Grass (*Pennisetum purpureum*, Schum). *Trop. Anim. Sci. J.* **2022**, *45*, 56–63. [CrossRef]
5. Miyaji, M.; Nonaka, K. Effects of Altering Total Mixed Ration Conservation Method When Feeding Dry-Rolled versus Steam-Flaked Hulled Rice on Lactation and Digestion in Dairy Cows. *J. Dairy Sci.* **2018**, *101*, 5092–5101. [CrossRef]
6. Mcguffey, R.K.; Richardson, L.F.; Wilkinson, J.I.D. Ionophores for Dairy Cattle: Current Status and Future Outlook. *J. Dairy Sci.* **2001**, *84*, E194–E203. [CrossRef]
7. Žižek, S.; Hrženjak, R.; Kalcher, G.T.; Šrimpf, K.; Šemrov, N.; Zidar, P. Does Monensin in Chicken Manure from Poultry Farms Pose a Threat to Soil Invertebrates? *Chemosphere* **2011**, *83*, 517–523. [CrossRef] [PubMed]
8. Calsamiglia, S.; Busquet, M.; Cardozo, P.W.; Castillejos, L.; Ferret, A. Invited Review: Essential Oils as Modifiers of Rumen Microbial Fermentation. *J. Dairy Sci.* **2007**, *90*, 2580–2595. [CrossRef] [PubMed]
9. Bassolé, I.H.N.; Juliani, H.R. Essential Oils in Combination and Their Antimicrobial Properties. *Molecules* **2012**, *17*, 3989–4006. [CrossRef]
10. Hyldgaard, M.; Mygind, T.; Meyer, R.L. Essential Oils in Food Preservation: Mode of Action, Synergies, and Interactions with Food Matrix Components. *Front. Microbiol.* **2012**, *3*, 12. [CrossRef]
11. Lazzari, G.; Poppi, A.C.O.; Machado, J.; Bueno, A.V.I.; Gomes, A.L.M.; Jobim, C.C.; Daniel, J.L.P. Effects of Protein Source and Lipid Supplementation on Conservation and Feed Value of Total Mixed Ration Silages for Finishing Beef Cattle. *J. Anim. Sci.* **2021**, *99*, skab032. [CrossRef] [PubMed]
12. Borreani, G.; Tabacco, E.; Schmidt, R.J.; Holmes, B.J.; Muck, R.E. Silage Review: Factors Affecting Dry Matter and Quality Losses in Silages. *J. Dairy Sci.* **2018**, *101*, 3952–3979. [CrossRef] [PubMed]
13. Muller, L.D.; Bargo, F.; Ipharraguerre, I.R. Review: Monensin for Grazing Dairy Cows. *Prof. Anim. Sci.* **2015**, *22*, 115–119. [CrossRef]
14. Chen, L.; Li, X.; Wang, Y.; Guo, Z.; Wang, G.; Zhang, Y. The Performance of Plant Essential Oils against Lactic Acid Bacteria and Adverse Microorganisms in Silage Production. *Front. Plant Sci.* **2023**, *14*, 1285722. [CrossRef] [PubMed]
15. National Research Council. *Nutrient Requirements of Small Ruminants: Sheep, Goats, Cervids, and New World Camelids*, 2nd ed.; National Academies Press: Washington, DC, USA, 2007.
16. Gupta, A.; Jeyakumar, E.; Lawrence, R. Journey of Limonene as an Antimicrobial Agent. *J. Pure Appl. Microbiol.* **2021**, *15*, 1094–1110. [CrossRef]
17. Belanche, A.; Newbold, C.J.; Morgavi, D.P.; Bach, A.; Zweifel, B.; David, R.Y. A Meta-Analysis Describing the Effects of the Essential Oils Blend Agolin Ruminant on Performance, Rumen Fermentation and Methane Emissions in Dairy Cows. *Animals* **2020**, *10*, 620. [CrossRef] [PubMed]
18. Susanto, I.; Rahmadani, M.; Wiryawan, K.G.; Laconi, E.B.; Jayanegara, A. Evaluation of Essential Oils as Additives during Fermentation of Feed Products: A Meta-Analysis. *Fermentation* **2023**, *9*, 583. [CrossRef]
19. National Research Council. *Nutrient Requirements of Dairy Cattle*, 7th ed.; National Academy Press: Washington, DC, USA, 2001.
20. Jobim, C.C.; Nussio, L.G.; Reis, R.A.; Schmidt, P. Avanços Metodológicos Na Avaliação da Qualidade Da Forragem Conservada Methodological Advances in Evaluation. *Rev. Bras. Zootec.* **2007**, *36*, 101–119. [CrossRef]
21. Playne, M.J.; McDonald, P. The Buffering Constituents of Herbage and of Silage. *J. Sci. Food Agric.* **1966**, *17*, 264–268. [CrossRef]

22. Pryce, J.D. A Modification of the Barker-Summerson Method for the Determination of Lactic Acid. *Analyst* **1969**, *94*, 1151–1152. [CrossRef]
23. Association of Official Analytical Chemists. *Official Methods of Analysis*, 18th ed.; Helrich, K., Ed.; AOAC International: Arlington, VA, USA, 2005.
24. Tilley, J.M.A.; Terry, R.A. A Two-Stage Technique for the In Vitro Digestion of Forage Crops. *Grass Forage Sci.* **1963**, *18*, 104–111. [CrossRef]
25. Kung, L., Jr.; Shaver, R.D.; Grant, R.J.; Schmidt, R.J. Silage Review: Interpretation of Chemical, Microbial, and Organoleptic Components of Silages. *J. Dairy Sci.* **2018**, *101*, 4020–4033. [CrossRef] [PubMed]
26. Scott, R.; Knott, M. A Cluster Analysis Method for Grouping Means in the Analysis of Variance. *Biometrics* **1974**, *30*, 507–512. [CrossRef]
27. Barmaki, S.; Alamouti, A.A.; Khadem, A.A.; Afzalzadeh, A. Effectiveness of Chopped Lucerne Hay as a Moisture Absorbent for Low Dry-Matter Maize Silage: Effluent Reduction, Fermentation Quality and Intake by Sheep. *Grass Forage Sci.* **2018**, *73*, 406–412. [CrossRef]
28. Kung Junior, L.; Stokes, M.R.; Lin, C.J. Silage Additives. In *Silage Science and Technology*; Buxton, D.R., Muck, R.E., Harrison, J.H., Eds.; American Society of Agronomy: Madison, WI, USA, 2003; pp. 305–360.
29. Muck, R.E. Silage Microbiology and Its Control through Additives. *Rev. Bras. Zootec.* **2010**, *39*, 183–191. [CrossRef]
30. Cunha, S.S.; Orrico Junior, M.A.P.; Reis, R.A.; Orrico, A.C.A.; Schwingel, A.W.; Reis, S.D.S.; Silva, M.S.J. Use of Crude Glycerine and Microbial Inoculants to Improve the Fermentation Process of Tifton 85 Haylages. *Trop. Anim. Health Prod.* **2020**, *52*, 871–879. [CrossRef] [PubMed]
31. Zhang, G.; Fang, X.; Feng, G.; Li, Y.; Zhang, Y. Silage Fermentation, Bacterial Community, and Aerobic Stability of Total Mixed Ration Containing Wet Corn Gluten Feed and Corn Stover Prepared with Different Additives. *Animals* **2020**, *10*, 1775. [CrossRef] [PubMed]
32. Turan, A.; Önenç, S.S. Effect of Cumin Essential Oil Usage on Fermentation Quality, Aerobic Stability and In Vitro Digetibility of Alfalfa Silage. *Asian Australas. J. Anim. Sci.* **2018**, *31*, 1252–1258. [CrossRef] [PubMed]
33. Aniceto, E.S.; Oliveira, T.S.; Meirelles, J.R.; Silva, I.N.; Filho, E.J.L.M.; Gomes, R.S.; Arévalo, J.P.; Moraes, P.R. Evaluation of Essential Oils and Their Blends on the Fermentative Profile, Microbial Count, and Aerobic Stability of Sorghum Silage. *Fermentation* **2024**, *10*, 335. [CrossRef]
34. Fuller, J.R.; Johnson, D.E. Monensin and Lasalocid Effects on Fermentation In Vitro. *J. Anim. Sci.* **1981**, *53*, 1574–1580. [CrossRef]
35. Cantoia Júnior, R.; Capucho, E.; Garcia, T.M.; Del Valle, T.A.; Campana, M.; Zilio, E.M.C.; Azevedo, E.B.; Morais, J.P.G. Lemongrass Essential Oil in Sugarcane Silage: Fermentative Profile, Losses, Chemical Composition, and Aerobic Stability. *Anim. Feed Sci. Technol.* **2020**, *260*, 114371. [CrossRef]
36. Brahmakshatriya, R.D.; Donker, J.D. Five Methods for Determination of Silage Dry Matter. *J. Dairy Sci.* **1971**, *54*, 1470–1474. [CrossRef]
37. Pahlow, G.; Muck, R.E.; Driehuis, F.; Oude Elferink, S.J.W.H.; Spoelstra, S.F. Microbiology of Ensiling. In *Silage Science and Technology*; Buxton, D.R., Muck, R.E., Harrison, J.H., Eds.; American Society of Agronomy: Madison, WI, USA, 2003; pp. 31–93.
38. Wilkinson, J.M.; Rinne, M. Highlights of Progress in Silage Conservation and Future Perspectives. *Grass Forage Sci.* **2018**, *73*, 40–52. [CrossRef]
39. Akinci, Y.; Soycan Önenç, S. The Effect of Cumin Essential Oil on the Fermentation Quality, Aerobic Stability, and In Vitro Digestibility of Vetch-Oat Silages. *J. Agric. Fac. Ege Univ.* **2021**, *58*, 217–228. [CrossRef]
40. Hall, M.B.; Arbaugh, J.; Binkerd, K.; Carlson, A.; Thi Doan, T.; Grant, T.; Heuer, C.; Inerowicz, H.D.; Jean-Louis, B.; Johnson, R.; et al. Determination of Dietary Starch in Animal Feeds and Pet Food by an Enzymatic-Colorimetric Method: Collaborative Study. *J. AOAC Int.* **2015**, *98*, 397–409. [CrossRef] [PubMed]
41. Jiang, H.; Lu, J. Using an Optimal CC-PLSR-RBFNN Model and NIR Spectroscopy for the Starch Content Determination in Corn. *Spectrochim. Acta Part A Mol. Biomol. Spectrosc.* **2018**, *196*, 131–140. [CrossRef] [PubMed]
42. Besharati, M.; Palangi, V.; Niazifar, M.; Nemati, Z. Comparison Study of Flaxseed, Cinnamon and Lemon Seed Essential Oils Additives on Quality and Fermentation Characteristics of Lucerne Silage. *Acta Agric. Slov.* **2020**, *115*, 455–462. [CrossRef]
43. Chaves, A.V.; Baah, J.; Wang, Y.; Mcallister, A.; Benchaar, C. Effects of Cinnamon Leaf, Oregano and Sweet Orange Essential Oils on Fermentation and Aerobic Stability of Barley Silage. *J. Food Agric.* **2012**, *92*, 906–915. [CrossRef] [PubMed]
44. Romero, S.; Merino, E.; Bolívar, F.; Gosset, G.; Martinez, A. Metabolic Engineering of Bacillus Subtilis for Ethanol Production: Lactate Dehydrogenase Plays a Key Role in Fermentative Metabolism. *Appl. Environ. Microbiol.* **2007**, *73*, 5190–5198. [CrossRef]
45. Ding, W.R.; Long, R.J.; Guo, X.S. Effects of Plant Enzyme Inactivation or Sterilization on Lipolysis and Proteolysis in Alfalfa Silage. *J. Dairy Sci.* **2013**, *96*, 2536–2543. [CrossRef]

agriculture

Article

Ensiling of Willow and Poplar Biomass Is Improved by Ensiling Additives

Søren Ugilt Larsen [1,2,*], Helle Hestbjerg [3], Uffe Jørgensen [1,2] and Anne Grete Kongsted [1]

1 Department of Agroecology, Aarhus University, Blichers Allé 20, DK-8830 Tjele, Denmark; uffe.jorgensen@agro.au.dk (U.J.)
2 CBIO—Aarhus University Centre for Circular Bioeconomy, Blichers Allé 20, DK-8830 Tjele, Denmark
3 Danish Technological Institute, Agro Food Park 15, DK-8200 Aarhus N, Denmark; helh@teknologisk.dk
* Correspondence: soren.ugilt.larsen@agro.au.dk

Abstract: Biomass from willow and poplar harvested for feed during the growing season may be preserved by ensiling; however, little research has focused on ensiling of these biomasses. This study focuses on the use of ensiling additives to reduce the pH to around 4.0 to secure stable storage. Lab-scale ensiling experiments were conducted with different willow and poplar clones, shoot ages, and harvest times (June or September). Ensiling without additives often resulted in limited pH reduction. The pH could be reduced in the biomass of both species by adding formic acid, and the required dose to reduce the pH to 4.0 (buffering capacity, BC) ranged significantly between biomass types but was in the range of 2–5 kg formic acid (78%) per ton fresh weight. BC decreased with increasing dry matter (DM) content and decreasing crude protein content. The pH could also be reduced during ensiling by applying molasses and/or lactic acid bacteria, although not sufficiently in poplar. Willow biomass was ensiled effectively at the pilot scale with less than 7% DM loss by adding formic acid or by mixing with grass biomass. Comparable pH results were obtained at the lab scale and pilot scale. The study demonstrates how willow and poplar can be ensiled; however, more research is needed on quality changes during ensiling.

Keywords: silage; lactic acid bacteria; molasses; formic acid; buffering capacity; pH; Salix; Populus; storage loss

Citation: Larsen, S.U.; Hestbjerg, H.; Jørgensen, U.; Kongsted, A.G. Ensiling of Willow and Poplar Biomass Is Improved by Ensiling Additives. *Agriculture* **2024**, *14*, 1853. https://doi.org/10.3390/agriculture14101853

Academic Editor: Siran Wang

Received: 6 September 2024
Revised: 2 October 2024
Accepted: 8 October 2024
Published: 21 October 2024

1. Introduction

Fast-growing tree species such as willow and poplar possess a range of advantages typical of many perennial crops compared to annual crops and which are desirable in relation to the green transition: high biomass yield [1,2], build-up of soil carbon [3–5], low use of pesticides except in the establishment phase [6] and generally low loss of nutrients from the rootzone [5,7].

A specific use of such tree species occurs in free-range pig and poultry production where fast-growing trees may both take up nutrients from nutrient hot-spots in the range area and provide shade and shelter for the animals and, hence providing environmental and animal welfare services [8–11]. There is also increasing interest in the use of trees and agroforestry concepts in relation to grazing ruminants, especially in organic production [12,13].

In addition to environmental and welfare aspects, the produced biomass from willow and poplar may also constitute a potential feed source of supplementary forage [14–18], particularly when harvested as 'green biomass' during the growing season, i.e., the harvest of leaves and new shoots. As reviewed by Vandermeulen et al. [19], willow is regarded as a valuable supplement to grass during dry periods for grazing ruminants, providing energy, protein, and micronutrients. Limited research exists on the use of green willow biomass as feed for monogastric livestock. However, its relatively high protein content [17] suggests that it could serve as a viable alternative to grass clover silage, a well-established amino acid source, particularly for gestating sows, which demonstrate a substantial capacity for roughage intake [20].

Use of tree biomass for feed may also be relevant due to the content of secondary compounds, such as condensed tannins and phenolic glycosides in willow and poplar [14], as well as polyphenols, salicin, and flavonoids in various parts of willow [21,22], including the shoots and leaves [23]. These bioactive compounds may have positive effects on livestock performance [14,24], and the content of tannins in plant biomass used as feed for ruminants may potentially reduce the emission of enteric methane [25] which has been demonstrated specifically for sheep grazing willow [26].

For year-round use of tree biomass for feed, the biomass needs to be preserved to reduce loss of dry matter (DM) and quality during storage. Ensiling is an ideal method for preserving wet biomass and has been used extensively for various types of forage [27].

A few studies have investigated ensiling of willow biomass [16,18,28–30]. Also, a couple of practical guides for low-tech and small-scale ensiling of willow and poplar have been published [31,32], and the experience with ensiling of biomass from browse trees in Africa has been reviewed [33]. Overall, only little research has been conducted on ensiling of green biomass from willow and poplar.

A good ensiling process is characterized by a rapid reduction in pH to a low level, preferably around 4.0, which will protect the biomass from spoilage microorganisms [34]. The fermentation loss is potentially very low but may vary considerably depending on the conditions [35]. The pH reduction during ensiling is driven by conversion of water-soluble carbohydrates (WSCs) to organic acids by lactic acid bacteria (LAB) [35] and, therefore, requires the presence of both WSCs and LAB at sufficient levels. However, the pH reduction during ensiling is also dependent on the buffering capacity (BC) of the biomass, which expresses the resistance of the biomass to pH lowering [36] and which can be defined as the quantity of acid required to reduce the pH to a certain level, e.g., 4.0. Biomasses with high BC, therefore, require higher levels of acids to reduce the pH to the desired level and, consequently, are more dependent on a high concentration of WSCs [36]. Moreover, a high DM content of the biomass may also enhance the ensiling process [36].

The few existing studies on ensiling of green willow biomass indicate a generally low ensilability of the biomass, expressed as a relatively high pH level after ensiling without any additives, e.g., pH levels of 4.7–6.3 [28], 5.8 [16], 5.1 [29] and 4.7–4.9 [18]. It has been suggested that the low ensilability of biomass from willow and other species may be due to the low content of WSCs [16,33].

Ensiling of biomass can be enhanced, e.g., by applying a source of WSCs such as molasses [36] and/or an ensiling additive that can be a biological additive such as an LAB inoculant, or a chemical additive, such as formic acid [37]. Depending on the dose, formic acid can cause direct acidification and suppress spoilage bacteria [37]. Another approach is to mix the poorly ensilable biomass with another crop biomass with a higher content of WSCs as suggested for tree biomass [33]. For instance, forage grasses often have high WSCs concentrations [38] and maize has been co-ensiled with willow [30]. Molasses and LAB have been investigated as additives for ensiling of willow biomass [29]; however, the use of formic acid as a chemical ensiling additive as well as the BC of willow and poplar biomass appear to have not been studied.

Lab-scale ensiling experiments can be performed using e.g., vacuum bags [39] or anaerobic jars [29], and such model systems are convenient for studying the effects of various factors, such as ensiling additives on qualitative silage parameters, including pH. However, for assessment of the mass balance during ensiling, better estimates can be obtained by using larger quantities of biomass, e.g., by ensiling in barrels [29,30].

In this study, we investigated ensiling of different types of green biomass from willow and poplar at the lab-scale as well as the pilot scale for selected treatments. The study consisted of four experiments, with Experiment 1 focusing on the buffering capacity of the biomasses and Experiments 2–4 focusing on ensiling of the biomasses. The main focus was on the reduction in pH during ensiling as a primary indicator of the ensiling process. The aim of the study was to investigate (i) the buffering capacity of green biomass harvested from willow and poplar during the growing season, (ii) the ensilability of the biomass, (iii) the requirement for and effects of various ensiling additives to reduce the pH of the biomass, and (iv) up-scaling from lab-scale ensiling to pilot-scale ensiling and to explore the feeding value of the ensiled willow. The main hypothesis is that the pH can be reduced significantly during the ensiling of willow and poplar biomass by the appropriate use of ensiling additives.

2. Materials and Methods

2.1. Types of Biomasses Harvested

All biomasses were produced and harvested by the company Ny Vraa (www.nyvraa.dk/en (accessed on 2 September 2024)) near Tylstrup, Denmark (57°10′25.4″ N, 9°55′10.5″ E) as part of field trials with two or three replicate blocks. Tree biomass was obtained from the willow clones Tordis ((*Salix schwerinii* × *S. viminalis*) × *S. viminalis*) [2] and Clone X (from breeding program, not yet commercially available) and from the poplar clone OP42 (*Populus triochocarpa* × *Populus maximowiczii*) [1]. The willow clone Tordis was established in 2010 on humus soil and had not been fertilized since its establishment. The willow Clone X was established in 2019 on coarse sandy soil and was fertilized with slurry in the establishment year. The poplar clone OP42 was established on coarse sandy soil in 2012 and had not been fertilized since its establishment. All trees had been harvested at least once before harvesting for this study.

Tree biomass was harvested in June and September, i.e., during the growing season, and shoots and leaves were harvested together, typically at a stubble height of 20 cm, which was just above the stool from the previous harvest. For the willow clone Tordis, biomass was harvested at two different 'shoot ages', where '$\frac{1}{2}$-year shoots' indicates that the harvested biomass originated solely from growth in the current growth year, i.e., the trees had been harvested in the preceding winter, and '$1\frac{1}{2}$-year shoots' originated from both the current and the preceding growth year. For willow Clone X and the poplar clone OP42, only $\frac{1}{2}$-year shoots were harvested. Details on biomass type, shoot age, harvest date, shoot length, and contents of dry matter (DM) and crude protein (CP) in the biomass batches are shown in Table 1, and the table also shows in which of the four experiments (explained in Sections 2.3 and 2.6–2.8) the various biomass batches were used.

In 2021, all tree shoots were harvested using a shrub cutter, whereas shoots harvested in 2022 were harvested by a machine (ZG80, Zero Grazer, Kilnaleck, County Cavan, Ireland), which picks up the biomass directly without chopping the biomass. In both years, the harvested tree shoots were subsequently chopped using a wood-chipper (JF40 Ensiling Machine, JF, Jardim Guarujá, Brazil). The chopper was adjusted to the finest chopping, resulting in a particle size of typically 1–2 cm in length.

In addition to tree biomass, herbaceous biomass was obtained in 2022 from a late first cut in a grass–clover field, dominated by perennial ryegrass and with some red clover and other broadleaved species. This biomass was harvested by machine and chopped as described for tree shoots.

The batches of chopped biomasses were used for the four ensiling experiments which were all initiated on the same day as the harvest. Subsamples were analyzed for DM content on the same day, and other subsamples were frozen for later analysis of CP content and pH.

Table 1. Overview of biomasses used in the four experiments. Shoot age indicates if the harvested biomass originates solely from the current growth year ('$\frac{1}{2}$-year shoots') or if it originates from both the current and the preceding growth year ('$1\frac{1}{2}$-year shoots'). Mean and standard deviation are given for dry matter (DM) content and crude protein (CP) content, based on either two or three replicate samples.

Biomass Batch No.	Species	Clone	Shoot Age	Previous Harvest	Harvest Date	Shoot Length (m)	DM Content (%)		CP Content (% in DM)		Used in Experiment			
											Exp. 1	Exp. 2	Exp. 3	Exp. 4
1	Willow	Tordis	$\frac{1}{2}$-year shoots	Winter 2020/2021	15 June 2021	1.0	19.3	±0.4	-		x	x		
2					24 June 2021	1.0–1.2	19.9	±0.5	17.5	±1.2			x	
3					9 September 2021	-	33.0	±0.2	-		x			
4					24 September 2021	3.5	37.7	±0.5	8.4	±0.3			x	
5				Winter 2021/2022	22 June 2022	-	25.1	±0.2	14.1	-				x
6	Willow	Tordis	$1\frac{1}{2}$-year shoots	Winter 2019/2020	15 June 2021	-	36.9	±0.1	-		x			
7					24 June 2021	3.0–3.3	36.9	±0.6	6.2	±0.1			x	
8					9 September 2021	-	41.4	±0.4	-		x			
9					24 September 2021	4.3	43.1	±1.5	6.3	±0.8			x	
10	Willow	Clone X	$\frac{1}{2}$-year shoots	Winter 2020/2021	24 June 2021	1.1–1.2	25.1	±0.4	13.6	±0.2			x	
11					9 September 2021	-	35.7	±0.2	-		x			
12					24 September 2021	2.7	40.8	±0.2	4.7	±0.2			x	
13	Poplar	OP42	$\frac{1}{2}$-year shoots	Winter 2020/2021	15 June 2021	-	26.4	-	-		x			
14					24 June 2021	0.7–0.8	26.2	±0.3	13.7	±0.5			x	
15					9 September 2021	-	35.0	±0.7	-		x			
16					24 September 2021	2.0–2.5	38.3	±0.5	5.9	±0.6			x	
17	Grass-clover mixture	-	1st cut	-	22 June 2022	-	29.9	±0.3	-					x
18	Willow + grass	50:50 mix of no. 5 and 17			22 June 2022	-	27.2	±0.1	10.8	-				x

2.2. Analysis of Dry Matter Content, Crude Protein Content and pH

The dry matter content was analyzed for each biomass batch by drying either two or three replicate subsamples of approx. 150 g at 60°C until constant weight. The content of N was analyzed for some of the biomass batches (Table 1) by Eurofins Agro Testing, Vejen, Denmark, via Dumas analysis (CEE22, Internal Method Dumas/combustion), with three replicate samples per batch. The CP content was calculated by multiplying the N content by 6.25 [40].

The pH was measured in various samples of fresh and ensiled biomass using the following procedure. Two subsamples of 10 g fresh weight (FW) each were taken from each batch. The samples were either taken directly from unfrozen samples or from frozen samples where the sample material was obtained using a drill. Each sample of 10 g FW was loaded into a PA/PE bag (120 × 250 × 0.09 mm, from LogiCon Nordic, Kolding, Denmark) and 100 g of demineralized water was added. The mixture of biomass and water was massaged initially and after 15 and 30 min, after which the pH was measured in the liquid. The pH was measured with a WTW pH 3310 SET 2 with a SenTix® 41 probe (WTW, Weilheim, Germany), which was calibrated with pH buffer solutions of pH 7.00 and 4.01.

The content of amino acids (Lysine, Methionine, Cystine, and Threonine) in the silage of willow and in the silage of willow + grass was analyzed in accordance with Commission Regulation EC [41], as detailed by Johannsen et al. [42]. The energy content, expressed in Danish Feed Units for sows (FUsow), was analyzed using in vitro enzymatic digestion, as described by Boisen [43]. The feed units were converted to net energy (NE, MJ) according to Theil et al. [44].

2.3. Experiment 1: Dose–Response Experiment for Determination of Buffering Capacity

The buffering capacity (BC) was determined in three types of tree biomass harvested 15 June 2021 and four types harvested 9 September 2021, representing both willow and poplar clones and different shoot ages (biomass batch no. 1, 3, 6, 8, 11, 13, 15, Table 1). BC was determined from a dose–response experiment by the addition of various doses of formic acid and measurement of the subsequent pH, where BC was estimated as the acid dose required to reduce pH to 4.0.

For each biomass type, 14 portions of chopped biomass of 20 g FW were loaded into PA/PE bags (120 × 250 × 0.09 mm) and 5 g of demineralized water was added. For each biomass type, 7 different doses of formic acid (78%) were applied to two replicate bags. The doses corresponded to 0, 0.5, 1, 2, 4, 8 and 12 kg formic acid (78%) per ton FW. After loading the acid into the bags, the bags were massaged to thoroughly mix the biomass, water and acid, and the bags were vacuum packed and stored at 5–10 °C for 20 h (June) or 44 h (September) to allow equilibration between the biomass and the acid. The bags were then opened, 100 g of demineralized water was added to each bag, and the pH was measured as described above.

2.4. General Procedure for Lab-Scale Ensiling Experiments

Lab-scale ensiling experiments were carried out using a vacuum bag system [28,45], using PA/PE bags (LogiCon Nordic, Kolding, Denmark). The general procedure included weighing 750 g FW of the given tree biomass, which was loaded into an inner bag (525 × 600 × 0.090 mm). Depending on the treatment, the ensiling additives were sprayed onto the biomass in the bag using spray bottles with known dose per push and with the additive dose being checked by weighing each bag during application. Afterwards, the biomass was mixed thoroughly with the ensiling additives within the bag, which was then placed into another bag (240 × 650 × 0.150 mm), and this bag was vacuum-packed by a Webomatic vacuum-packing machine (M. Type C 15-HL, M. No. 0310TB1000. Webomatic, Hansastr. 119, D-44866 Bochum, Germany). The silage bags were then stored in an unheated barn (temperature typically in the range of 5–15 °C) for the preplanned duration, after which the bags were frozen at −18 °C until further analysis.

2.5. Ensiling Additives

The additives used in the ensiling experiments included formic acid, molasses and LAB, and all were applied in an aqueous solution to ensure a good distribution on the biomass. For all three additives, the dose of water corresponded to 20 kg water per ton of FW biomass. This addition of water reduced the DM content of the biomass slightly, e.g., with a reduction from 20.0 to 19.6% DM or from 40.0 to 39.2% DM. This reduction in DM content was expected to have little impact on the ensiling process.

Formic acid: Formic acid of 78% concentration was used (Brenntag Nordic, Vejle, Denmark), with a density of 1.18 kg per liter. In the ensiling experiments, the applied dose of formic acid was determined from the analysis of BC, i.e., the required dose of formic acid to reduce the pH to 4.0 for each individual type of biomass (see Section 2.3), typically in the range of 3 to 6 kg formic acid (78%) per ton FW. With a BC of 6 kg per ton FW, for instance, an aqueous solution of formic acid was prepared by dissolving 100 g of formic acid (78%) in 333 g of water. This solution was then applied with a spray bottle at a dose of 19.5 g for a biomass sample of 750 g FW, corresponding to the addition of 4.5 g formic acid (78%) + 15 g water per sample.

Molasses: The applied molasses was standardized sugar beet molasses (Amequ by Dangro flydende melasse, produced by Nordic Sugar, distributed by Amequ and Dangro Nordic A/S, Videbæk, Denmark), with a density of 1.35 kg per liter. According to the declaration, the sucrose content was minimum 43%, the water content was maximum 26%, and the maximum content of microorganisms was 10,000 CFU g^{-1} for both yeasts and molds and 1000 CFU g^{-1} for both *Bacillus* species and *Clostridium* species. Based on previous experiments, a standard dose of 20 kg molasses per ton of FW biomass was used, corresponding to at least 8.6 kg sucrose per ton of FW biomass. Molasses was applied in an aqueous solution which was prepared by dissolving 100 g of molasses in 100 g of tap water. Hence, the molasses treatment included the addition of 20 kg molasses + 20 kg water per ton of FW biomass. The solution was applied with a spray bottle at a dose of 30 g for a biomass sample of 750 g FW, resulting in the addition of 15 g molasses + 15 g water per sample.

Lactic acid bacteria: The applied LAB was the product TOPSIL Max ØKO (SH Ensilering, Højslev, Denmark). This is the same product as TOPSIL Max but it uses organic sugar and is, therefore, allowed for use in organic farming in Denmark. The product consists of three homofermentative LAB species:

- *Lactobacillus plantarum* DSMZ 1,6627/1k20749: 75.0 $\times$ 10^9 CFU g^{-1}.
- *Pediococcus acidilactici* NCIMB 3,0005/1k21013: 37.5 $\times$ 10^9 CFU g^{-1}.
- *Lactobacillus paracasei* NCIMB 3,0151/1k20748: 37.5 $\times$ 10^9 CFU g^{-1}.

The total content of LAB is 150 $\times$ 10^9 CFU g^{-1}. The recommended dose is 2 g per ton of FW of maize, grass and alfalfa (typical DM content of 30–35%), and this dose was also applied in these experiments. An aqueous solution was prepared by dissolving 0.5 g of TOPSIL Max ØKO in 5 kg of cold water. This solution was applied with a spray bottle at a dose of 15 g for a biomass sample of 750 g FW, resulting in the addition of 0.0015 g TOPSIL + 15 g water per sample.

2.6. Experiment 2, Lab-Scale-Ensiling: Development of pH in Willow Silage over Time

A lab-scale experiment was performed to study the development of pH in willow biomass during ensiling with or without ensiling additives. A series of 32 silage bags were prepared with willow biomass from $\frac{1}{2}$-year shoots from the clone Tordis, harvested 15 June 2021 with a DM content of 19.3% (biomass batch no. 1, Table 1). Sixteen bags were prepared without any ensiling additives, and 16 bags were prepared with a combination of molasses (20 kg per ton FW) and LAB TOPSIL Max Øko (2 g per ton FW). Both molasses and LAB were applied as aqueous solutions, corresponding to the addition of 40 kg water per ton FW. The biomass was vacuum packed within 8 h from harvest. Two replicate bags of each of the two additive treatments were frozen after the following ensiling durations: 0, 1, 2, 5, 10, 20, 48 and 85 days. Samples were later used for analysis of pH.

2.7. Experiment 3, Lab-Scale Ensiling: Effects of Additives on Ensiling of Willow and Poplar Harvested in June or September

A lab-scale experiment was performed to study the effect of different ensiling additives on ensiling of four types of green tree biomass harvested in June or September 2021. The experiment included four factors: biomass type (4 levels), harvest time (2 levels), ensiling additive (5 levels) and ensiling duration (2 levels, with or without ensiling). The four biomass types represented both willow and poplar clones, different shoot ages, biomass harvested 24 June or 24 September 2021.The DM content ranged from 19.9 to 43.1% (biomass batch no. 2, 4, 7, 9, 10, 12, 14, 16, Table 1).

For each of the four biomass types and two harvest times, five different additive treatments were applied, as described in Table 2. The treatments included a control without additive and the addition of either formic acid, molasses, LAB, or molasses + LAB. The dose of formic acid (78%) was determined from the analysis of BC within two weeks prior to the ensiling experiment (see Section 2.3) and ranged between 2.6 and 6.0 kg per ton FW (Table 2). The doses of molasses and LAB were based on previous experiences and recommendations, respectively. All additives were applied as aqueous solutions, and a total of either 20 or 40 kg water was added per ton FW (Table 2).

Table 2. Overview of additive treatments to the four biomass types and the two harvest dates in ensiling Experiment 3.

Species	Clone	Shoot Age	Harvest Date	Treatment No. and Ensiling Additive				
				1. Untreated	2. Formic Acid	3. Molasses	4. Lactic Acid Bacteria	5. Molasses + Lactic Acid Bacteria
				Dose				
				-	kg 78% Acid + kg Water per ton FW	kg Molasses + kg Water per ton FW	g Topsil Max ØKO + kg Water per ton FW	kg Molasses + g Topsil Max ØKO + kg Water per ton FW
Willow	Tordis	$\frac{1}{2}$-year shoots	24 June 2021	-	6.0 kg + 20 kg	20 kg + 20 kg	2 g + 20 kg	20 kg + 2 g + 40 kg
			24 September 2021	-	3.0 kg + 20 kg	20 kg + 20 kg	2 g + 20 kg	20 kg + 2 g + 40 kg
Willow	Tordis	$1\frac{1}{2}$-year shoots	24 June 2021	-	3.0 kg + 20 kg	20 kg + 20 kg	2 g + 20 kg	20 kg + 2 g + 40 kg
			24 September 2021	-	2.6 kg + 20 kg	20 kg + 20 kg	2 g + 20 kg	20 kg + 2 g + 40 kg
Willow	Clone X	$\frac{1}{2}$-year shoots	24 June 2021	-	6.0 kg + 20 kg	20 kg + 20 kg	2 g + 20 kg	20 kg + 2 g + 40 kg
			24 September 2021	-	5.0 kg + 20 kg	20 kg + 20 kg	2 g + 20 kg	20 kg + 2 g + 40 kg
Poplar	OP42	$\frac{1}{2}$-year shoots	24 June 2021	-	5.0 kg + 20 kg	20 kg + 20 kg	2 g + 20 kg	20 kg + 2 g + 40 kg
			24 September 2021	-	2.6 kg + 20 kg	20 kg + 20 kg	2 g + 20 kg	20 kg + 2 g + 40 kg

For each combination of biomass type, harvest time, and additive treatment, three replicate vacuum bags were prepared with 750 g FW, resulting in a total of 120 bags. The bags were vacuum packed within 10 h from harvest of the biomass. One replicate bag was frozen at −18 °C within 12 h after vacuum packing, corresponding to an unensiled control (0 days of ensiling). The other two replicate bags (vacuum packed with two layers of 0.150 mm PA/PE, i.e., first vacuum-packing in one bag and then in another bag) were stored for ensiling for 75 days at ambient temperature in an unheated barn. The pH was measured as described above in duplicate subsamples from the unensiled bag and from the two ensiled bags per combination.

2.8. Experiment 4, Lab-Scale and Pilot-Scale Ensiling: Ensiling of Willow and Willow + Grass Mixture

Based on results from Experiments 2 and 3 in 2021, a combined lab-scale and pilot-scale experiment was initiated in 2022. The aim was to test upscaling of the ensiling procedure for willow with formic acid and to test grass as an 'ensiling additive' in mixture with willow to improve the ensiling of willow biomass. Also, the aim was to estimate mass loss during ensiling at the pilot scale.

The tree biomass represented $\frac{1}{2}$-year shoots of the willow clone Tordis with a DM content of 25.1% (biomass batch no. 5, Table 1), which was ensiled with the addition of 5 kg formic acid (78%) per ton FW. In addition, willow biomass was co-ensiled with grass–clover biomass (biomass batch no. 17) in a ratio of approx. 50:50 % FW basis, resulting in a willow + grass mixture with 27.2% DM (biomass batch no. 18).

Both biomasses were harvested 22 June 2022 and chopped. Formic acid was applied to willow biomass as an aqueous solution of 15.6% formic acid applied at a dose of 25 kg per ton FW. Approx. 180 kg FW of chopped willow was distributed in a 15 cm layer on a concrete floor, and formic acid solution was sprayed onto the biomass using a garden sprayer. The biomass was turned and mixed during spraying to obtain a uniform distribution. The mixture of willow + grass was obtained by weighing 80 kg FW of each biomass and mixing thoroughly by turning on a concrete floor.

Both willow biomasses with formic acid and willow + grass were loaded into two 60 L open top barrels (Plastic Blue UN Approved Open Top Drum in HDPE). The biomass was continuously compressed during filling of the barrels by a person trampling on the biomass layer by layer. The barrels were filled and closed with as much biomass as possible to diminish the air content. After filling, the barrels were weighed to determine the initial biomass weight. The barrels were stored in an unheated barn for 257 days to mimic long-term storage for year-round supply, then weighed to determine the final biomass weight, and opened for visual evaluation. Biomass samples were taken before ensiling and after ensiling (at 10–15 cm distance from the surface after opening the barrels) for duplicate analysis of DM content and pH as described above.

In addition to pilot-scale ensiling, lab-scale ensiling was performed by vacuum packing four bags of approx. 700 g of willow biomass with formic acid and four bags with willow + grass, and the bags were stored together with the barrels and opened after 257 days of ensiling for duplicate analysis of pH. In addition, samples of fresh frozen willow biomass without and with formic acid and willow + grass was analyzed for pH as unensiled references.

2.9. Data Analyses

2.9.1. Experiment 1

The BC of willow and poplar biomasses was estimated individually for the different types of tree biomass and the two harvest times (Table 1), and both on a FW and a DM basis by fitting the relationship between the dose of formic acid and pH by Equation (1):

$$y = a + b \times e^{(-c \times x)} \quad (1)$$

where y is the pH value, x is the dose of 78% formic acid (kg ton^{-1} FW or DM), a is the lower asymptote for pH at infinitely high doses, b is the difference between the lower asymptote and the intercept with the y axis, and c is a constant expressing the rate at which pH decreases with increasing dose. The relationship was fitted with SAS software release 9.3 [46] using the proc nlmixed procedure, and estimates and 95% confidence limits were calculated for pH without formic acid addition, and the required formic acid dose to reduce the pH to 4.0, which corresponds to BC, was calculated by rewriting Equation (1) as Equation (2):

$$x = \ln((y - a)/b) - c \quad (2)$$

Significant differences in estimates of BC were evaluated from the 95% confidence limits.

The relationships between measured DM and estimated BC and between measured CP (% of DM) and estimated BC were analyzed by linear regression, with separate analyses for BC on FW and DM basis. In both models, the response variable was BC, and the explanatory variables were either DM (%) or CP (% in DM). The analyses were performed with R software version 4.1.3 using the function lm.

2.9.2. Experiment 2

The data for pH development in willow silage over time were analyzed with R software, both as analysis of variance and non-linear regression. The analysis of variance was performed using a model that included treatment (class variable, with or without ensiling additives), ensiling duration (class variable, 0, 1, 5, 10, 20, 48 and 85 days) and their interacting effects, as well as replicate vacuum bags (class variable, two bags per combination of treatment and ensiling duration). The mean value of the two replicate pH measurements per vacuum bag was used in the analysis. A total of 64 observations were included. The analysis was carried out using the R functions lm and annova. The functions emmeans and multcomp were used to estimate the expected means and LSD groups (compact letter display).

The non-linear regression analysis was performed separately for the treatments without and with ensiling additive using a three-parameter model, as described in Equation (1), where y is the pH value, x is the ensiling duration (numerical variable), and a, b, and c are parameters. The analysis was performed using the nlme package in R and the nls function.

2.9.3. Experiment 3

The data for pH before and after ensiling of different types and harvest times of tree biomass with different additive treatments were all analyzed in one analysis using the optimx package in R software version 4.1.3., with pH as the response variable and with four explanatory variables, all as class variables: biomass type (4 levels), harvest time (2 levels), ensiling additive (5 levels) and ensiling duration (with or without ensiling). All possible interactions between the main factors were included in the model, and the model was not reduced, since all interactions were significant. The two replicate silage bags per combination of biomass type × harvest time × ensiling additive were included as the random effect. The model fit was evaluated by plotting residuals versus predicted values. LSD groups were calculated within each combination of biomass type and harvest time, i.e., for comparison of the effects of ensiling additives and ensiling duration.

2.9.4. Experiment 4

The pH data from the lab-scale and pilot-scale experiment were analyzed together using the lmer function of the lm4 package in R software version 2022.07.2 Build 576. The model included pH as the response variable and a treatment factor with seven treatments: (1) willow fresh frozen without formic acid, (2) willow fresh frozen with formic acid, (3) willow ensiled with formic acid in vacuum bags, (4) willow ensiled with formic acid in barrels, (5) willow + grass fresh frozen, (6) willow + grass ensiled in vacuum bags and (7) willow + grass ensiled in barrels. The two replicate vacuum bags and two replicate barrels within each biomass type were included as the random effect. The distribution of the data was tested using the Shapiro–Wilk test and the data fit was evaluated from a residual plot. Estimates of expected marginal means and compact letter display were calculated using the em_res function.

FW and DM mass losses during ensiling in barrels were calculated based on the initial and final weight and initial and final DM content of the biomass in each barrel. Mean and standard deviation were calculated across two barrels per biomass type.

3. Results

3.1. Buffering Capacity in Willow and Poplar Harvested in June or September

The relationships between the dose of formic acid and the response pH for various types of biomasses are shown in Figure 1 along with the estimates of BC, and in general, the observed relationship was described well by the function. The estimated BC required to reduce the pH to 4.0 varied between 2.5 and 4.8 kg formic acid per ton FW and between 6.1 and 24.9 kg per ton DM. For $\frac{1}{2}$-year shoots of willow clone Tordis and poplar clone OP42, BC was significantly lower in September than in June, with reductions of 58 and 34% on a FW basis, respectively, and 75 and 50% on a DM basis, respectively. BC was not estimated for willow Clone X in June, but in September BC was significantly higher for $\frac{1}{2}$-year shoots of Clone X than for $\frac{1}{2}$-year shoots of Tordis, demonstrating that differences in BC may occur between willow clones. For $1\frac{1}{2}$-year shoots of the willow clone Tordis, BC was relatively low in both June and September, with no significant difference between the two harvest times.

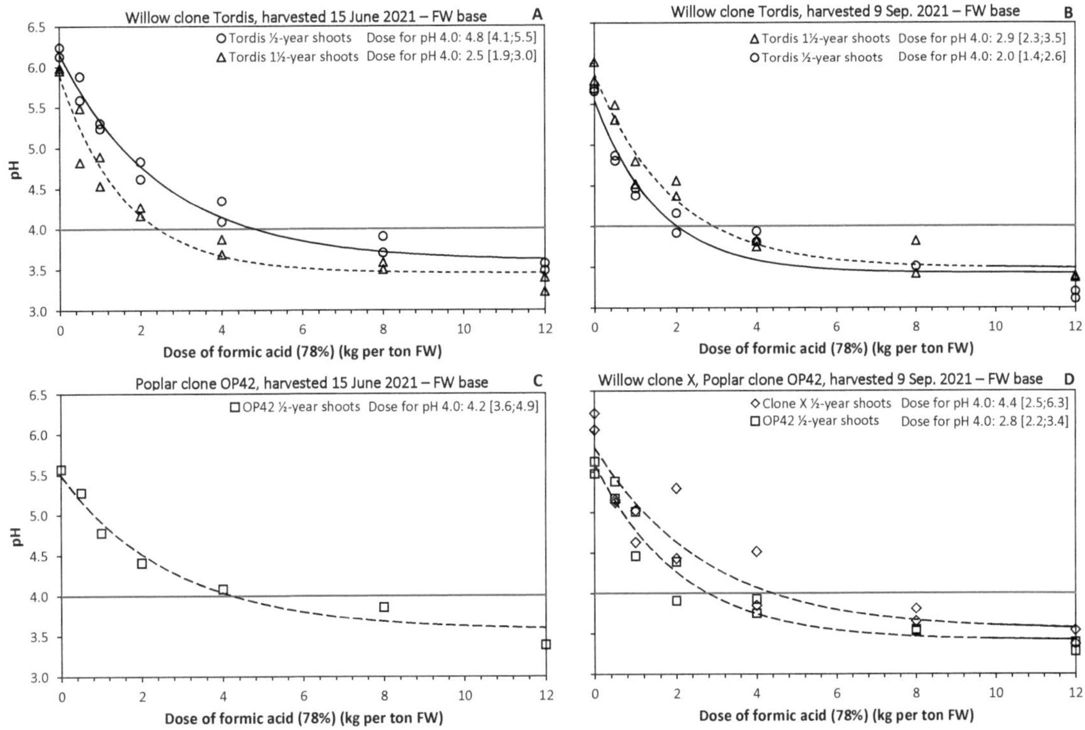

Figure 1. *Cont.*

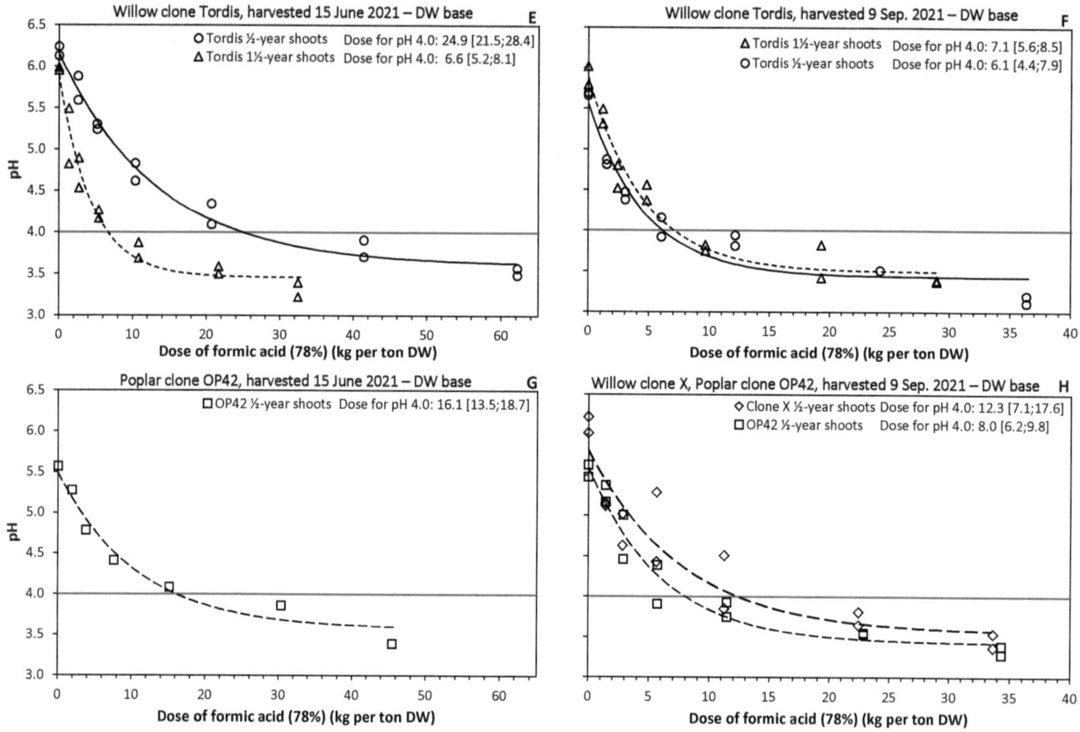

Figure 1. Dose–response relationship between added formic acid and pH in biomass from willow and poplar clones, harvested either (**A**,**C**,**E**,**G**) 15 June 2021 or (**B**,**D**,**F**,**H**) or 9 September 2021. Figures show relationships on (**A**–**D**) a fresh weight (FW) basis and (**E**–**H**) a dry weight (DW) basis, respectively. Shoots were either $\frac{1}{2}$-year shoots or $1\frac{1}{2}$-year shoots, see Table 1 for details. Symbols indicate measured values (with two replicates per dose and biomass type), and lines indicate the predicted relationship. The horizontal line indicates pH 4.0. The estimated dose to achieve pH 4.0 and 95% confidence limits are given for each biomass type, corresponding to the buffering capacity.

When analyzing across biomass types, shoot ages, and harvest times, BC was negatively correlated with DM content and positively correlated with CP content (Figure 2). However, the effect was only significant for BC calculated on a DM basis. Hence, the predicted BC decreased significantly ($p = 0.006$) from 22.1 to 5.3 kg formic acid (78%) per ton DM, when the DM content of the biomass increased from 20 to 40% (Figure 2A). Likewise, BC increased significantly ($p = 0.012$) from 6.7 to 19.2 kg formic acid per ton DM when the CP content increased from 5 to 15% (Figure 2B).

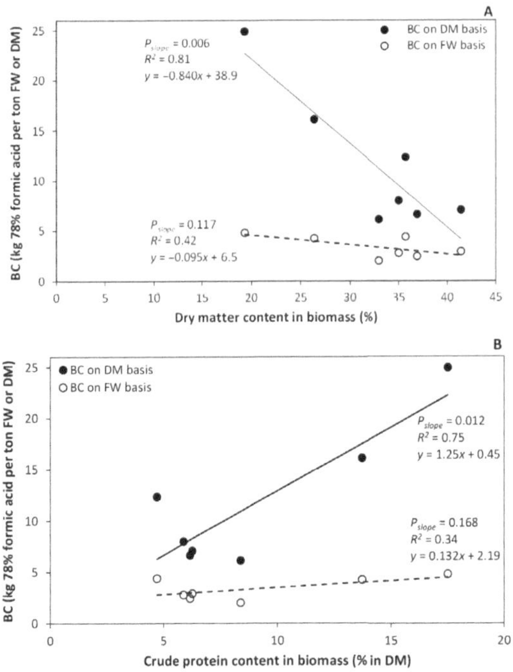

Figure 2. Relationship between (**A**) dry matter (DM) content and buffering capacity (BC) and (**B**) crude protein (CP) content and BC in tree biomass. BC is estimated both on a fresh weight (FW) and a DM basis. The tree biomass covers different clones and shoot ages of willow and poplar harvested in either June or September 2021. Note that CP was analyzed in biomass that was harvested 9–15 days later than the biomass used for the analysis of BC. See Table 1 for details. Symbols indicate the measured values of DM and CP and the estimated value of BC based on dose–response relationships in Figure 1. The full and dashed lines indicate significant and non-significant linear relationship, respectively.

3.2. Development of pH in Willow Silage over Time

The development of pH in willow silage over 85 days with and without ensiling additives in ensiling Experiment 2 is shown in Figure 3. The fitted non-linear functions clearly illustrate that the pH stabilized within a few days but at a much lower level when applying molasses + LAB compared to no additive application. In the analysis of variance, there was highly significant effect of both additive treatment, ensiling duration, and their interacting effects ($p < 0.001$ for all). For biomass ensiled without additives, the pH decreased within the first 5–10 days and stabilized at a level of around 5.18–5.50. For biomass ensiled with molasses and LAB, the pH decreased more rapidly and to a lower level, and the pH was reduced to 4.19 after 5 days of ensiling. There was a further slight decline in the pH to 4.10 after 90 days, but there were no significant differences between ensiling durations from 5 to 90 days, indicating that the pH stabilized within 5 days. When comparing the two ensiling additive treatments within each ensiling duration, the treatment with ensiling additives resulted in a significantly lower pH compared to no additive for all ensiling durations from 2 days up to 90 days.

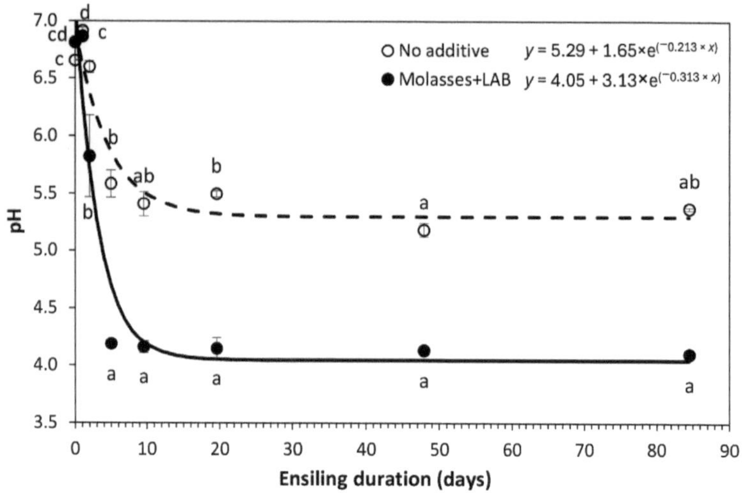

Figure 3. Development of pH during ensiling of biomass from $\frac{1}{2}$-year shoots of the willow clone Tordis in ensiling Experiment 2, harvested 15 June 2021. Ensiling was performed either without any additives or with the addition of a combination of sugar beet molasses and lactic acid bacteria (LAB). Symbols and error bars represent the mean value and standard deviation of two pH measurements in each of two individual vacuum bags. Letters indicate significant differences; values within additive treatments followed by the same letter are not significantly different ($p = 0.05$). Lines indicate the predicted non-linear relationship.

3.3. Effects of Additives on Ensiling of Willow and Poplar Harvested in June or September

The effects of biomass type, harvest time, ensiling additives, and ensiling duration on pH in Experiment 3 are shown in Figure 4. There were significant effects of all main factors as well as all their interactions on pH ($p < 0.001$ for all).

In general, there was only a limited reduction in pH during ensiling of both willow and poplar when no ensiling additive was applied ('untreated'). This was very pronounced for biomass harvested in September (Figure 4B,D,F,H) with only a significant but small reduction in pH for poplar OP42 biomass (Figure 4H). For biomass harvested in June, the pH was significantly reduced for all four biomass types (Figure 4A,C,E,G), but the reduction was very small in willow Tordis biomass from $\frac{1}{2}$-year shoots (Figure 4A) and moderate in poplar OP42 biomass (Figure 4G).

The addition of formic acid reduced the pH immediately after application, and since the acid dose was adjusted specifically to each biomass type and harvest time, it is not surprising that the pH was relatively close to pH 4.0 for unensiled biomass, except for poplar OP42 biomass harvested in September, where the applied acid dose appeared to have been too low (Figure 4H). Moreover, the pH remained at the same level or slightly lower after 75 days of ensiling for all combinations of biomass type and harvest time.

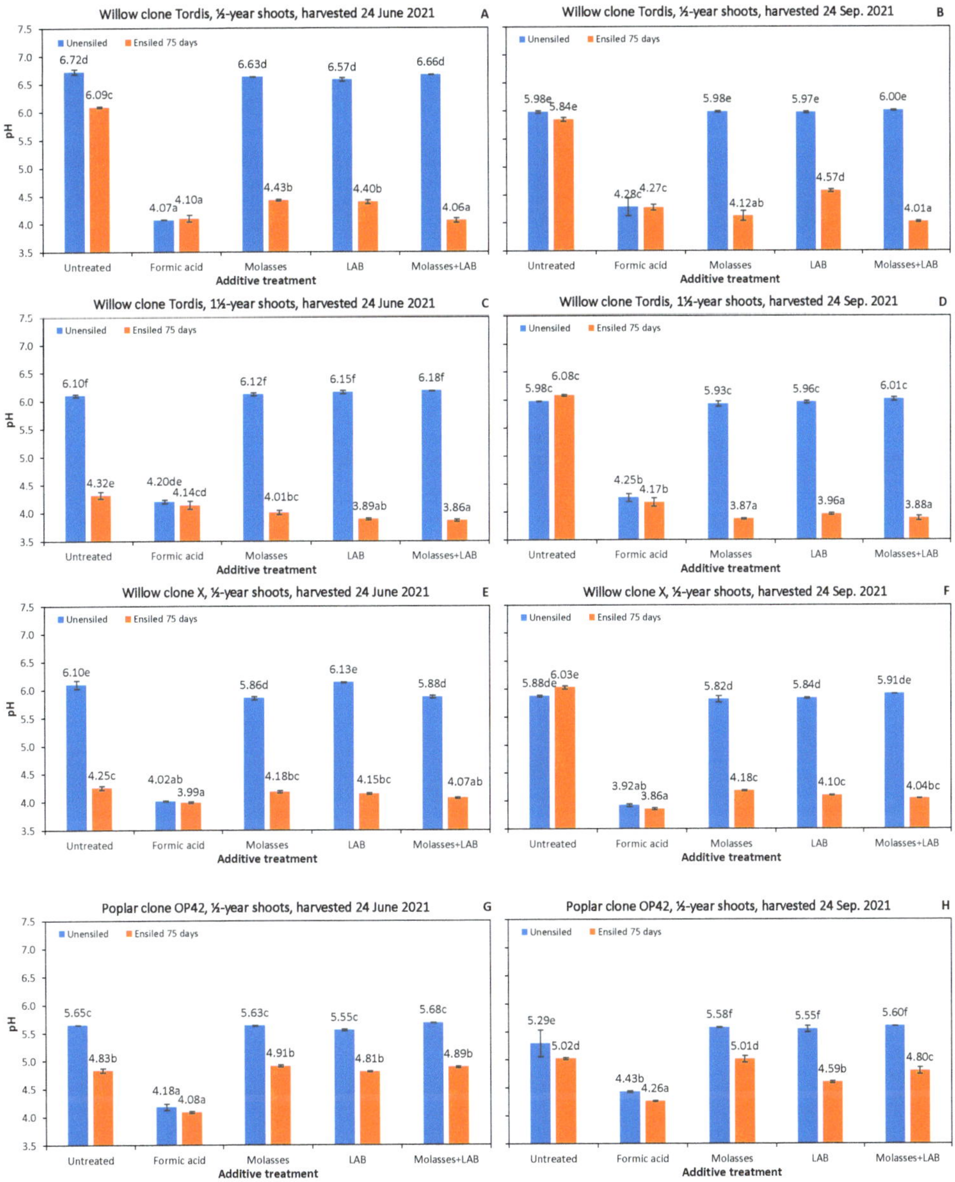

Figure 4. pH in willow and poplar biomass before and after ensiling for 75 days in ensiling Experiment 3. Biomass from the willow clones (**A–D**) Tordis and (**E**,**F**) Clone X and (**G**,**H**) the poplar clone OP42 was harvested either (**A**,**C**,**E**,**G**) 24 June 2021 or (**B**,**D**,**F**,**H**) 24 September 2021. The biomass was harvested from plants that had previously been harvested during the winter 2020/2021 ($\frac{1}{2}$-year shoots) or during the winter 2019/2020 ($1\frac{1}{2}$-year shoots), see Table 1 for details. The biomass was ensiled either without any additives (untreated) or with formic acid, molasses, lactic acid bacteria (LAB) or molasses + LAB, see Table 2 for details. The formic acid dose was adjusted specifically to each biomass type to obtain an initial pH of approx. 4.0. Columns and error bars indicate the mean and standard deviation of two replicate samples from one silage bag for unensiled biomass and from two silage bags for ensiled biomass. Letters indicate LSD groups; within each figure, columns with the same letter are not significantly different.

As an alternative to chemical ensiling, the application of molasses and/or LAB for biological ensiling also resulted in a significant reduction in pH compared to untreated ensiled biomass for most combinations of willow biomass type and harvest time (Figure 4A–F), except for the biomass from $\frac{1}{2}$-year shoots of Clone X harvested in June, where the pH was at a relatively low level after ensiling of untreated biomass (Figure 4E).

For poplar, there was either little or no significant effect on pH of applying molasses and/or LAB compared to the untreated, ensiled biomass (Figure 4G,H), and the pH was only reduced to a pH level in the range 4.6–5.0 after biological ensiling of poplar biomass.

3.4. Ensiling of Willow and Willow + Grass Mixture at Pilot-Scale

When opening the barrels after 257 days of ensiling in Experiment 4, there was considerable occurrence of mold on the surface and to approx. 5–8 cm depth of the willow silage but only little mold on the surface of the willow + grass silage. From a subjective evaluation, the silage in all four barrels smelled fresh and good, with willow + grass silage smelling more aromatic than pure willow silage. The willow silage and willow + grass silage in the vacuum bags also smelled fresh and good; however, the color of the silage appeared slightly darker compared to the silage in the barrels.

The pH differed significantly between the seven biomasses/treatments ($p < 0.001$), and the mean values are shown in Figure 5. For fresh willow, the pH was reduced from 6.22 to 4.35 when adding 5 kg of formic acid (78%) per ton FW. After ensiling, the pH was reduced to around 4.2 in willow silage, both in bags and in barrels. Hence, the addition of formic acid may have enhanced subsequent fermentation and further pH reduction in willow biomass.

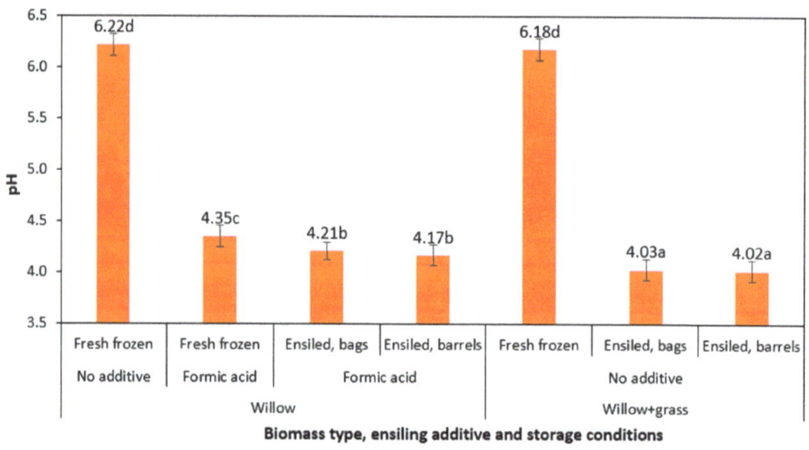

Figure 5. pH in willow and willow + grass before and after ensiling in either vacuum bags or 60 L barrels for 257 days in ensiling Experiment 4, harvested 22 June 2022. For pure willow, formic acid was added as an ensiling additive at a dose of 5 kg (78%) per ton fresh weight. For fresh frozen willow biomass, pH was measured both before and after addition of formic acid. Columns indicate the mean value; for fresh frozen biomass, the columns represent two replicate samples, and for ensiled biomass, the columns represent two replicate samples from each of two vacuum bags or two barrels. Letters indicate LSD groups; columns with the same letter are not significantly different. Error bars indicate the 95% confidence limits as calculated in the analysis of variance.

For willow + grass, the pH was reduced from 6.18 in fresh biomass to 4.02–4.03 in silage. There was no significant difference in pH between ensiling in vacuum bags and ensiling in barrels, neither in willow nor in willow + grass (Figure 5).

The initial density and initial and final DM content and biomass weight in the barrels are shown in Table 3. On a FW basis, there were only minor mass losses of 0.3 and 1.2% during ensiling of willow and willow + grass, respectively. However, due to a slight reduction in DM content in the biomass, there were DM losses of 6.5 and 6.7%, respectively (Table 3).

Table 3. Mass balance for ensiling of willow and willow + grass at pilot-scale in ensiling Experiment 4, harvested 22 June 2022. Silage samples were taken before and after 257 days of ensiling to obtain initial and final values and mean and standard deviation were calculated across two barrels per biomass type.

Parameter	Time	Willow Mean ± Std.dev.	Willow + Grass Mean ± Std.dev.
Density, kg FM m^{-3}	Initial	605 ± 14	547 ± 1
DM content, %	Initial	24.9 ± 0.27	27.1 ± 0.04
	Final	23.2 ± 0.00	25.6 ± 0.10
CP content, %	Initial	14.1 [1]	10.8 [1]
	Final	13.6 ± 0.5	11.6 ± 0.0
Biomass fresh weight, kg FW	Initial	36.3 ± 0.8	32.8 ± 0.1
	Final	36.2 ± 0.9	32.5 ± 0.1
Biomass dry weight, kg DM	Initial	9.0 ± 0.2	8.9 ± 0.02
	Final	8.4 ± 0.2	8.3 ± 0.03
FW mass loss, % of initial weight	-	0.3 ± 0.3	1.2 ± 0.1
DM mass loss, % of initial weight	-	6.5 ± 0.8	6.7 ± 0.1

[1] Only one observation per biomass type.

For both the fresh and ensiled biomasses, the CP content was numerically higher in willow than in the mixture of willow and grass (Table 4). The ensiling process reduced the CP content in willow by 4% and increased the CP content in the willow + grass biomasses by 7%. The energy content decreased in the ensiled biomasses compared to their fresh counterparts, by approximately 37% for willow and 27% for willow + grass. The in vitro organic matter digestibility (EDOM and EDOMi) of the ensiled biomasses similarly decreased by 19% to 31% compared to that of the fresh biomasses.

Table 4. Analysis of feed value of freshly harvested willow and willow + grass harvested on 22 June 2022 and ensiled willow and willow + grass sampled from barrels on 6 March 2023. For fresh biomass, values are based on one sample per biomass type. For silage, the mean and standard deviation is based on two samples per biomass type, except analyses of amino acids, which were based on one sample per type.

Parameter	Unit	Fresh Willow	Fresh Willow + Grass	Silage of Willow	Silage of Willow + Grass
		Mean	Mean	Mean Std. dev.	Mean Std. dev.
DM	%	24.6	27.1	23.2 ± 0.0	25.6 ± 0.1
Crude ash	% in DM	7.7	6.9	7.9 ± 0.7	7.6 ± 0.2
Fat	% in DM	2.1	2	1.9 ± 0.1	2.4 ± 0.2
CP	% in DM	14.1	10.8	13.6 ± 0.5	11.6 ± 0.0
NE [1]	FUsow [2]/kg DM	0.35	0.45	0.22 ± 0.03	0.33 ± 0.02
NE [1]	MJ/kg DM	2.70	3.48	1.71 ± 0.20	2.56 ± 0.13
EDOM [3]	%	38.2	43.0	30.4 ± 0.5	34.7 ± 1.0
EDOMi [4]	%	26.9	33.7	18.7 ± 2.2	26.1 ± 0.7
Amino acids					
Lysine	g/kg DM	-	-	9.2	5.3
Methionine	g/kg DM	-	-	3.0	1.8
Cysteine + Cystine	g/kg DM	-	-	1.4	0.8
Threonine	g/kg DM	-	-	6.8	4.3

[1] Net energy (NE). [2] Danish Feed Units for sows corresponding to 7.7 MJ [43,44]. [3] Enzyme digestible organic matter. [4] Enzyme digestible organic matter at ileal level.

4. Discussion

4.1. Buffering Capacity in Willow and Poplar Harvested in June and September

The results from Experiment 1 in 2021 show that the required dose of formic acid (78%) for reducing the pH in the willow and poplar biomass is generally in the range of 2–5 kg per ton FW, corresponding to 6–25 kg per ton DW (Figure 1). However, both the harvest time and shoot age may affect the BC of tree biomass; the BC appears to decrease from June to September for the $\frac{1}{2}$-year shoots, whereas the BC for the $1\frac{1}{2}$-year shoots is low already in June. For comparison, the BC for wheat and barley straw has been estimated to be 9.9 and 20.3 kg formic acid per kg DM [45].

The BC appears to be negatively correlated with the DM content and positively correlated with the CP content in the tree biomass (Figure 2). It should be mentioned that the DM and CP content in the biomass are confounded, and from the present data, the causality of these two factors cannot be determined. The BC of herbage biomass has been found to be mainly related to the anion fraction of the plant material, whereas proteins contributed to a lesser extent [47]. Therefore, the variation in the BC in the tree biomass may partly be related to the CP content but also to other constituents which were not analyzed.

It must also be noted that the CP content was analyzed in the biomass that was harvested 9 or 15 days later than the biomass used for the BC analysis (see Table 1), and the CP may have changed slightly during this period. Nevertheless, the predicted relationship between the DM content and BC provides a practical and useful indication of the required dose of formic acid to apply to reduce the pH to 4.0 in willow or poplar biomass, given that the DM content is known.

In the pilot-scale ensiling of willow in June 2022 in Experiment 4, a dose of 5 kg of formic acid was applied, based on the results obtained in June 2021. The pH results from Experiment 4 indicate that a higher acid dose may have been required for the biomass harvested in June 2022 (Figure 5), and this illustrates that some variation in the BC may occur also between biomasses even of the same cultivar and shoot age.

4.2. Ensiling of Willow and Poplar without Ensiling Additives

The results from Experiment 2 and Experiment 3 clearly show that the willow biomass harvested in June or September ensiled poorly without any additives, whereas the addition of molasses + LAB enhanced the ensiling process, both in terms of the more rapid pH decrease (Figure 3) and the pH reduction to a lower final level (Figures 3 and 4). The poor ensilability of the willow harvested during the growing season is consistent with previous studies of ensiling biomass from $\frac{1}{2}$-year-old willow shoots without additives, which also resulted in a high pH [16,18,28,29]. Our results are somehow in contrast to those from the study by Larsen et al. [28], which indicated a poorer ensilability of willow when harvesting in June–July compared to harvesting in August or September. There seems to be very little knowledge on the ensilability of poplar; however, a study reported a relatively low pH of 4.2 in willow silage and a higher pH of 4.6 in poplar silage [48].

Our results and previous studies strongly suggest that there is often a need for the application of ensiling additives to green willow and poplar biomass to ensure a good ensiling process with a sufficiently low pH in the range of 4.0, and the need seems to be more pronounced when harvesting biomass in September rather than in June.

4.3. Effect of Additives on Ensiling of Willow and Poplar

The results from Experiment 3 demonstrate that chemical ensiling by the application of formic acid in the appropriate dose can ensure a low and stable pH during the ensiling of both willow and poplar biomass (Figure 4). Formic acid is frequently used as an ensiling additive due to its low production costs and high acidification capacity [49].

The addition of molasses and LAB also had a significant effect on the pH reduction during the ensiling of willow (Figure 4). Interestingly, the separate addition of molasses and LAB had very similar effects on the pH after ensiling, with no significant differences in the pH between these two treatments in six of the eight combinations of biomass and harvest time (all except Figure 4B,H). Also, there seems to be a very small additive effect of applying a combination of molasses and LAB compared to molasses or LAB alone, since the combination only reduced the pH slightly more and only significantly for the biomass of $\frac{1}{2}$-year-old shoots of the willow clone Tordis (Figure 4A). Therefore, ensiling willow biomass may be enhanced by either applying molasses or LAB, whereas the combined treatment only provides little extra effect. For comparison, Muklada et al. [29] found that LAB significantly reduced the pH to 4.2 when ensiling willow biomass, whereas molasses had no significant effect (pH 4.9) compared to with no additive (pH 5.1). The lacking effect of molasses in their study may have been due to a relatively high concentration of WSCs of 107 g kg^{-1} DM, which is much higher than the 35 g kg^{-1} DM reported by Smith et al. [16]. Heubeck [30] reported a concentration of WSCs of 122 g kg^{-1} DM in willow biomass; however, the pH in the resulting silage was not reported. Unfortunately, WSCs in the biomasses were not analyzed in our study.

Since the addition of molasses and LAB to the poplar biomass did not decrease or only slightly decreased the pH further compared to no treatment (Figure 4G,H), the ensiling process in poplar OP42 biomass may be inhibited by other factors and, hence, chemical ensiling with formic acid may be more relevant for this type of biomass.

Overall, the results show that there is often a need for applying ensiling additives to achieve a good ensiling process with a low pH when ensiling the green biomass of willow and poplar. For willow, similar effects on the pH can be obtained by applying an appropriate dose of formic acid or by applying either molasses or LAB, and additional benefit seems to be obtained only occasionally by applying a combination of molasses and LAB. For poplar, there seems to be little effect of molasses and LAB on the pH, and the application of formic acid appears to be the most relevant ensiling additive for poplar biomass.

4.4. Ensiling of Willow and Willow + Grass Mixture on Pilot Scale

The occurrence of mold primarily in the barrels with the silage of willow in Experiment 4 indicates that the ingression of some oxygen was enabled between the barrel and the lid. However, it also shows that the pure willow silage is more vulnerable to secondary fermentation than the willow + grass silage, and this risk needs consideration when ensiling willow commercially, not least the risk of quality loss during feed-out from a silage silo. Heubeck [30] also reported the occurrence of mold in the upper 10 cm layer when ensiling willow in barrels but concluded that "this problem does not occur at field scale, where flexible plastic covers allow fermentation gases to escape from the silage stack, while generally excluding oxygen from the silage".

The pH dropped to a low level during the ensiling of willow + grass (Figure 5), which demonstrates that the grass can serve as an ensiling additive when mixed with the willow biomass and ensure a sufficiently low pH for good preservation of the biomass mixture. The positive effect of grass may be due to a higher content of WSCs, which is often observed in grass species [38].

Since there was no difference in the pH between the biomass ensiled in vacuum bags and barrels (Figure 5), the ensiling process appears to be comparable on these two scales and, therefore, ensiling in vacuum bags can serve to mimic ensiling on a larger scale. Hence, our results support those from previous studies showing that ensiling in vacuum bags can serve as a realistic model system for studying the effects of various factors on ensiling [39].

The DM loss of 6.5–6.7% during the ensiling of willow and willow + grass on the pilot scale (Table 3) is slightly higher than the DM loss 2.4–3.3% found during the pilot-scale ensiling of grass pulp after the extraction of protein-rich juice from grass [50]. The DM loss during ensiling may partly be due to the reduction in the FW mass and partly due to the reduction in the DM content (Table 3). Hatt and Clauss [32] also reported a slight reduction in the DM content from 47.8 to 46.8% after ensiling the biomass of willow and other tree species; however, DM mass loss was not reported.

4.5. Feeding Value

Identifying local protein sources to reduce the dependence on imported soybeans is a key objective in livestock production. Willow silage has been recognized as a potential source of crude protein (CP), with reported CP levels of 18% in DM, and up to 22% in DM in silage made from leaves only [16]. These values are significantly higher than the CP content observed in ensiled willow and willow + grass (11.6–13.6% of DM) in the current study (Table 4). As the ensiling process had minimal effect on the CP content, in accordance with earlier studies [16,24], no clear explanation exists for the relatively low CP levels in the willow silage tested. From a pig nutritional standpoint, the high lysine and methionine levels in the willow silage (Table 4) are advantageous and comparable to, or even exceed, those found in fresh and ensiled grass-clover [20]. Nevertheless, the potential of ensiled willow biomass to provide energy for pigs appears to be low, with substantially lower net energy levels (0.22 to 0.33 FUsow/kg DM) compared to those of ensiled grass-clover (0.65 FUsow/kg DM) [42].

4.6. Practical Perspectives and Future Research

Overall, the ensiling experiments demonstrate that it is possible to preserve green willow and poplar biomass effectively by ensiling, provided an appropriate ensiling additive is applied that can ensure rapid and sufficient pH reduction and minimize DM loss. A low pH is generally a good indicator of a successful ensiling process, and the pH reduction facilitated by ensiling additives is generally expected to improve the preservation of the feed quality when ensiling willow biomass. Future research should investigate the potential changes in qualitative parameters during the ensiling of tree biomass, including the production of different volatile fatty acids, which may affect the palatability of the silage. Also, it is important to consider what the willow silage is to be used for, since quality requirements may differ between different applications. For instance, the tree silage may be used as (1) possible supplementary forage for large and small ruminants [14,16,18,24,26], or (2) with the specific purpose of reducing enteric methane production from ruminants in general [25,26] or (3) even as feed for browsing ungulates such as giraffes in zoos [33]. The pilot testing of the pig nutritional value of willow biomass in this study indicates its low energy content and low organic matter digestibility. These outcomes could potentially be improved by ensiling only willow leaves to reduce the fiber content from the shoots or by harvesting at earlier stages of maturity. Future research should investigate the potential for improving the pig nutritional value while transferring bioactive compounds from the willow to sow milk, and thereby support piglet health.

Our study does not reveal the mechanisms restricting the ensiling of willow and poplar biomass without additives. Further investigations are required to illuminate and distinguish between the potential effects of a low DM content, a low content of WSCs, and high BC as well as other potentially inhibitory factors for the ensiling process. For example, it is relevant to analyze the content of WSCs in various types of tree biomass. Nevertheless, this study provides valuable knowledge in a poorly studied area; it documents a frequently occurring need for the application of additives when ensiling willow and poplar biomass, and it provides practically applicable knowledge regarding the choice of ensiling additive as well as guidelines for the required dose of formic acid.

5. Conclusions

This study is concerned with preserving green tree biomass harvested from willow and poplar during the growing season by ensiling, i.e., by reducing the pH to a low level around 4.0. The lab-scale ensiling experiments demonstrate that pH is often not reduced sufficiently in willow and poplar biomass during ensiling without applying an ensiling additive. However, a low pH level can be obtained in willow biomass either by applying a dose of 2–5 kg per ton FW of formic acid (78%) or by adding molasses and/or LAB, which can ensure a rapid and adequate pH reduction. For poplar biomass, molasses and LAB seem less effective, and formic acid appears to be more suitable as an ensiling additive.

The buffering capacity of tree biomass, defined as the required dose of formic acid to reduce the pH to 4.0, differed between biomass types and harvest dates but typically ranged between 2 and 5 kg formic acid (78%) per ton FW. The buffering capacity decreased with increasing DM content of the biomass.

The pilot-scale experiment demonstrates that comparable pH results can be obtained when ensiling at the lab-scale in vacuum bags as when ensiling at the pilot-scale in 60 L barrels. The experiment also showed that willow biomass can be ensiled either by the application of formic acid or by mixing with grass biomass, with a mass loss of 6–7%.

Overall, the study provides guidelines for how to ensure a low pH during ensiling of willow and poplar biomass. Further research is needed to investigate the mechanisms involved in the ensiling process as well as the effect of ensiling on silage quality parameters.

Author Contributions: Conceptualization, S.U.L. and A.G.K.; methodology, S.U.L. and A.G.K.; software, S.U.L.; validation, S.U.L.; formal analysis, S.U.L.; investigation, S.U.L.; resources, S.U.L. and A.G.K.; data curation, S.U.L.; writing—original draft preparation, S.U.L.; writing—review and editing, S.U.L., H.H., U.J. and A.G.K.; visualization, S.U.L.; project administration, S.U.L. and A.G.K.; funding acquisition, S.U.L. and A.G.K. All authors have read and agreed to the published version of the manuscript.

Funding: This work was carried out as part of the Danish project 'Outdoor sows in novel concepts to benefit the environment–OUTFIT'. The project was supported by the Green Development and Demonstration Program (GUDP) in Denmark through the Organic Research, Development, and Demonstration Program (Organic RDD6), coordinated by International Centre for Research in Organic Food Systems (ICROFS), grant number 34009-20-1701. Also, the work was supported by the MIXED project (Grant agreement No. 862357), funded by the Horizon Europe Framework Program (HORIZON 2020).

Institutional Review Board Statement: Not applicable.

Data Availability Statement: The data presented in this study are available on request from the corresponding author.

Acknowledgments: Technical support from the company Ny Vraa, Tylstrup, Denmark is greatly appreciated. The support included production of willow and poplar biomass and assistance related to harvest and processing of the biomass. Valuable insights from Jakob C. Johannsen and Maria Eskildsen, Department of Animal and Veterinary Sciences, Aarhus University, to the sections relating to the feeding value of willow biomasses are highly appreciated.

Conflicts of Interest: The authors declare no conflicts of interest.

References

1. Lærke, P.E.; Larsen, S.U.; Nielsen, A.T.; Stupak, I.; Raulund-Rasmussen, K.; Jørgensen, U. Yield and quality of poplar clones grown as SRC at three sites with different soil types in Denmark. *Asp. Appl. Biol.* **2015**, *131*, 269–274.
2. Larsen, S.U.; Jørgensen, U.; Kjeldsen, J.B.; Lærke, P.E. Long-term yield effects of establishment method and weed control in willow for short rotation coppice (SRC). *Biomass Bioenergy* **2014**, *71*, 266–274. [CrossRef]
3. Rytter, R.-M. The potential of willow and poplar plantations as carbon sinks in Sweden. *Biomass Bioenergy* **2012**, *36*, 86–95. [CrossRef]
4. Dimitriou, I.; Mola-Yudego, B.; Aronsson, P.; Eriksson, J. Changes in organic carbon and trace elements in the soil of willow short-rotation coppice plantations. *Bioenergy Res.* **2012**, *5*, 563–572. [CrossRef]
5. Pugesgaard, S.; Schelde, K.; Larsen, S.U.; Lærke, P.E.; Jørgensen, U. Comparing annual and perennial crops for bioenergy production—Influence on nitrate leaching and energy balance. *GCB Bioenergy* **2014**, *7*, 1136–1149. [CrossRef]
6. Verwijst, T.; Lundkvist, A.; Edelfeldt, S.; Albertsson, J. Development of Sustainable Willow Short Rotation Forestry in Northern Europe. In *Biomass Now—Sustainable Growth and Use*; Matovic, M.D., Ed.; INTECH: London, UK, 2013; pp. 479–502.
7. Schmidt-Walter, P.; Lamersdorf, N.P. Biomass production with willow and poplar short rotation coppices on sensitive areas—The impact on nitrate leaching and groundwater recharge in a drinking water catchment near Hanover, Germany. *Bioenergy Res.* **2012**, *5*, 546–562. [CrossRef]
8. Jørgensen, U.; Thuesen, J.; Eriksen, J.; Horsted, K.; Hermansen, J.E.; Kristensen, K.; Kongsted, A.G. Nitrogen distribution as affected by stocking density in a combined production system of energy crops and free-range pigs. *Agrofor. Syst.* **2018**, *92*, 987–999. [CrossRef]

9. Jakobsen, M. Integrating Foraging and Agroforestry into Organic Pig Production—Environmental and Animal Benefits. Ph.D. Thesis, Aarhus University, Tjele, Denmark, 2018.
10. Kongsted, A.G.; Lambertsen, L.; Jakobsen, M.; Hermansen, J.E.; Andersen, H.M.-L.; Markussen, C.; Kyed, S.; Eriksen, S.; Serup, T.; Jørgensen, U. *Træer i Svinefolde [Trees in Pig Paddocks]*; Aarhus Universitet: Aarhus, Denmark, 2018; p. 37.
11. Manevski, K.; Steenfeldt, S.; Hellwing, A.L.F.; Andersen, H.M.-L.; Jørgensen, U. Nitrate leaching and nitrogen balances for integrated willow-poultry organic systems in Denmark. *Agric. Syst.* **2024**, Submitted AGSY-D-24-01349.
12. Smith, J. Combining Organic Livestock and Bioenergy Production. Leaflet Produced in the AGFORWARD Project. Available online: https://euraf.isa.utl.pt/files/pub/46_combining_organic_livestock_and_bioenergy_production.pdf (accessed on 2 September 2024).
13. García de Jalón, S.; Burgess, P.J.; Graves, A.; Moreno, G.; McAdam, J.; Pottier, E.; Novak, S.; Bondesan, V.; Mosquera-Losada, R.; Crous-Durán, J.; et al. How is agroforestry perceived in Europe? An assessment of positive and negative aspects by stakeholders. *Agrofor. Syst.* **2018**, *92*, 829–848. [CrossRef]
14. Kemp, K.; Barry, T.N.; Douglas, G.B. Edible forage yield and nutritive value of poplar and willow. *Grassl. Res. Pract. Ser.* **2003**, *2003*, 53–64. [CrossRef]
15. Kemp, P.D.; Mackay, A.D.; Matheson, L.A.; Timmins, M.E. The forage value of poplars and willows. *Proc. N. Z. Grassl. Assoc.* **2001**, *63*, 115–119. [CrossRef]
16. Smith, J.; Kuoppala, K.; Yáñez-Ruiz, D.; Leach, K.; Rinne, M. Nutritional and fermentation quality of ensiled willow from an integrated feed and bioenergy agroforestry system in UK. *Suom. Maatal. Seuran Tied.* **2014**, *30*, 1–9. [CrossRef]
17. Luske, B.; van Eekeren, N. Nutritional potential of fodder trees on clay and sandy soils. *Agrofor. Syst.* **2018**, *92*, 975–986. [CrossRef]
18. Steffen, K.; Muro, M.; Trigo, M.; Rodrigues-Matias, L.; Cordiviola, C.; Arias, R. Willow leaf silage: Evaluation of quality, organoleptic characteristics, acceptability, and rumen degradability in goat feed. *Chil. J. Agric. Anim. Sci.* **2024**, *40*, 312–321.
19. Vandermeulen, S.; Ramírez-Restrepo, C.A.; Beckers, Y.; Claessens, H.; Bindelle, J. Agroforestry for ruminants: A review of trees and shrubs as fodder in silvopastoral temperate and tropical production systems. *Anim. Prod. Sci.* **2018**, *58*, 767–777. [CrossRef]
20. Kongsted, A.G.; van der Heide, M.E.; Stødkilde, L.; Johannsen, J.C.; Eskildsen, M. Green feed in organic pig farming. Advances in pig nutrition. *Burleigh Dodds Series in Agricultural Science*; Wiseman, J., Ed.; Burleigh Dodds: Cambridge, UK, 2024. Available online: https://shop.bdspublishing.com/store/bds/detail/workgroup/3-190-136515 (accessed on 3 September 2024).
21. Nahrstedt, A.; Schmidt, M.; Jaggi, R.; Metz, J.; Khayyal, M.T. Willow bark extract: The contribution of polyphenols to the overall effect. *Wien. Med. Wochenschr.* **2007**, *157*, 348–351. [CrossRef]
22. Brereton, N.J.B.; Berthod, N.; Lafleur, B.; Pedneault, K.; Pitre, F.E.; Labrecque, M. Extractable phenolic yield variation in five cultivars of mature short rotation coppice willow from four plantations in Quebec. *Ind. Crop. Prod.* **2017**, *97*, 525–535. [CrossRef]
23. Wiesneth, S.; Aas, G.; Heilmann, J.; Jürgenliemk, G. Investigation of the flavan-3-ol patterns in willow species during one growing-season. *Phytochemistry* **2018**, *145*, 26–39. [CrossRef]
24. Muklada, H.; Klein, J.D.; Glasser, T.A.; Dvash, L.; Azaizeh, H.; Halabi, N.; Davidovich-Rikanati, R.; Lewinsohn, E.; Landau, S.Y. Initial evaluation of willow (*Salix acmophylla*) irrigated with treated wastewater as a fodder crop for dairy goats. *Small Rumin. Res.* **2018**, *163*, 76–83. [CrossRef]
25. Jayanegara, A.; Togtokhbayar, N.; Makkar, H.P.S.; Becker, K. Tannins determined by various methods as predictors of methane production reduction potential of plants by an in vitro rumen fermentation system. *Anim. Feed Sci. Technol.* **2009**, *150*, 230–237. [CrossRef]
26. Ramírez-Restrepo, C.A.; Barry, T.N.; Marriner, A.; López-Villalobos, N.; McWilliam, E.L.; Lassey, K.R.; Clark, H. Effects of grazing willow fodder blocks upon methane production and blood composition in young sheep. *Anim. Feed Sci. Technol.* **2010**, *155*, 33–43. [CrossRef]
27. Wilkinson, J.M. *Silage*, 1st ed.; Chalcombe: London, UK, 2005; p. 254.
28. Larsen, S.U.; Lærke, P.E.; Jørgensen, U. Harvest of green willow biomass for feed—Effects of harvest time and frequency on yield, nutrient concentration, silage quality and regrowth. *Acta Agric. Scand. Sect. B Soil Plant Sci.* **2020**, *70*, 532–540. [CrossRef]
29. Muklada, H.; Davidovich-Rikanati, R.; Awabdeh, S.; Weinberg, Z.G.; Hen, Y.; Deutch, T.; Klein, J.D.; Voet, H.; Lewinsohn, E.; Landau, S.Y. Ensiling willow (*Salix acmophylla*) fodder modifies the contents of plant specialized metabolites, but not nutritional attributes. *Anim. Feed Sci. Technol.* **2021**, *278*, 115019. [CrossRef]
30. Heubeck, S. Willow biomass ensiling trials. In *Milestone Report for the Productive Riparian Buffer Project*; National Institute of Water & Atmospheric Research Ltd.: Hamilton, New Zealand, 2020; p. 19.
31. Ghalley, K.B. Willow silage: An alternative to winter fodder. In *Potential Good Practice Note*; SA PPLPP: Delhi, India, 2009; BHGP13, 1-6.
32. Hatt, J.M.; Clauss, M. Browse silage in zoo animal nutrition: Feeding enrichment of browsers during winter. *Zoo Anim. Nutr.* **2001**, *2*, 201–204.
33. Mbatha, K.R.; Bakare, A.G. Browse silage as potential feed for captive wild ungulates in southern Africa: A review. *Anim. Nutr.* **2018**, *4*, 1–10. [CrossRef]
34. Mohd-Setapar, S.H.; Abd-Talib, N.; Aziz, R. Review on Crucial Parameters of Silage Quality. *APCBEE Procedia* **2012**, *3*, 99–103. [CrossRef]
35. Borreani, G.; Tabacco, E.; Schmidt, R.J.; Holmes, B.J.; Muck, R.E. Silage review: Factors affecting dry matter and quality losses in silages. *J. Dairy Sci.* **2018**, *101*, 3952–3979. [CrossRef]

36. da Silva, T.C.; da Silva, L.D.; Santos, E.M.; Oliveira, J.S.; Perazzo, A.F. Importance of the Fermentation to Produce High-Quality Silage. *Ferment. Process.* **2017**, *8*, 1–20.
37. Muck, R.E.; Nadeau, E.M.G.; McAllister, T.A.; Contreras-Govea, F.E.; Santos, M.C.; Kung, L., Jr. Silage review: Recent advances and future uses of silage additives. *J. Dairy Sci.* **2018**, *101*, 3980–4000. [CrossRef]
38. King, C.; McEniry, J.; Richardson, M.; O'Kiely, P. Yield and chemical composition of five common grassland species in response to nitrogen fertiliser application and phenological growth stage. *Acta Agric. Scand. Sect. B Soil Plant Sci.* **2012**, *62*, 644–658. [CrossRef]
39. Johnson, H.E.; Merry, R.J.; Davies, D.R.; Kell, D.B.; Theodorou, M.K.; Griffith, G.W. Vacuum packing: A model system for laboratory-scale silage fermentations. *J. Appl. Microbiol.* **2005**, *98*, 106–113. [CrossRef] [PubMed]
40. Solati, Z.; Manevski, K.; Jørgensen, U.; Labouriau, R.; Shahbazi, S.; Lærke, P.E. Crude protein yield and theoretical extractable true protein of potential biorefinery feedstocks. *Ind. Crop. Prod.* **2018**, *115*, 214–226. [CrossRef]
41. Official Journal European Union. Commission Regulation (EC) No 152/2009 of 27 January 2009 laying down the methods of sampling and analysis for the official control of feed. *Off. J. Eur. Union* **2009**, *54*, 2–52.
42. Johannsen, J.C.; Eskildsen, M.; Kongsted, A.G.; Theil, P.K. Effect of reduced dietary protein on productivity and plasma, urine, and milk metabolites in organic sows during winter conditions. *Livest. Sci.* **2022**, *263*, 105019. [CrossRef]
43. Boisen, S.; A New Concept for Practical Feed Evaluation Systems. DJF Animal Science No. 79, Faculty of Agricultural Sciences, Aarhus University, 2007. p. 70. Available online: https://dcapub.au.dk/djfpublikation/djfpdf/djfhus79.pdf (accessed on 2 September 2024).
44. Theil, P.K.; Chwalibog, A.; Jørgensen, H. Energy for pigs: Metabolism, requirement, utilisation and prediction of dietary content. In *Text Book in Physiology [Lærebog i Fysiologi]*; SEGES, Ed.; SEGES: Skejby, Denmark, 2020; pp. 1–106.
45. Larsen, S.U.; Ma, N.; Nielsen, S.V.; Ward, A.J.; Møller, H.B. The impact of water content and additives on ensiling and methane yield of cereal straw. *Bioresour. Technol. Rep.* **2023**, *24*, 101672. [CrossRef]
46. SAS, I. *SAS/STAT, Release 9.3*; SAS Institute: Cary, NC, USA, 2010.
47. Playne, M.J.; McDonald, P. The buffering constituents of herbage and of silage. *J. Sci. Food Agric.* **1966**, *17*, 264–268. [CrossRef]
48. Kleinmans, J. *Investigating Silage Quality and Feed Value of Chopped Willow and Poplar Material in the Waikato in Autumn 2021*; NutriAssist Ltd.: Aukland, New Zealand, 2021; p. 7.
49. Fijałkowska, M.; Pysera, B.; Lipiński, K.; Strusińska, D. Changes of nitrogen compounds during ensiling of high protein herbages—A review. *Ann. Anim. Sci.* **2015**, *15*, 289–305. [CrossRef]
50. Larsen, S.U.; Ambye-Jensen, M.; Jørgensen, H.; Jørgensen, U. Ensiling of the pulp fraction after biorefining of grass into pulp and protein juice. *Ind. Crop. Prod.* **2019**, *139*, 111576. [CrossRef]

agriculture

MDPI

Article

Effects of Biological Additives on the Fermentation Quality and Microbial Community of High-Moisture Oat Silage

Yifei Huang , Siling Zhang, Maoling Hu, Shuyuan Yang, Ailing Hui, Wencheng Zhang and Zeyu Wu *

Engineering Research Center of Bio-Process of Ministry of Education, School of Food and Biological Engineering, Hefei University of Technology, Hefei 230601, China
* Correspondence: wuzeyu@hfut.edu.cn; Tel.: +86-138-5518-4210

Abstract: The primary objective of this study was to explore the effects of biological additives, including *Streptococcus bovis* (SB), *Bacillus subtilis* (BS), xylanase (XT), and their combined treatments, including SB + BS (SBBS), SB + XT (SBXT), and BS + XT (BSXT), on the chemical composition, fermentation characteristics, and microbial community of high-moisture oat silage. Compared with the CK group (control group without additives), SB and SBBS treatments increased the lactic acid content ($p < 0.05$) and reduced the contents of acetic acid, propionic acid, butyric acid, and ammonia nitrogen in silage ($p < 0.05$). XT, SBXT, and BSXT treatments decreased the neutral detergent fiber and acid detergent fiber contents ($p < 0.05$), increasing the water-soluble carbohydrate content ($p < 0.05$). The SB, SBBS, and SBXT treatments increased the abundance of *Lactiplantibacillus* ($p < 0.05$) and significantly decreased microbial richness with diversity ($p < 0.05$), improving the microbial community structure in silage. The addition of XT increased the relative abundance of *Clostridium* and *Enterobacteriaceae*, but its combination with SB and BS increased the abundance of *Lactiplantibacillus* and inhibited the development of undesirable bacteria. Moreover, different additives changed the metabolism of carbohydrates, amino acids, energy, cofactors and vitamins of bacterial communities during ensiling. In summary, the addition of SB and SBBS was more conducive to improving the fermentation characteristics of oat, while XT, SBXT, and BSXT performed better in degrading lignocellulose in plants.

Keywords: oat ensiling; lactic acid bacteria; exogenous fibrolytic enzymes; fermentation quality; bacterial diversity

Citation: Huang, Y.; Zhang, S.; Hu, M.; Yang, S.; Hui, A.; Zhang, W.; Wu, Z. Effects of Biological Additives on the Fermentation Quality and Microbial Community of High-Moisture Oat Silage. *Agriculture* **2024**, *14*, 2275. https://doi.org/10.3390/agriculture14122275

Academic Editor: Siran Wang

Received: 18 October 2024
Revised: 8 December 2024
Accepted: 8 December 2024
Published: 11 December 2024

1. Introduction

Forage oat (*Avena sativa* L.) is considered as a primary feed resource and food crop globally [1]. Oat is rich in dietary fiber, vitamins, minerals, high-quality protein, and carbohydrates [2], making it an ideal choice for livestock feed. However, the seasonal production of oat seriously disrupts the consistent year-round supply of forage, and inadequate preservation methods also cause substantial losses of fresh forage. Thus, there is an urgent need to develop suitable preservation techniques for oats.

Ensiling has emerged as a viable solution for preserving fresh forage. Silage making involves the anaerobic fermentation of lactic acid bacteria (LAB), which produces lactic acid to inhibit the growth of undesirable microorganisms, thereby preserving fresh crops [3]. During this process, high-moisture-content ensiled crops often produce effluent or leachate. However, the development and multiplication of LAB require an appropriate moisture content (MC) [4]. The high-humidity environment often promotes *Clostridium* and mold growth in silage, which is not conducive to LAB growth [5]. Furthermore, effluent release during silage production results in the loss of nutritional and feeding value, thereby threatening the preservation quality of oats [6].

To address the above challenges, microbial additives have gained more and more attention. LAB metabolize water-soluble carbohydrates (WSCs) in silage under anaerobic conditions, producing abundant organic acids that rapidly decrease the pH value [7]. *Streptococcus bovis* (*S. bovis*), a type of LAB isolated from rumen, has a growth rate that is 30% higher than other commonly used LAB in silage. This accelerated growth indicates that *S. bovis* could serve as an effective starter strain in silage fermentation, promoting a rapid pH reduction in silage [8]. *Bacillus subtilis* (*B. subtilis*) is characterized by its strong resistance to harsh environments. Additionally, *B. subtilis* produces broad-spectrum lignocellulose, facilitating lignocellulose degradation and enhancing the nutritional quality of lignocelluloses [9]. Furthermore, *B. subtilis* can improve aerobic stability by producing antifungal substances and bacteriocins, which suppress the development of undesirable organisms, such as yeast and filamentous fungi [10].

Furthermore, oat contains significant levels of both neutral detergent fiber (NDF) and acid detergent fiber (ADF) [11]. These fibers often pose a challenge for degradation by intestinal microorganisms [12], adversely affecting both the ensiling process and the digestibility of ruminant feed. To mitigate this issue, the use of exogenous fibrolytic enzymes has received increasing attention in recent years. These enzymes facilitate the degradation of fiber components, converting them into WSCs during ensiling [13]. For example, cellulase and xylanase target the cellulose and hemicellulose fractions of plants, respectively. Xylanase catalyzes xylan hydrolysis [14], with commercial sources mainly derived from filamentous fungi and certain bacteria like *Bacillus stearothermophilus* and *B. subtilis* [15]. Therefore, the addition of xylanase could improve the fermentation quality of oat silage at different stages of ensiling [16,17].

However, current research indicates that the effectiveness of xylanase is closely related to MC [18]. While a higher MC can facilitate the diffusion rate of enzyme molecules to some extent, it may also lead to excessive dilution of the enzyme solution, subsequently reducing enzyme activity and exerting a negative impact on the ensiling process. Thus, a new method for the combined treatment of xylanase and microbial inoculants is urgently needed to maximize the fermentation profile and microbial community structure of high-moisture oat silage. This approach would effectively compensate for the potential deficiencies associated with the addition of xylanase alone.

Therefore, we utilized high-moisture fresh oat as the substrate to examine the influence of biological additives, including *S. bovis*, *B. subtilis* and xylanase, on the fermentation characteristics and microbial community of oat silage. In this study, we hypothesized that the application of microbial additives (*S. bovis* and *B. subtilis*) can effectively improve the fermentation profile of high-moisture oat silage, and the utilization of xylanase can markedly enhance the digestibility of fibrous materials during ensiling.

2. Materials and Methods

2.1. Materials and Reagents

The oat forage was harvested at its heading stage from Chengdu City, Sichuan Province, China (104°07′ E, 30°57′ N), on 11 April 2023 and immediately transported to the laboratory. *S. bovis* (strain number CICC21604) was purchased from China Center of Industrial Culture Collection (CICC). *B. subtilis* was isolated from fermented bean curd, and its DNA sequence was determined by Sangon Biotech (Shanghai) Co., Ltd. (Shanghai, China). The nucleic acid BLAST analysis conducted through the NCBI database identified the strain as *Bacillus subtilis* CP026662.1. Xylanase, with an enzyme activity of 50,000 U/g, was obtained from Shanghai Macklin Biochemical Co., Ltd. (Shanghai, China).

2.2. Silage Preparation

The oat raw materials were chopped to an average length of 2.50 ± 0.07 cm and divided into seven equal portions: (1) a control group without additives (CK), (2) a group treated with *S. bovis* (SB, 1×10^6 CFU g^{-1} FW), (3) a group treated with *B. subtilis* (BS, 1×10^6 CFU g^{-1} FW), (4) a group treated with xylanase (XT, 50 U g^{-1} FW), (5) a group treated with a combination of *S. bovis* and *B. subtilis* (SBBS, 1×10^6 CFU g^{-1} FW + 1×10^6 CFU g^{-1} FW), (6) a group treated with a combination of *S. bovis* and xylanase

(SBXT, 1 × 10^6 CFU g^{-1} FW + 50 U g^{-1} FW), and (7) a group treated with a combination of *B. subtilis* and xylanase (BSXT, 1 × 10^6 CFU g^{-1} FW + 50 U g^{-1} FW). These additives were dissolved in a certain volume of sterile water and then evenly sprayed onto the raw materials. The CK group received an equivalent volume of sterile water. Each group was thoroughly mixed and prepared in triplicate. Approximately 200 g of chopped oat was packed into each 250 mm × 300 mm polyethylene bag and vacuum-sealed immediately. A total of 63 experimental bags (7 treatments × 3 ensiling days × 3 replicates) were stored in a constant temperature incubator at 28 °C. Chemical composition and fermentation characteristics were analyzed after 7, 30, and 60 days of ensiling. The microbial community was assessed using samples collected after 7 and 60 days of ensiling.

2.3. Analysis of Fermentation Characteristics and Chemical Composition

Ten grams of fresh silage sample was mixed with 90 mL of distilled water and stored at 4 °C for 24 h. The mixture was filtered through four layers of cheesecloth, and its pH value was measured using a portable pH meter (PHB-1, Hangzhou Qiwei Instrument Co., Ltd., Hangzhou, China). Ammonia nitrogen (NH_3-N) content was determined by phenol-sodium hypochlorite colorimetry [19]. Concentrations of organic acids, including lactic acid (LA), acetic acid (AA), propionic acid (PA) and butyric acid (BA), were determined using high-performance liquid chromatography (Agilent 1260 infinity, Agilent Technologies, Santa Clara, CA, USA). Chromatographic conditions included a ZORBAX SB-Aq chromatographic column, with 0.1% H_3PO_4 as the mobile phase, a column temperature of 30 °C, a flow rate of 1 mL/min, an injection volume of 20 µL, and a detection wavelength of 210 nm. In addition, fresh forage and silage samples were dried in an oven at 65 °C for 48 h to determine dry matter (DM) content. Dried samples were ground using a 1 mm screen by a plant sample mill (BJ-800A, Deqing Baijie Electric Appliance Co., Ltd., Huzhou, China) for chemical composition analysis. WSC content was quantified using the anthrone colorimetric method [20]. Total nitrogen (TN) and crude protein (CP) contents were determined using the Kjeldahl method [21]. NDF and ADF contents were analyzed using the VanSoest method [22], with hemicellulose (HC) content calculated by subtracting ADF from NDF. The fermentation quality of the oat silage was evaluated using Flieg's point index, as detailed in reference [23,24].

$$\text{Flieg's point} = 220 + [(2 \times \%DM) \times 15] - 40 \times pH \tag{1}$$

2.4. Microbiological Analysis

The microbial populations were quantified using the plate count method [25]. De Man–Rogosa–Sharpe agar (AoBoxing Bio-Technology Co., Ltd., Beijing, China) was used to count the numbers of LAB after incubation at 37 °C for 48 h. Potato dextrose agar (Hope Bio-Technology Co., Ltd., Qingdao, China) was used to count the numbers of molds and yeasts following incubation at 30 °C for 72 h.

2.5. Bacterial Community Sequencing Analysis

The bacterial community of silage samples was analyzed through high-throughput sequencing. Briefly, 20 g of fresh oat or oat silage sample from each bag was mixed with 180 mL of sterile PBS solution and centrifuged at 12,000 rpm for 2 min. The sediment obtained was reserved for DNA extraction. Following the manufacturer's protocol, total DNA was extracted using the E.Z.N.A™ Mag-Bind Soil DNA Kit (Omega Bio-Tek, Norcross, GA, USA).

The 16S rRNA V3-V4 region of the eukaryotic ribosomal RNA gene was amplified by two rounds of PCR using the primers 341F: CCTACGGGNGGCWGCAG and 805R: GACTACHVGGGTATCTAATCC. Subsequently, the PCR amplicons were sequenced on an Illumina MiSeq PE300 platform (Sangon Biotech Co., Ltd., Shanghai, China). Bioinformatics analysis and graphic drawing were analyzed via the Sangon Biotech Cloud (https://ngs.sangon.com (accessed on 10 May 2024)).

2.6. Statistical Analysis

The fixed effects of treatment, ensiling day, and their interactions were analyzed using two-way analysis of variance with SPSS 26.0 software (IBM Corp., Armonk, NY, USA). Significant differences were compared using Duncan's multiple range tests, with $p < 0.05$ as the level of statistical significance. Sequence alignment against the RDP database was performed using the RDP classifier (version 2.12) to identify the best alignment results for each sequence. Spearman correlation analysis was completed with R. Differences in bacterial community structures between various treatments were evaluated using ANOSIM analysis at a significance level of $p < 0.05$.

3. Results

3.1. Characteristics of the Fresh Material Before Ensiling

The chemical and microbial compositions of oat are presented in Table 1. The DM content of oat was 227.19 g kg^{-1} FM, which meets the standard for high-moisture silage. Adequate levels of WSC and CP indicate good quality of oats. However, fresh oats contain relatively high amounts of structural carbohydrates, which are detrimental to silage fermentation. This results in decreased feed intake and digestibility in the rumen due to the rigid structure of fibrous materials, which are not easily degraded by microorganisms and enzymes. The number of epiphytic LAB was 3.11 log $cfug^{-1}$ FW, which may prevent them from becoming the dominant species. In contrast, the counts of molds and yeasts were 6.05 log $cfug^{-1}$ FW.

Table 1. Chemical and microbial compositions of oat before ensiling (Mean ± SD, $n = 3$).

Item	Oat
DM (g kg^{-1} FM)	227.19 ± 3.74
CP (g kg^{-1} DM)	149.80 ± 2.65
WSC (g kg^{-1} DM)	73.73 ± 1.55
NDF (g kg^{-1} DM)	565.73 ± 2.01
ADF (g kg^{-1} DM)	381.50 ± 4.84
HC (g kg^{-1} DM)	184.23 ± 5.01
pH	6.02 ± 0.05
LAB (log $cfug^{-1}$ FM)	3.11 ± 0.06
Molds and yeasts (log $cfug^{-1}$ FM)	6.05 ± 0.10

Data are means of three replicates. Oat, fresh material; FM, fresh matter; cfu, colony-forming unit; SD, standard deviation.

3.2. Chemical Composition of Oat Silage

The nutritional composition of oat silage is detailed in Table 2. As the ensiling period extended, the DM content displayed a declining trend. When compared to the CK treatment at different stages of ensiling, except for BS, XT, and BSXT treatments, the DM content was significantly increased in other additive treatments ($p < 0.05$). Meanwhile, the CP content in the XT treatment decreased faster than that in the CK group during the early stage of ensiling but remained relatively stable in the later stage. Moreover, the SBXT and BSXT treatments exhibited a higher CP content after 30 days of ensiling compared with the addition of XT alone ($p < 0.05$). The WSC content gradually decreased throughout the ensiling process. Compared with the CK group, all additive treatments significantly increased the WSC content ($p < 0.05$), with the combined treatments showing a higher WSC content than individual treatments. As fermentation time progressed, the NDF, ADF, and HC contents of all treatments exhibited a downward trend. The contents of NDF, ADF, and HC were significantly decreased in all treatments compared with CK treatment ($p < 0.05$). Furthermore, the NDF, ADF, and HC contents in the XT and all combined treatments were significantly lower than those treated with the addition of SB and BS alone ($p < 0.05$).

Table 2. Chemical compositions of oat silage (Mean ± SEM, n = 3).

Item		Treatment								*p*-Value		
	Day	CK	SB	BS	XT	SBBS	SBXT	BSXT	SEM	T	D	TxD
DM ($g\ kg^{-1}$ FM)	7	216.86 ^{Ca}	228.39 ^{ABCa}	228.71 ^{ABCa}	215.79 ^{Ca}	237.05 ^{Aa}	231.57 ^{ABa}	219.32 ^{BCa}				
	30	209.38 ^{Ba}	219.31 ^{ABa}	219.61 ^{ABa}	212.50 ^{Ba}	226.64 ^{Aa}	221.32 ^{ABa}	218.25 ^{ABa}	1.17	<0.001	0.005	0.977
	60	214.68 ^{CDa}	227.16 ^{ABa}	224.51 ^{ABCa}	209.84 ^{Da}	231.47 ^{Aa}	225.74 ^{ABCa}	217.38 ^{BCDa}				
CP ($g\ kg^{-1}$ DM)	7	140.28 ^{Aa}	151.92 ^{Aa}	147.35 ^{Aa}	140.14 ^{Aa}	150.06 ^{Aa}	148.85 ^{Aa}	145.66 ^{Aa}				
	30	136.05 ^{CDa}	152.63 ^{ABa}	144.07 ^{BCa}	130.18 ^{Da}	155.73 ^{Aa}	146.49 ^{ABCa}	142.27 ^{BCa}	1.05	<0.001	0.455	0.898
	60	138.62 ^{Ca}	150.73 ^{ABa}	145.88 ^{ABCa}	137.73 ^{Ca}	153.29 ^{Aa}	148.16 ^{ABCa}	142.13 ^{ABCa}				
WSC ($g\ kg^{-1}$ DM)	7	52.02 ^{BCa}	52.95 ^{ABCa}	50.87 ^{Ca}	52.63 ^{ABCa}	54.26 ^{ABCa}	56.46 ^{Aa}	55.50 ^{ABa}				
	30	35.14 ^{Db}	38.15 ^{BCb}	37.91 ^{BCb}	40.96 ^{ABb}	40.59 ^{ABb}	42.73 ^{Ab}	40.15 ^{ABb}	0.99	<0.001	<0.001	0.675
	60	32.08 ^{Cb}	36.15 ^{BCb}	35.28 ^{BCb}	38.48 ^{ABb}	38.72 ^{ABb}	41.16 ^{Ab}	40.68 ^{Ab}				
NDF ($g\ kg^{-1}$ DM)	7	557.53 ^{Aa}	542.22 ^{Aa}	549.47 ^{Aa}	508.44 ^{Ca}	526.61 ^{Ba}	503.57 ^{Ca}	503.50 ^{Ca}				
	30	496.38 ^{Ab}	464.25 ^{BCb}	473.16 ^{Bb}	413.57 ^{Db}	455.27 ^{Cb}	405.29 ^{Db}	408.74 ^{Db}	6.48	<0.001	<0.001	0.005
	60	485.82 ^{Ab}	446.37 ^{BCb}	457.53 ^{Bb}	429.05 ^{Db}	436.17 ^{CDb}	404.13 ^{Eb}	402.47 ^{Eb}				
ADF ($g\ kg^{-1}$ DM)	7	362.64 ^{A}	355.62 ^{A}	363.85 ^{A}	358.32 ^{A}	353.18 ^{A}	355.26 ^{A}	359.38 ^{A}				
	30	335.36 ^{A}	307.29 ^{BC}	308.41 ^{B}	287.02 ^{BCD}	295.62 ^{BC}	268.93 ^{D}	285.74 ^{CD}	4.46	<0.001	<0.001	0.001
	60	307.73 ^{Ab}	282.48 ^{Bb}	288.04 ^{Bb}	285.39 ^{Bb}	275.49 ^{Bb}	275.16 ^{Bb}	282.36 ^{Bb}				
HC ($g\ kg^{-1}$ DM)	7	194.89 ^{Aa}	186.60 ^{Ba}	185.62 ^{Ba}	150.12 ^{Da}	173.43 ^{Ca}	148.31 ^{Da}	144.12 ^{Da}				
	30	161.02 ^{Ac}	156.96 ^{Ab}	164.75 ^{Ab}	126.55 ^{Cb}	159.65 ^{Ab}	136.36 ^{Bab}	123.00 ^{Cb}	2.71	<0.001	<0.001	<0.001
	60	178.09 ^{Ab}	163.89 ^{BCb}	169.49 ^{Bb}	143.66 ^{Da}	160.68 ^{Cb}	128.97 ^{Eb}	120.11 ^{Fb}				

Values in the same row (A–F) or column (a–c) followed by different letters differ at $p < 0.05$. CK: control; SB: *S. bovis*; BS: *B. subtilis*; XT: xylanase; SBBS: combination of *S. bovis* and *B. subtilis*; SBXT: combination of *S. bovis* and xylanase; BSXT: combination of *B. subtilis* and xylanase; SEM, standard error of means.

3.3. Fermentation Characteristics of Oat Silage

The fermentation characteristics of oat silage are summarized in Table 3. All treatments exhibited a gradual diminution in pH levels as silage fermentation progressed. After 60 days of fermentation, the pH in all additive treatments was significantly lower than that in the CK treatment ($p < 0.05$). Notably, SB and SBBS treatments achieved relatively low pH levels after 7 days of ensiling, considerably lower than other treatments ($p < 0.05$). In contrast, the XT treatment decreased the pH of silage at a slower rate. The LA content in all treatments gradually increased throughout the fermentation period, with the SB, SBBS, and SBXT treatments maintaining higher levels. However, compared to BS alone, the pH value of the combined treatment of BS and XT was significantly higher after 7 and 30 days of ensiling ($p < 0.05$), and its LA content was also significantly lower ($p < 0.05$). As expected, the AA content in the SB, SBBS, and SBXT treatments was significantly lower than that of other treatments after 7, 30, and 60 days of fermentation ($p < 0.05$). During the 60-day ensiling period, the PA content of all treatments exhibited an upward trend. The PA content in additive treatments was decreased compared with the CK treatment, but the XT treatment exhibited significantly higher PA content after 30 and 60 days of ensiling ($p < 0.05$). The BA content remained low in all treatments, but NH_3-N content steadily increased throughout the ensiling process. The SB, SBBS, and SBXT treatments had significantly lower NH_3-N content than other treatments ($p < 0.05$). However, after 60 days of silage fermentation, the BS, XT, and BSXT treatments led to a significant elevation in NH_3-N and BA contents compared to the other additive treatments. Overall, the SB, SBBS, and SBXT groups achieved higher Flieg's points than the others.

Table 3. Fermentation characteristics of oat silage (Mean ± SEM, n = 3).

Item		Treatment								*p*-Value		
	Day	CK	SB	BS	XT	SBBS	SBXT	BSXT	SEM	T	D	TxD
pH	7	4.24 ^{Aa}	3.83 ^{Ca}	4.04 ^{Ba}	4.33 ^{Aa}	3.82 ^{Ca}	3.94 ^{BCa}	4.20 ^{Aa}				
	30	4.16 ^{Aab}	3.79 ^{Ca}	3.83 ^{Cab}	4.12 ^{ABb}	3.77 ^{Ca}	3.78 ^{Cb}	4.02 ^{Bb}	0.02	<0.001	<0.001	<0.001
	60	4.07 ^{Ab}	3.82 ^{Ca}	3.86 ^{BCb}	3.88 ^{BCc}	3.83 ^{Ca}	3.80 ^{Cb}	3.95 ^{Bb}				
LA ($g\ kg^{-1}$ DM)	7	37.68 ^{Eb}	64.64 ^{Ab}	54.36 ^{BCb}	44.86 ^{DEb}	67.39 ^{Ab}	59.52 ^{ABb}	49.32 ^{CDb}				
	30	43.43 ^{Db}	75.11 ^{ABa}	68.84 ^{BCa}	59.77 ^{Ca}	80.51 ^{Aa}	73.53 ^{ABab}	61.57 ^{Ca}	1.71	<0.001	<0.001	0.757
	60	55.48 ^{Ca}	78.18 ^{Aa}	76.43 ^{Aa}	65.46 ^{Ba}	82.79 ^{Aa}	77.31 ^{Aa}	66.38 ^{Ba}				

Table 3. *Cont.*

Item		Treatment								*p*-Value		
	Day	CK	SB	BS	XT	SBBS	SBXT	BSXT	SEM	T	D	TxD
AA ($g\ kg^{-1}$ DM)	7	14.43 ^{Ab}	8.41B ^{Cb}	14.14 ^{Aa}	13.83 ^{Ab}	6.66 ^{Cb}	9.78 ^{BCb}	11.66B ^{Cb}	0.62	<0.001	<0.001	0.396
	30	21.40 ^{Aa}	10.19 ^{Cb}	16.60 ^{Ba}	17.75 ^{Bab}	8.98 ^{Cb}	10.36 ^{Cb}	17.01 ^{Ba}				
	60	24.81 ^{Aa}	15.17B ^{CDa}	19.27 ^{BCa}	19.53 ^{Ba}	13.43 ^{Da}	14.45 ^{CDa}	19.05 ^{BCa}				
PA ($g\ kg^{-1}$ DM)	7	3.03 ^{Ac}	1.47 ^{Dc}	2.57 ^{Bc}	2.09 ^{Cc}	1.98 ^{Cc}	1.35 ^{Db}	2.72 ^{ABc}	0.18	<0.001	<0.001	<0.001
	30	4.03 ^{Bb}	3.05 ^{Db}	3.50 ^{Cb}	4.69 ^{Ab}	2.38 ^{Eb}	2.27 ^{Ea}	4.13 ^{Bb}				
	60	5.91 ^{Ba}	3.77 ^{Ea}	4.40 ^{Da}	6.87 ^{Aa}	2.63 ^{Fa}	2.46 ^{Fa}	5.07 ^{Ca}				
BA ($g\ kg^{-1}$ DM)	7	1.56 ^{Bc}	0.77 ^{DEa}	1.05 ^{CDb}	1.38 ^{ABb}	0.57 ^{Eb}	0.79 ^{DEb}	1.13 ^{BCb}	0.08	<0.001	<0.001	<0.001
	30	2.25 ^{Ab}	0.83 ^{Da}	1.37 ^{Ca}	1.76 ^{Ba}	0.68 ^{Da}	0.95 ^{Dab}	1.69 ^{Ba}				
	60	3.25 ^{Aa}	0.88 ^{Ea}	1.54 ^{Ca}	1.97 ^{Ba}	0.75 ^{Ea}	1.15 ^{Da}	1.85 ^{Ba}				
NH_3-N ($g\ kg^{-1}$ TN)	7	27.36 ^{Bc}	19.86 ^{DEc}	22.93 ^{CDc}	31.97 ^{Ac}	18.41 ^{Ec}	20.83 ^{DEc}	25.62 ^{BCc}	2.27	<0.001	<0.001	<0.001
	30	37.52 ^{Ab}	27.72 ^{CDb}	31.62 ^{BCb}	40.39 ^{Ab}	26.49 ^{Db}	28.24 ^{CDb}	33.10 ^{Bb}				
	60	74.56 ^{Aa}	55.07 ^{Ca}	63.95 ^{Ba}	66.21 ^{Ba}	48.63 ^{Da}	58.44 ^{Ca}	66.44 ^{Ba}				
Flieg's point	7	56.77 ^{Cb}	73.78 ^{Aa}	65.26 ^{Bb}	53.41 ^{Cc}	74.18 ^{Aa}	69.35 ^{ABb}	58.45 ^{Cb}	0.93	<0.001	<0.001	<0.001
	30	60.01 ^{Cab}	75.11 ^{Aa}	73.52 ^{Aa}	61.44 ^{BCb}	75.87 ^{Aa}	75.31 ^{Aa}	65.61 ^{Ba}				
	60	63.64 ^{Ca}	74.01 ^{Aa}	72.3 ^{ABa}	71.10 ^{ABa}	73.74 ^{Aa}	74.77 ^{Aa}	68.52 ^{Ba}				

Values in the same row (A–F) or column (a–c) followed by different letters differ at $p < 0.05$. CK: control; SB: *S. bovis*; BS: *B. subtilis*; XT: xylanase; SBBS: combination of *S. bovis* and *B. subtilis*; SBXT: combination of *S. bovis* and xylanase; BSXT: combination of *B. subtilis* and xylanase; SEM, standard error of means.

3.4. Effects of Additives on the Bacterial Community of Oat Silage

The alterations in the bacterial communities of oat silage at the phylum and genus levels are shown in Figure 1. Before ensiling, *Cyanobacteria* (65.60%) and *Proteobacteria* (30.26%) dominated bacterial phyla in fresh oat. Throughout different treatments and fermentation durations, the primary regulators of oat silage were *Firmicutes* and *Proteobacteria*, with *Firmicutes* exhibiting the highest relative abundance. As ensiling processed, regular changes in bacterial community composition of all treatments occurred. Throughout the entire fermentation process, the abundance of *Lactiplantibacillus* in all treatments increased rapidly. The abundance of *Rahnella1* in the SB, BS, SBBS, and SBXT treatments, *unclassified_Enterobacteriaceae* in the CK, SB, XT, SBXT, and BSXT treatments, and *Clostridium_sensu_stricto_1* in the XT treatment all decreased. *Lactiplantibacillus* dominated the fermentation process from 7 to 60 days post-ensiling, with higher abundance in the SB and SBBS treatments than in other treatments.

The beta-diversity was analyzed through the principal component analysis (PCA) and the results were depicted in Figure 2. The findings showed that there was a notable succession of bacterial communities throughout the fermentation process. The bacterial communities in the CK and XT treatments exhibited distinct separation and differences during different ensiling stages, while the BS, BSXT, and SBXT treatments showed less variation in bacterial communities over time. Notably, no significant variations were detected in the SB and SBBS treatments during different ensiling periods, which were mainly clustered in the second quadrant.

As shown in Figure 3, both the Chao1 and Sobs indices of bacterial communities were significantly affected by the duration of fermentation time ($p < 0.001$). Specifically, the Chao1 and Sobs indices in the CK, XT, SBBS, and SBXT treatments significantly decreased after 60 days of ensiling ($p < 0.05$). In contrast, the Shannon and Simpson indices exhibited opposite trends as ensiling time increased. Compared to the early stage of ensiling, the Shannon index in the SB, XT, SBBS, SBXT, and BSXT treatments in the later stage of ensiling significantly decreased ($p < 0.05$), while the Simpson index in the SB, XT, SBBS, and SBXT treatments in the later stage of ensiling significantly increased ($p < 0.05$).

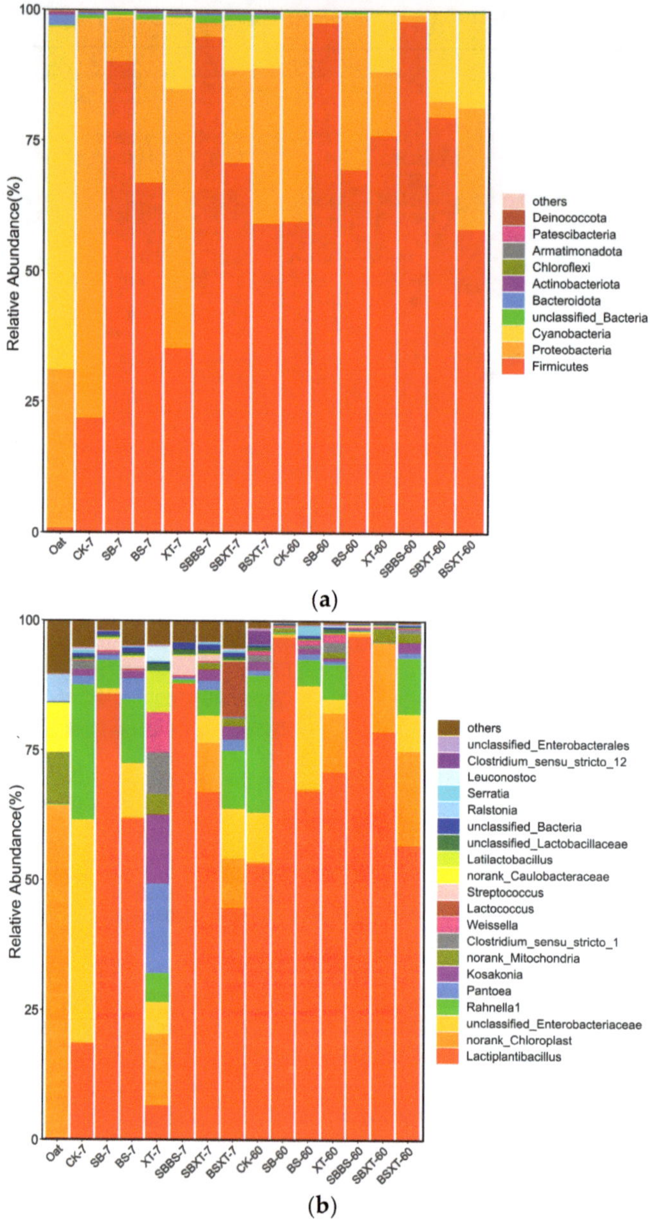

Figure 1. Relative abundances of oat silage bacterial at the phylum level (**a**) and genus level (**b**), respectively, during different treatments and periods of ensiling. Oat: fresh material; CK: control; SB: *S. bovis*; BS: *B. subtilis*; XT: xylanase; SBBS: combination of *S. bovis* and *B. subtilis*; SBXT: combination of *S. bovis* and xylanase; BSXT: combination of *B. subtilis* and xylanase; 7, 60: 7 and 60 days of ensiling.

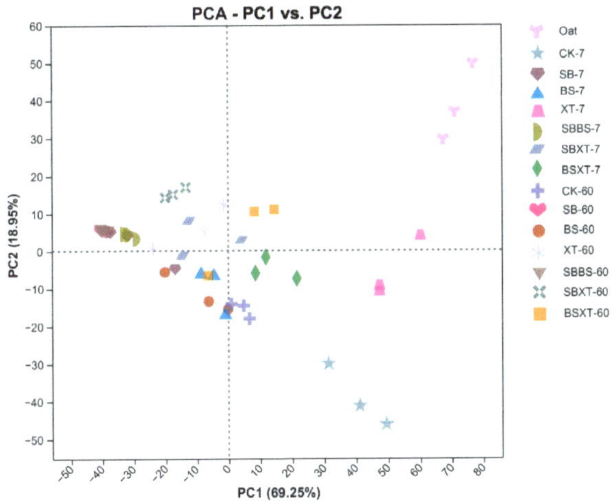

Figure 2. Principal component analysis of the bacterial communities in oat silage. Oat: fresh material; CK: control; SB: *S. bovis*; BS: *B. subtilis*; XT: xylanase; SBBS: combination of *S. bovis* and *B. subtilis*; SBXT: combination of *S. bovis* and xylanase; BSXT: combination of *B. subtilis* and xylanase; 7, 60: 7 and 60 days of ensiling.

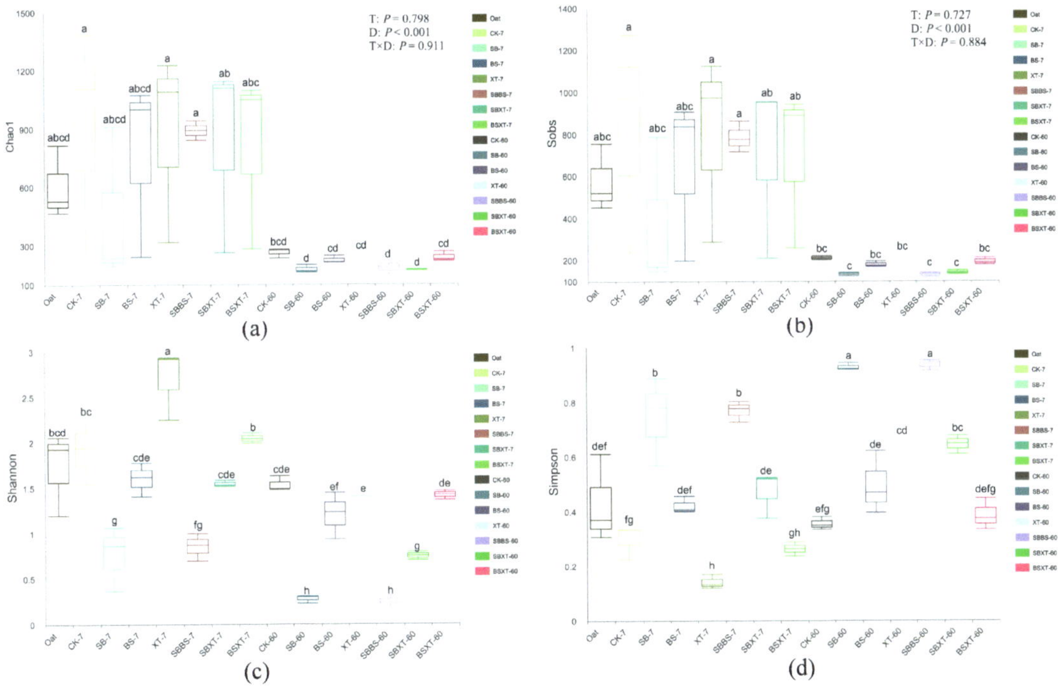

Figure 3. The variations in alpha-diversity of bacterial communities (**a**–**d**) of oat silage during different treatments and periods of ensiling. Different lowercase letters indicate significant differences at $p < 0.05$. (**a**) Chao1 index; (**b**) Sobs index; (**c**) Shannon index; (**d**) Simpson index. Oat: fresh material; CK: control; SB: *S. bovis*; BS: *B. subtilis*; XT: xylanase; SBBS: combination of *S. bovis* and *B. subtilis*; SBXT: combination of *S. bovis* and xylanase; BSXT: combination of *B. subtilis* and xylanase; 7, 60: 7 and 60 days of ensiling.

A correlation heatmap was employed to evaluate the relationship between dominant bacterial genera and the fermentation profile of oat silage, utilizing the Spearman correlation coefficient for analysis. As shown in Figure 4, the NDF and ADF contents displayed a notably negative relationship with *Lactiplantibacillus* (NDF: $R = -0.322$, $p < 0.05$; ADF: $R = -0.519$, $p < 0.001$). The WSC content showed strong negative correlations with *Weissella* ($R = -0.534$, $p < 0.001$) and *Clostridium_sensu_stricto_1* ($R = -0.371$, $p < 0.05$), while its correlation with *Lactiplantibacillus* was weaker. The result indicated that the pH value had a significantly negative correlation with *Lactiplantibacillus* ($R = -0.838$, $p < 0.001$). Meanwhile, the LA content had a significantly positive correlation with *Lactiplantibacillus* ($R = 0.764$, $p < 0.001$) and a negative correlations with *Rahnella1* ($R = -0.628$, $p < 0.001$), *unclassified_Enterobacteriaceae* ($R = -0.496$, $p < 0.001$), and *Clostridium_sensu_stricto_1* ($R = -0.386$, $p < 0.05$). However, the NH_3-N and AA contents exhibited a substantial negative correlation with *Lactiplantibacillus* (NH_3-N: $R = -0.643$, $p < 0.001$; AA: $R = -0.384$, $p < 0.05$) and a positive correlation with *Rahnella1* (NH_3-N: $R = 0.514$, $p < 0.001$; AA: $R = 0.423$, $p < 0.01$), *unclassified_Enterobacteriaceae* (NH_3-N: $R = 0.396$, $p < 0.01$; AA: $R = 0.556$, $p < 0.001$), and *Clostridium_sensu_stricto_1* (NH_3-N: $R = 0.320$, $p < 0.05$; AA: $R = 0.659$, $p < 0.001$).

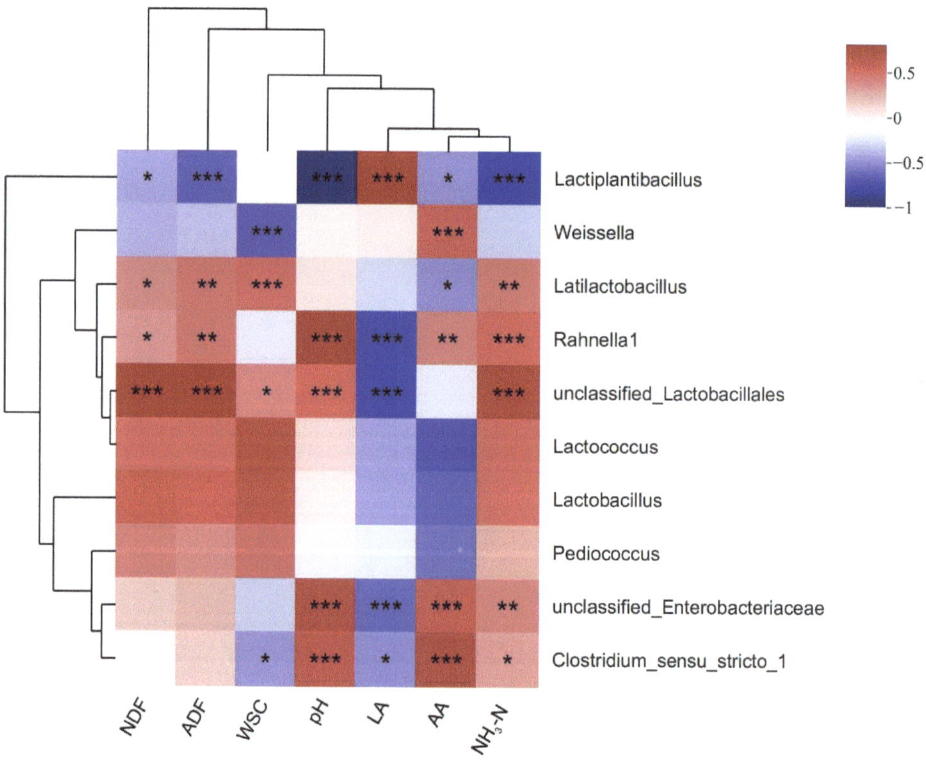

Figure 4. The Spearman correlation coefficients between dominant bacterial genera and the fermentation quality of oat silage. * Significant correlation at the $p < 0.05$ level, ** significant correlation at the $p < 0.01$ level, *** significant correlation at the $p < 0.001$ level.

The metabolic functions were predicted by PICRUSt2 based on the Kyoto Encyclopedia of Genes and Genomes (KEGG) pathway database. Figure 5 depicts the relative abundance of dominant metabolic pathways from levels 1 to 3 of oat silage. Metabolism emerged as the primary metabolic pathway at level 1 in oat silage, with carbohydrate metabolism and amino acid metabolism emerging as essential metabolic pathways at level 2. During the early stages of ensiling, except for the XT treatment, the majority of the metabolic pathways involved in carbohydrate metabolism, such as the glycolysis/gluconeogenesis, pentose phosphate pathway, and citrate cycle (TCA cycle), as well as some metabolic pathways during lipid metabolism like the secondary bile acid biosynthesis and fatty acid biosynthesis, all increased. This was particularly notable in the SB and SBBS treatments, exhibiting a higher relative abundance. Intriguingly, with the extension of ensiling time, excluding the SB, SBBS, and SBXT treatments, carbohydrate metabolism underwent a certain degree of increase in other treatments. In the later stages of ensiling, the BS, XT, and BSXT treatments exhibited some increase in amino acid metabolism, energy metabolism, and specific pathways of metabolism of cofactors and vitamins.

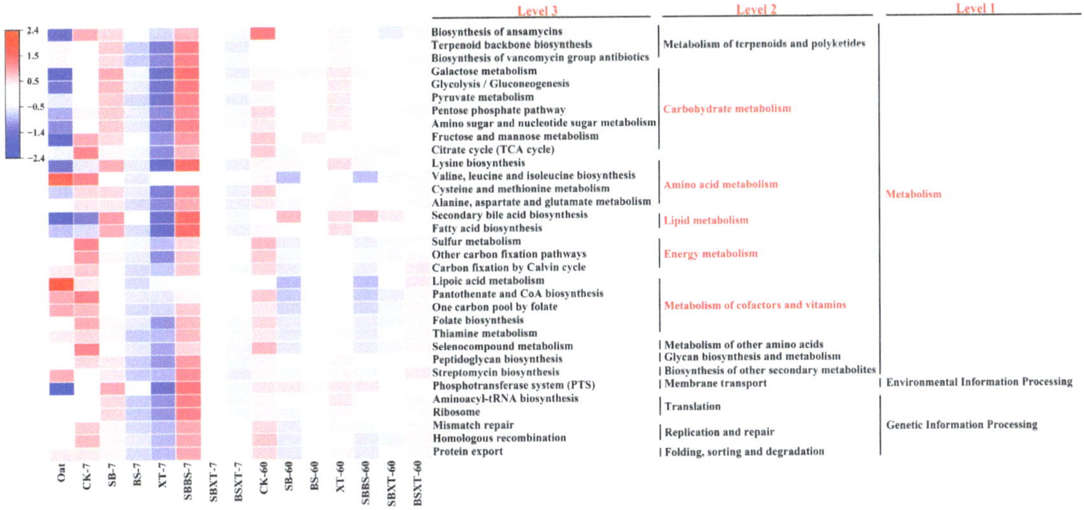

Figure 5. Dynamics of predicted microbial functional profiles of oat silage analyzed by PICRUSt2 for bacteria. Level 3 KEGG orthologue functional predictions of the relative abundances of the dominant metabolic functions. Oat: fresh material; CK: control; SB: *S. bovis*; BS: *B. subtilis*; XT: xylanase; SBBS: combination of *S. bovis* and *B. subtilis*; SBXT: combination of *S. bovis* and xylanase; BSXT: combination of *B. subtilis* and xylanase; 7, 60: 7 and 60 days of ensiling.

4. Discussion

4.1. Analysis of Fermentation Characteristics and Chemical Compositions of Oat Silage

The SB and SBBS treatments significantly increased the DM content in oat silage ($p < 0.05$), which may be related to the pH decreasing. The lower pH level inhibited the multiplication of most bacteria, consequently conserving a greater amount of DM. However, the addition of XT seemed to have no notable influence on the preservation of DM content, which may be associated with the mechanism of xylanase [26]. Xylan, an important component of the plant cell wall, can be decomposed and converted into xylose and other small-molecule substances by xylanase, which are then available for microbial utilization, potentially resulting in a reduction in DM content. Additionally, for oat with high MC, the disruption of the plant cell wall by xylanase leads to the leakage of a large amount of plant cell sap, thereby increasing the MC in silage. While the high-moisture environment is conducive to the propagation of a large amount of undesirable microorganisms, thus increasing the DM loss rate [27]. Compared with fresh oats, the decrease in WSC content during the early stage of fermentation serves a crucial function in the rapid acidification of the silage environment, the suppression of undesirable microorganisms, and the preservation of nutrients. After 7 days of fermentation, the pH values of all treatments decreased to varying degrees, accompanied by an elevation in LA and AA contents. The pH value of silage serves as a crucial indicator for evaluating the quality of ensiling. The decline in pH is directly correlated to the production of large amounts of organic acids by LAB from utilizing WSCs during fermentation. Notably, the pH in the XT treatment declined at a slower rate, which may be related to the time required for xylanase to take effect [28], as well as the diluting effect of high MC on enzyme concentration [18]. Therefore, addressing the aforementioned issues arising from the addition of XT during ensiling and conducting further research on the optimal dosage of enzymes in high-moisture oat silage are essential to enhance enzyme efficiency during the early stage of ensiling while simultaneously reducing silage effluent production, which is still a future challenge. During fermentation, the pH levels in the SB, SBBS, and SBXT treatments decreased more rapidly, accompanied by elevated LA concentrations, while the AA and PA concentrations were higher in the BS, XT, and BSXT treatments. These discrepancies might be related to the unique fermentation patterns exhibited by LAB during ensiling. Previous studies suggested that homofermentative LAB possess the capacity to produce LA from glucose, pentose, and D-xylose through various metabolic pathways, including the glycolysis/gluconeogenesis, phosphoketolase, and pentosephosphate pathways [29,30]. Homofermentation converts glucose into LA with an efficiency that surpasses 80% of the theoretical maximum. On the other hand, heterofermentative LAB, such as *Lactobacillus hilgardii* and *Lactobacillus buchneri*, could increase the concentrations of AA and PA via the pentosephosphate pathway, thus enhancing the aerobic stability of silage [31]. *S. bovis*, as a homofermentative LAB isolated from the rumen, could grow at an extremely fast rate to achieve doubling in 20 min, outpacing other LAB commonly used in silage [8]. Therefore, SB can rapidly proliferate during the early stage of fermentation, producing a large amount of LA and quickly reducing the pH of silage. This was likely the main reason for the insignificant variation in pH between the SB and SBBS treatments in silage. The addition of BS to silage can rapidly consume oxygen in the ensiling environment during the early stage of fermentation [32] and produce antimicrobial peptides to accelerate the proliferation of homofermentative LAB [33]. Nevertheless, this effect was relatively minor for SB, which could rapidly proliferate in the early stage of fermentation. BA, typically produced by *Clostridium* [34], could degrade the crude protein to cause nutrient loss. The lower pH level may help inhibit the development and metabolism of undesirable microorganisms, such as *Clostridium* and *Aspergillus*, thereby reducing protein degradation and BA production [35]. Therefore, compared with other treatments, the CP content in the SB, SBBS, and SBXT treatments was higher, while NH_3-N and BA concentrations were lower. However, the CP content in the XT treatment was notably lower compared to that of other treatments in the early stage of ensiling ($p < 0.05$), and the NH_3-N and BA contents were significantly higher compared with other treatments ($p < 0.05$). As mentioned earlier, the addition of XT could increase the MC in the ensiling environment, and the growth of *Clostridium* and other aerobic microorganisms in silage often tends to be promoted by the high humidity environment [36], thereby accelerating CP degradation. However, as the ensiling process progressed, xylanase could hydrolyze the plant cell walls to produce more WSC, which promotes the growth and reproduction of LAB, thus inhibiting CP degradation by undesirable bacteria [37]. Notably, compared with the addition of XT alone, the combined treatments of SB and BS with XT seemed to improve the preservation rate of CP. This may be attributed to the lower pH provided by SB and the antimicrobial properties provided by BS.

Xylanase is capable of hydrolyzing plant cell walls to produce WSC, thereby reducing the NDF and ADF content in oat silage [38]. HC in silage can be partially hydrolyzed into pentoses by *S. Bovis*, such as xylose and arabinose, during the ensiling process, thereby reducing the degree of lignification of structural carbohydrates in the silage [39]. In this study, the addition of XT and SB exhibited good effects on the decomposition and transformation of structural carbohydrates, which significantly reduced NDF, ADF, and HC contents in silage ($p < 0.05$), and the combined treatment of SB and XT may have a synergistic effect on improving the utilization of fibrous substances. However, BS supplementation also significantly reduced the contents of NDF and ADF ($p < 0.05$), possibly due to the production of cellulase, α-amylase, feruloyl esterase, and other enzymes by *Bacillus* during the silage fermentation process, degrading the structural carbohydrates of plants [40]. Similarly, compared to BS supplementation alone, the combined treatment of SB and BS had a better effect on degrading the lignocellulose of the plant. The WSC content in silage on the 30th day was generally lower than that on the 7th day and decreased as the ensiling process progressed, which was a typical occurrence resulting from the utilization of substrate by microorganisms during ensiling. During the 60-day fermentation process, the WSC content in all treatments was remarkably higher than that in the CK ($p < 0.05$). As mentioned before, the elevation in WSC content observed in the XT-treated samples was primarily attributed to the transformation of fiber components. In contrast, the increase in WSC content in the SB treatment may be related to the fermentation pattern of homofermentative LAB, which results in a lower pH level that often leads to an increase in acidification of the silage environment, which may help to preserve more WSCs [41]. The WSC content in the combined treatments was relatively higher than that in the individual treatments to a certain extent, further corroborating the synergistic effect among SB, BS, and XT.

4.2. Analysis of Microbial Community of Oat Silage

The ensiling process involves complex interactions among various microorganisms, and the microbial community structure influences the silage quality [42]. *Cyanobacteria* and *Proteobacteria* were dominant bacterial phyla in fresh oats, but throughout the fermentation process, *Firmicutes* and *Proteobacteria* took over the leading position at the phylum level, which was in line with the result of the previous research [16]. Additionally, an intriguing phenomenon was observed. In the early stages of silage fermentation, the abundance of *Bacteroidota* in the XT treatment (0.40%) was relatively high, while it was lower in the SB (0.14%) and BS (0.35%) treatments. Studies have shown that both *Firmicutes* and *Bacteroidota* are the dominant phyla in the rumen, playing crucial roles in degrading plant cellulose and hemicellulose, respectively [43]. Therefore, the lower abundance of *Bacteroidota* may also contribute to the reduced fiber digestibility of silage in the SB and BS treatments compared to the XT treatment ($p < 0.05$). Furthermore, as ensiling time progressed, the abundance of *Bacteroidota* decreased sharply, with reductions ranging from 93.19% to 99.32%, and the dominant bacterial phylum gradually transitioned from *Proteobacteria* to *Firmicutes*. This shift could be attributed to the relatively stable microecological environment in the rumen, where pH and temperature conditions favor the growth of *Bacteroidota*. However, the anaerobic and acidic conditions during silage fermentation promote the growth and proliferation of *Firmicutes*, while this acidic environment has a certain inhibitory effect on the growth of *Bacteroidota* [44]. Consequently, the lower pH values in the later stages of ensiling led to a significant reduction in the abundance of *Bacteroidota*. In summary, the formation of microbial fermentation niches depends on the types of available substrates and growth environments. The significant differences in the types of available substrates and ecological conditions between the rumen and silage result in notable diversities and distinct community structures among different microbial assemblages.

Notably, in the early stage of ensiling, the relative abundance of *Lactiplantibacillus* in the XT treatment was relatively lower (6.81%), while the relative abundances of *Pantoea* and *Clostridium_sensu_stricto_1* reached 24.33% and 8.33%, respectively. The abundance of *Lactiplantibacillus* increased significantly in the later stage of ensiling. As mentioned above, xylanse can increase the MC in silage, and a high-moisture environment may inhibit the proliferation of LAB, but favor the growth of *Pantoea* and *Clostridium* [45]. Therefore, the abundance of *Lactiplantibacillus* in the XT treatment was the lowest among all treatments. However, as ensiling time progressed, xylanase promoted the production of WSCs by hydrolyzing plant cell walls, which stimulated LAB proliferation. Consequently, the abundance of *Lactiplantibacillus* increased significantly in the later stage of ensiling, thus inhibiting the development of undesirable bacteria. Nevertheless, compared with the individual treatments of BS and XT, the combined treatment of BS and XT seemed to inhibit LAB growth in the later stage of ensiling, which was consistent with the change trends in pH value and LA content of oat silage. This could be attributed to the variety of hydrolytic enzymes produced by BS, just like with cellulose [46], which accelerated the hydrolysis of plant cell walls when acting together with xylanase, further increasing the MC in oat silage and thus inhibiting the growth and reproduction of LAB. This may also be one of the reasons for the higher NH_3-N and BA contents in the later stage of ensiling. Moreover, compared with the addition of SB alone, the combined treatment of SB and XT also led to a reduction in the abundance of *Lactiplantibacillus* during the entire fermentation process, which further validated the aforementioned speculation.

The PCA clearly demonstrated the differences in bacterial communities among various treatments during different ensiling periods. In particular, significant differences were observed between the silage treated with XT and BS compared to those treated with SB, indicating that the choice of additives significantly influenced the microbial community structure of oat silage. Furthermore, the CK and XT treatments underwent distinct separation during different ensiling periods, while no obvious changes were observed in the SB and SBBS treatments. This suggested that ensiling time had a substantial effect on the microbial community structure of the CK and XT treatments, whereas the SB and SBBS treatments showed no significant difference between the early and later stage of ensiling. This could be attributed to the unique growth characteristics of *S. bovis*, which enable it to proliferate quickly in the initial phase of ensiling [8], thus becoming the dominant species in silage.

Ensiling time had a significant effect on the Chao1 and Sobs indices ($p < 0.001$). As shown in Figure 3, the Chao1 and Sobs indices of all treatments were significantly lower in the later stage of ensiling compared to the early stage, indicating that the prolongation of ensiling time altered the silage environment, leading to a substantial reduction in bacterial diversity. The Shannon and Simpson indices revealed that the additive treatments altered the diversity of bacterial communities in silage, with the Shannon and Simpson indices of SB, XT, SBBS, SBXT, and BSXT treatments showing significant differences between the early and later stage of ensiling ($p < 0.05$). Additionally, the α-diversity of the SB, SBBS, and SBXT treatments was notably lower than that of other treatments ($p < 0.05$), indicating that *S. bovis* has a prominent effect in rapidly reducing the pH of silage, thereby exercising inhibition on the development of undesirable microorganisms.

4.3. Correlation Analysis of Microbial Community and Fermentation Quality of Oat Silage

As shown in Figure 1b, during the initial phase of silage fermentation, the abundance of *Lactiplantibacillus* in the CK and XT treatments was relatively lower, while the abundance of *unclassified_Enterobacteriaceae*, *Rahnella1*, and *Pantoea* was higher in CK, BS, and XT treatments. These three genera belong to the *Enterobacteriaceae*, capable of converting LA into AA and PA in silage [42]. According to Figure 4, *Enterobacteriaceae* was positively correlated with pH value and AA content but negatively correlated with LA. Throughout the fermentation period, the relative abundance of *Clostridium_sensu_stricto_1* remained relatively higher in the XT treatment. It belongs to the *Clostridiaceae*, which is typically associated with protein degradation and butyric acid production [45]. Therefore, it is positively correlated with the NH_3-N content. In the later stage of ensiling, the relative abundance of *Lactiplantibacillus* was high in all treatments, and the pH value showed a significantly negative correlation with *Lactiplantibacillus* ($p < 0.001$). This finding aligned with the results reported by Mu et al. [41], mainly due to the rapid acidification ability of *Lactiplantibacillus*. Furthermore, a weak correlation was observed between *Lactiplantibacillus* and WSC content, potentially related to the addition of BS and XT in this study. The degradation of fiber components by BS and XT may be the main reason for the WSC content increase.

4.4. Predicted Functions of the Microbial Community in Oat Silage

Based on the KEGG pathway database, PICRUSt2 was utilized to evaluate the predicted functions of bacterial communities from levels 1 to 3. The metabolism of carbohydrates, amino acids, energy, cofactors, and vitamins was primarily associated with silage fermentation. In the early stage of ensiling, except for the XT treatment, the relative abundance of most metabolic pathways in carbohydrate metabolism gradually increased, aligning with findings by Bai et al. [47]. Furthermore, SB and SBBS treatments displayed significantly higher relative abundance compared to other treatments, which may be attributed to the unique growth characteristics of *S. bovis*. This rapid increase in LAB abundance enhanced the carbohydrate metabolism capacity during ensiling. However, in the later fermentation stage, the abundance of carbohydrate metabolism in the SB, SBBS, and SBXT treatments declined, likely owing to the accumulation of a substantial quantity of lactic acid over time, which inhibited LAB growth and consequently slowed their metabolism. Notably, the BS, XT, and BSXT treatments exhibited a rising trend in the relative abundance of amino acid metabolism, such as valine, leucine, and isoleucine biosynthesis, along with alanine, aspartate, and glutamate metabolism, which was consistent with higher pH values and NH_3-N contents in silage. Additionally, energy metabolism pathways, including carbon fixation in photosynthetic organisms and sulfur metabolism, were also enhanced. This enhancement may be attributed to the presence of certain undesirable microorganisms in silage, such as *Clostridiaceae* and *Enterobacteriaceae*, typically leading to increased metabolism of amino acids and energy [48,49].

5. Conclusions

The study indicated that the application of different types of biological additives for high-moisture oat had significant effects on silage quality by altering the microbial community composition during the ensiling process. The SB and SBBS treatments had significant effects on enhancing silage quality by increasing the abundance of *Lactiplantibacillus*, outperforming other treatments in terms of increasing the LA content and reducing the pH, BA, and NH_3-N contents. XT had a better effect in the later stage of ensiling, but its combination with SB and BS had certain inhibitory effects on fermentation quality compared with SB and BS alone. The combined treatments had better effects on degrading the lignocellulose of the plant compared with individual treatments. In summary, in the practical applications of the silage industry, SB or SBBS demonstrates significant advantages in enhancing silage quality and reducing production costs. Therefore, it is recommended that the fermentation period for these two additives be 60 days or longer. If there is a need to prioritize the improvement of fiber digestibility in the rumen while simultaneously enhancing silage quality and microbial community structure, the combined treatment of SBXT could be a better choice. Additionally, to minimize the production of silage effluent, it is advisable to use this additive with an ensiling period of 30 to 60 days.

Author Contributions: Conceptualization, Z.W. and W.Z.; methodology, Z.W. and W.Z.; software, Y.H., S.Z. and S.Y.; validation, Y.H. and M.H.; formal analysis, Y.H. and S.Z.; investigation, Y.H., S.Z. and M.H.; resources, Z.W., A.H. and W.Z.; data curation, Y.H., S.Z. and M.H.; writing—original draft preparation, Y.H. and Z.W.; writing—review and editing, Y.H., S.Z. and Z.W.; visualization, Y.H.; supervision, Z.W., A.H. and W.Z.; project administration, Z.W. and W.Z.; funding acquisition, Z.W. All authors have read and agreed to the published version of the manuscript.

Funding: This work was financially supported by the Key Project of Science and Technology Research and Development of Anhui Province (No. 2022j11020005) and the Fundamental Research Funds for the Central Universities of China (No. PA2024GDGP0037).

Institutional Review Board Statement: Not applicable.

Data Availability Statement: The sequence data reported in this study have been submitted to the NCBI Sequence Read Archive (SRA) database under the accession number PRJNA1111289.

Acknowledgments: We are deeply grateful to Gengcai Mima from Agriculture and Animal Husbandry Comprehensive Service Center of Langka County, Shannan Prefecture, Tibet Autonomous Region, for his invaluable contributions to this research.

Conflicts of Interest: The authors declare no conflicts of interest.

References

1. Menon, R.; Gonzalez, T.; Ferruzzi, M.; Jackson, E.; Winderl, D.; Watson, J. Oats-from farm to fork. *Adv. Food Nutr. Res.* **2016**, *77*, 1–55. [PubMed]
2. Zhang, K.; Dong, R.; Hu, X.; Ren, C.; Li, Y. Oat-Based Foods: Chemical Constituents, Glycemic Index, and the Effect of Processing. *Foods* **2021**, *10*, 1304. [CrossRef] [PubMed]
3. Mason, P.A. Dealing with silage effluent: Technical opportunities. In *Silage Effluent: Problems and Solutions*; Stark, B.A., Wilkinson, J.M., Eds.; Chalcombe Publications: Marlow, UK, 1988; pp. 21–36.
4. Troller, J.A.; Stinson, J.V. Moisture Requirements for Growth and Metabolite Production by Lactic Acid Bacteria. *Appl. Environ. Microbiol.* **1981**, *42*, 682–687. [CrossRef] [PubMed]
5. Li, P.; Liao, C.; Yan, L.; Zhang, C.; Chen, L.; You, M.; Cheng, Q.; Chen, C. Effects of Small-Scale Silo Types and Additives on Silage Fermentation and Bacterial Community of High Moisture Alfalfa on the Qinghai-Tibetan Plateau. *Anim. Feed. Sci. Technol.* **2023**, *299*, 115594. [CrossRef]
6. Gebrehanna, M.M.; Gordon, R.J.; Madani, A.; VanderZaag, A.C.; Wood, J.D. Silage Effluent Management: A Review. *J. Environ. Manag.* **2014**, *143*, 113–122. [CrossRef]
7. Zi, X.; Li, M.; Chen, Y.; Lv, R.; Zhou, H.; Tang, J. Effects of Citric Acid and Lactobacillus Plantarum on Silage Quality and Bacterial Diversity of King Grass Silage. *Front. Microbiol.* **2021**, *12*, 631096. [CrossRef]
8. Jones, B.A.; Muck, R.E.; Ricke, S.C. Selection and Application of Streptococcus Bovis as a Silage Inoculant. *Appl. Environ. Microbiol.* **1991**, *57*, 3000–3005. [CrossRef]
9. Guo, X.; Guo, W.; Yang, M.; Sun, Y.; Wang, Y.; Yan, Y.; Zhu, B. Effect of Bacillus Additives on Fermentation Quality and Bacterial Community during the Ensiling Process of Whole-Plant Corn Silage. *Processes* **2022**, *10*, 978. [CrossRef]
10. Lowes, K.F.; Shearman, C.A.; Payne, J.; MacKenzie, D.; Archer, D.B.; Merry, R.J.; Gasson, M.J. Prevention of Yeast Spoilage in Feed and Food by the Yeast Mycocin HMK. *Appl. Environ. Microbiol.* **2000**, *66*, 1066–1076. [CrossRef]
11. Wang, S.; Zhao, J.; Dong, Z.; Li, J.; Kak, N.A.; Shao, T. Sequencing and Microbiota Transplantation to Determine the Role of Microbiota on the Fermentation Type of Oat Silage. *Bioresour. Technol.* **2020**, *309*, 123371. [CrossRef]
12. Knudsen, K.E.B. Fiber and Nonstarch Polysaccharide Content and Variation in Common Crops Used in Broiler Diets. *Poult. Sci.* **2014**, *93*, 2380–2393. [CrossRef] [PubMed]
13. Wang, Q.; Wang, R.; Wang, C.; Dong, W.; Zhang, Z.; Zhao, L.; Zhang, X. Effects of Cellulase and Lactobacillus Plantarum on Fermentation Quality, Chemical Composition, and Microbial Community of Mixed Silage of Whole-Plant Corn and Peanut Vines. *Appl. Biochem. Biotechnol.* **2022**, *194*, 2465–2480. [CrossRef] [PubMed]
14. Barbosa, F.C.; Silvello, M.A.; Goldbeck, R. Cellulase and Oxidative Enzymes: New Approaches, Challenges and Perspectives on Cellulose Degradation for Bioethanol Production. *Biotechnol. Lett.* **2020**, *42*, 875–884. [CrossRef] [PubMed]
15. Polizeli, M.; Rizzatti, A.C.S.; Monti, R.; Terenzi, H.F.; Jorge, J.A.; Amorim, D.D.S. Xylanases from fungi: Properties and industrial applications. *Appl. Microbiol. Biotechnol.* **2005**, *67*, 577–591. [CrossRef]
16. Liu, W.; Si, Q.; Sun, L.; Wang, Z.; Liu, M.; Du, S.; Ge, G.; Jia, Y. Effects of Cellulase and Xylanase Addition on Fermentation Quality, Aerobic Stability, and Bacteria Composition of Low Water-Soluble Carbohydrates Oat Silage. *Fermentation* **2023**, *9*, 638. [CrossRef]
17. Liu, W.; Du, S.; Sun, L.; Wang, Z.; Ge, G.; Jia, Y. Study on Dynamic Fermentation of Oat Silage Assisted by Exogenous Fibrolytic Enzymes. *Plants* **2023**, *13*, 6. [CrossRef]
18. Xu, J.; Zhang, K.; Lin, Y.; Li, M.; Wang, X.; Yu, Q.; Sun, H.; Cheng, Q.; Xie, Y.; Wang, C.; et al. Effect of Cellulase and Lactic Acid Bacteria on the Fermentation Quality, Carbohydrate Conversion, and Microbial Community of Ensiling Oat with Different Moisture Contents. *Front. Microbiol.* **2022**, *13*, 1013258. [CrossRef]
19. Broderick, G.A.; Kang, J.H. Automated Simultaneous Determination of Ammonia and Total Amino Acids in Ruminal Fluid and in Vitro Media. *J. Dairy. Sci.* **1980**, *63*, 64–75. [CrossRef]
20. Arthur Thomas, T. An Automated Procedure for the Determination of Soluble Carbohydrates in Herbage. *J. Sci. Food Agric.* **1977**, *28*, 639–642. [CrossRef]
21. Association of Official Agricultural Chemists [AOAC]. *Official Methods of Analysis of the AOAC International*, 16th ed.; AOAC International: Washington, MD, USA, 1995.
22. Van Soest, P.J.; Robertson, J.B.; Lewis, B.A. Methods for Dietary Fiber, Neutral Detergent Fiber, and Nonstarch Polysaccharides in Relation to Animal Nutrition. *J. Dairy. Sci.* **1991**, *74*, 3583–3597. [CrossRef]
23. McDonald, P.; Henderson, A.R.; Heron, S.J.E. *The Biochemistry of Silage*; Chalcombe Publications: Marlow, UK, 1991; p. 340.
24. Xiong, Y.; Xu, J.; Guo, L.; Chen, F.; Jiang, D.; Lin, Y.; Guo, C.; Li, X.; Chen, Y.; Ni, K.; et al. Exploring the Effects of Different Bacteria Additives on Fermentation Quality, Microbial Community and in Vitro Gas Production of Forage Oat Silage. *Animals* **2022**, *12*, 1122. [CrossRef] [PubMed]
25. Tahir, M.; Li, J.; Xin, Y.; Wang, T.; Chen, C.; Zhong, Y.; Zhang, L.; Liu, H.; He, Y.; Wen, X.; et al. Response of Fermentation Quality and Microbial Community of Oat Silage to Homofermentative Lactic Acid Bacteria Inoculation. *Front. Microbiol.* **2023**, *13*, 1091394. [CrossRef] [PubMed]
26. Tian, J.; Xu, N.; Liu, B.; Huan, H.; Gu, H.; Chu, D.; Ding, C. Interaction Effect of Silo Density and Additives on the Fermentation Quality, Microbial Counts, Chemical Composition and in Vitro Degradability of Rice Straw Silage. *Bioresour. Technol.* **2020**, *297*, 122412. [CrossRef] [PubMed]

27. da Silva, T.C.; da Silva, L.D.; Santos, E.M.; Oliveira, J.S.; Perazzo, A.F. Importance of the Fermentation to Produce High-Quality Silage. *Ferment. Process* **2017**, *8*, 1–20.
28. Li, X.-M.; Chen, F.; Wang, X.; Sun, L.; Guo, L.; Xiong, Y.; Wang, Y.; Zhou, H.-Z.; Jia, S.; Yang, F.; et al. Impacts of Low Temperature and Ensiling Period on the Bacterial Community of Oat Silage by SMRT. *Microorganisms* **2021**, *9*, 274. [CrossRef]
29. Gänzle, M.G. Lactic Metabolism Revisited: Metabolism of Lactic Acid Bacteria in Food Fermentations and Food Spoilage. *Curr. Opin. Food Sci.* **2015**, *2*, 106–117. [CrossRef]
30. Valk, L.C.; Luttik, M.A.H.; de Ram, C.; Pabst, M.; van den Broek, M.; van Loosdrecht, M.C.M.; Pronk, J.T. A Novel D-Galacturonate Fermentation Pathway in Lactobacillus Suebicus Links Initial Reactions of the Galacturonate-Isomerase Route with the Phosphoketolase Pathway. *Front. Microbiol.* **2020**, *10*, 3027. [CrossRef]
31. Kim, D.; Dong Lee, K.; Choon Choi, K. Role of LAB in Silage Fermentation: Effect on Nutritional Quality and Organic Acid Production-an Overview. *Aims Agric. Food* **2021**, *6*, 216–234. [CrossRef]
32. Cruz Ramos, H.; Hoffmann, T.; Marino, M.; Nedjari, H.; Presecan-Siedel, E.; Dreesen, O.; Glaser, P.; Jahn, D. Fermentative Metabolism of Bacillus Subtilis: Physiology and Regulation of Gene Expression. *J. Bacteriol.* **2000**, *182*, 3072–3080. [CrossRef]
33. Wang, Y.; He, L.; Xing, Y.; Zheng, Y.; Zhou, W.; Pian, R.; Yang, F.; Chen, X.; Zhang, Q. Dynamics of Bacterial Community and Fermentation Quality during Ensiling of Wilted and Unwilted Moringa Oleifera Leaf Silage with or without Lactic Acid Bacterial Inoculants. *mSphere* **2019**, *4*, e00341-19. [CrossRef]
34. Wang, C.; He, L.; Xing, Y.; Zhou, W.; Yang, F.; Chen, X.; Zhang, Q. Fermentation Quality and Microbial Community of Alfalfa and Stylo Silage Mixed with Moringa Oleifera Leaves. *Bioresour. Technol.* **2019**, *284*, 240–247. [CrossRef] [PubMed]
35. Kung, L.; Shaver, R.D.; Grant, R.J.; Schmidt, R.J. Silage Review: Interpretation of Chemical, Microbial, and Organoleptic Components of Silages. *J. Dairy. Sci.* **2018**, *101*, 4020–4033. [CrossRef] [PubMed]
36. Zheng, M.; Niu, D.; Zuo, S.; Mao, P.; Meng, L.; Xu, C. The Effect of Cultivar, Wilting and Storage Period on Fermentation and the Clostridial Community of Alfalfa Silage. *Ital. J. Anim. Sci.* **2017**, *17*, 336–346. [CrossRef]
37. Qiu, C.; Liu, N.; Diao, X.; He, L.; Zhou, H.; Zhang, W. Effects of Cellulase and Xylanase on Fermentation Characteristics, Chemical Composition and Bacterial Community of the Mixed Silage of King Grass and Rice Straw. *Microorganisms* **2024**, *12*, 561. [CrossRef] [PubMed]
38. Liu, Q.; Zong, C.; Dong, Z.; Wu, J.; Zhu, J.; Li, J.; Zhang, J.; Shao, T. Effects of Cellulolytic Lactic Acid Bacteria on the Lignocellulose Degradation, Sugar Profile and Lactic Acid Fermentation of High-Moisture Alfalfa Ensiled in Low-Temperature Seasons. *Cellulose* **2020**, *27*, 7955–7965. [CrossRef]
39. Tanizawa, Y.; Tohno, M.; Kaminuma, E.; Nakamura, Y.; Arita, M. Complete Genome Sequence and Analysis of Lactobacillus Hokkaidonensis LOOC260T, a Psychrotrophic Lactic Acid Bacterium Isolated from Silage. *BMC Genom.* **2015**, *16*, 240. [CrossRef]
40. Donaghy, J.; McKay, A.M. Purification and characterization of a feruloyl esterase from the fungus Penicilliumexpansum. *J. Appl. Microbiol.* **1997**, *83*, 718–726. [CrossRef]
41. Mu, L.; Xie, Z.; Hu, L.; Chen, G.; Zhang, Z. Cellulase Interacts with Lactobacillus Plantarum to Affect Chemical Composition, Bacterial Communities, and Aerobic Stability in Mixed Silage of High-Moisture Amaranth and Rice Straw. *Bioresour. Technol.* **2020**, *315*, 123772. [CrossRef]
42. Ni, K.; Wang, F.; Zhu, B.; Yang, J.; Zhou, G.; Pan, Y.; Tao, Y.; Zhong, J. Effects of Lactic Acid Bacteria and Molasses Additives on the Microbial Community and Fermentation Quality of Soybean Silage. *Bioresour. Technol.* **2017**, *238*, 706–715. [CrossRef]
43. Wang, M.; Wu, L.; Guo, Y.; Sun, J.; Deng, M.; Liu, G.; Li, Y.; Sun, B. Effects of Fermented Herbal Tea Residue on Meat Quality, Rumen Fermentation Parameters and Microbes of Black Goats. *AMB Express* **2023**, *13*, 106. [CrossRef]
44. Xiao, Y.; Sun, L.; Xin, X.; Xu, L.; Du, S. Physicochemical Characteristics and Microbial Community Succession during Oat Silage Prepared without or with Lactiplantibacillus Plantarum or Lentilactobacillus Buchneri. *Microbiol. Spectr.* **2023**, *11*, e0222823. [CrossRef] [PubMed]
45. Jiang, H.; Wang, H.; Bao, B.; Qu, H.; Wang, J.; Sun, L.; Liu, B.; Gao, F. Effect of Compound Additives on Nutritional Composition, Fermentation Quality, and Bacterial Community of High-Moisture Alfalfa Silage. *Fermentation* **2023**, *9*, 453. [CrossRef]
46. Li, D.; Ni, K.; Zhang, Y.; Lin, Y.; Yang, F. Influence of Lactic Acid Bacteria, Cellulase, Cellulase-Producing Bacillus Pumilus and Their Combinations on Alfalfa Silage Quality. *J. Integr. Agric.* **2018**, *17*, 2768–2782. [CrossRef]
47. Bai, J.; Ding, Z.; Ke, W.; Xu, D.; Wang, M.; Huang, W.; Zhang, Y.; Liu, F.; Guo, X. Different Lactic Acid Bacteria and Their Combinations Regulated the Fermentation Process of Ensiled Alfalfa: Ensiling Characteristics, Dynamics of Bacterial Community and Their Functional Shifts. *Microb. Biotechnol.* **2021**, *14*, 1171–1182. [CrossRef]
48. Flythe, M.D.; Russell, J.B. The Effect of PH and a Bacteriocin (Bovicin HC5) on Clostridium Sporogenes MD1, a Bacterium That Has the Ability to Degrade Amino Acids in Ensiled Plant Materials1. *FEMS Microbiol. Ecol.* **2004**, *47*, 215–222. [CrossRef]
49. Yuan, X.; Li, J.; Dong, Z.; Shao, T. The Reconstitution Mechanism of Napier Grass Microiota during the Ensiling of Alfalfa and Their Contributions to Fermentation Quality of Silage. *Bioresour. Technol.* **2020**, *297*, 122391. [CrossRef]

agriculture

MDPI

Article

Identification and Correlation Analysis of Key Clostridia and LAB Species in Alfalfa Silages Prepared with Different Cultivars and Additives

Mingli Zheng [1], Qin Li [1,2], Peichun Mao [1], Xiaoxia Tian [1], Yuxia Guo [2] and Lin Meng [1,*]

1 Institute of Grassland, Flowers and Ecology, Beijing Academy of Agriculture and Forestry Sciences, Beijing 100097, China; zhengml@cau.edu.cn (M.Z.); 19503820416@163.com (Q.L.); mmpch@163.com (P.M.); tianxi8002@126.com (X.T.)

2 College of Animal Science and Technology, Henan Agricultural University, Zhengzhou 450046, China; yuxiaguo@163.com

* Correspondence: menglin9599@sina.com

Abstract: Clostridial fermentation is the determining process causing the spoilage of direct-cut alfalfa silage, and the application of lactic acid bacteria (LAB) inoculant is considered as the most promising technology for inhibiting clostridial fermentation. In order to screen target-based LAB strains, identification and correlation analysis of key Clostridia and LAB species in alfalfa silage were conducted in this study. Three alfalfa cultivars (Sanditi, SD; Celsius, CE; SW5909, SW) were harvested at the early bloom stage and ensiled without (CK) or with LAB inoculant (LB) and sucrose (SC) for 60 d. Single-molecule real-time sequencing was used to identify dominant Clostridia and LAB species, and LAB with significant inhibitory effects on dominant Clostridia was screened via correlation network analysis. The results showed that silages CK and LB encountered severe clostridial fermentation as indicated by large amounts of butyric acid (BA) and ammoniacal nitrogen (NH_3-N) production. Compared to silages CK and LB, SC treatment decreased ($p < 0.05$) BA and NH_3-N concentrations, as well as decreasing ($p < 0.05$) the bacterial community indexes of Shannon and Chao1. *Lactiplantibacillus pentosus* was the first dominant LAB in silage CK of alfalfa SD and CE. The first dominant LAB in silage LB was also identified as *L. pentosus*, rather than *Lentilactobacillus buchneri* and *Lactiplantibacillus plantarum* in the used inoculant. *L. buchneri* became more abundant in silage SC of alfalfa SD and CE, accounting for the high fermentation quality of these silages. *Clostridium tyrobutyricum*, *Clostridium luticellarii*, *Garciella* sp._GK3, *Clostridium sporogenes*, *Clostridium perfringens*, and *Clostridium* sp._BTY5 were the most dominant Clostridia species in alfalfa silage. Furthermore, *Enterococcus faecalis*, *L. buchneri*, and *L. pentosus* exhibited significant ($p < 0.05$) inhibitory effects on *C. tyrobutyricum*, *C. luticellarii*, and *Garciella* sp._GK3, respectively, which were the top three Clostridia species associated with clostridial fermentation. In conclusion, *E. faecalis*, *L. buchneri*, and *L. pentosus* were screened and can be used as potential LAB inoculants for the targeted inhibition of clostridial fermentation.

Keywords: clostridia; lactic acid bacteria; alfalfa silage; fermentation quality; bacterial community

Citation: Zheng, M.; Li, Q.; Mao, P.; Tian, X.; Guo, Y.; Meng, L. Identification and Correlation Analysis of Key Clostridia and LAB Species in Alfalfa Silages Prepared with Different Cultivars and Additives. *Agriculture* **2024**, *14*, 1963. https://doi.org/10.3390/agriculture14111963

Academic Editor: Zoltán Győri

Received: 10 September 2024
Revised: 25 October 2024
Accepted: 31 October 2024
Published: 1 November 2024

1. Introduction

Alfalfa (*Medicago sativa* L.) is the most widely cultivated legume in the world, with approximately 450 million tons grown on 30 million hectares worldwide, mainly in the U.S., Europe, and Argentina [1]. In recent years, under the policy of switching from grain crops to feed crops by the Ministry of Agriculture and Rural Affairs of China, alfalfa production in China has been growing year by year, with 546,700 hectares planted in 2020 [2]. Compared to hay preparation, ensiling has become an effective method to preserve alfalfa's nutritive value and improve animal performance [3]. Ensiling fermentation depends on lactic acid bacteria (LAB) to produce organic acids from water-soluble carbohydrates (WSCs) under anaerobic–acidic conditions, inhibiting the growth of harmful bacteria [2]. As is well known, however, direct-cut alfalfa is hard to ensile due to undesirable clostridial fermentation. In addition to causing reductions in nutritional value, clostridial fermentation affects the hygienic quality of silage by creating conditions that permit an increase in the number of pathogenic organisms and their toxins [4].

Many approaches have been proposed to prevent clostridial fermentation in alfalfa silage: wilting [5], LAB inoculants [6,7], sugar [3], and chemical additives [3]. Among these technologies, LAB inoculants are commonly used to accelerate lactic acid (LA) production as other technical methods have some inevitable defects. Wilting is affected by weather and causes severe damage, sugar is costly, and chemical additives can corrode machinery and pollute the environment. However, it is worth noting that general LAB inoculants do not improve alfalfa silage quality in some cases [8], which may be due to the poor adaptability to various ensiling microenvironments and insufficient targeted inhibition ability against key harmful microorganisms of some LAB inoculants. Li et al. [3] found that an LAB inoculant (a mixture of *Lactiplantibacillus plantarum*, *Bifidobacterium brevis*, and *Pediococcus pentosaceus*) did not inhibit the clostridial fermentation of alfalfa silage prepared in plastic bucket silos due to the growth of Clostridia at the early stage of ensiling because of the low silage density and high moisture and buffer capacity. Therefore, it is necessary to screen for LAB that are efficient and have stable usage effects, as the instability of general LAB inoculants limits their application in silage.

The complex interactions between different microorganisms (such as quorum sensing, mutual promotion, or hostility) collectively drive biochemical changes during microbial fermentation and are accompanied by microbial community dynamics, and these microbial interactions generally exhibit severe strain-dependent behaviors [9]. Therefore, using key Clostridia species that cause clostridial fermentation as targets provides a new approach for efficient LAB strain screening in alfalfa silage. Our preliminary research found the succession of Clostridial communities in alfalfa silage using high-throughput sequencing technology: Clostridia attached to alfalfa raw materials were mainly *Clostridium tyrobutyricum* and *Clostridium perfringens*; *C. perfringens*, *Garciella* sp., and *Clostridium barati* rapidly proliferated and initiated clostridial fermentation after three days of fermentation; *C. perfringens* and *C. barati* became dominant Clostridia subsequently; and *C. tyrobutyricum* played a dominant role in the clostridial fermentation process after 14 days [4,10]. However, research on Clostridia in alfalfa silage is still in its early stages, and the key Clostridia species contributing to clostridial fermentation are still unclear. We hypothesis that common Clostridia species that cause clostridial fermentation exist in different alfalfa silages, and these key Clostridia species can be used as targets for screening LAB inoculants. Therefore, this study was conducted to reveal these common Clostridia species and targeted LAB. In addition, changes in chemical and microbial compositions were also monitored to thoroughly understand and improve the ensiling process [11].

2. Materials and Methods

2.1. Alfalfa Cultivar and Additive and Silage Preparation

On 18 April 2022, three alfalfa cultivars, including 'Sanditi' (SD), 'Celsius' (CE), and 'SW5909' (SW), were grown in experimental plots of the Beijing Academy of Agricultural and Forestry Sciences (39°34′ N, 116°28′ E), Beijing, China. The fall dormancy level of alfalfa SD, CE, and SW is 5, 6, and 5, belonging to the moderate fall dormancy type. These three alfalfa cultivars are one of the main alfalfa cultivars planted in North China due to their excellent cold resistance. Triplicate experimental plots were used for each alfalfa cultivar, each plot with an area of 3 × 10 m. Approximately 60 kg per hectare of chicken manure was applied prior to sowing, with no additional fertilizer being applied during crop growth. Organic matter and total N, P, and K in chicken manure were 223.67, 21.11, 27.54, and 12.41 g/kg, respectively. On 16 May 2023, first-cut alfalfa was randomly harvested artificially at the early bloom stage from each of the three experimental plots, leaving a stubble of 10 cm. The commercial LAB inoculant used in this study was composed of *L.*

plantarum ($\geq$1.3 $\times$ 10^{10} cfu/g) and *Lentilactobacillus buchneri* ($\geq$7.0 $\times$ 10^9 cfu/g). Sucrose was used as an analytical reagent.

Direct-cut alfalfa samples were chopped to about 1–2 cm using a forage cutter and then treated separately with distilled water (control, CK), LAB inoculant (LB), and sucrose (SC). The LAB inoculant and sucrose were dissolved in the same volume of distilled water before silage preparation, and the application rates were 1 $\times$ 10^6 cfu/g FM and 20 g/kg FM, respectively. Approximately 500 g for each of the three replicates of pre-ensiled material was packed into a plastic film bag silo (Hiryu KN type, 180 $\times$ 260 mm; Asahikasei, Tokyo, Japan), sealed with a vacuum sealer (BH 950; Matsushita, Tokyo, Japan), and stored at ambient temperature (20–35 °C) for 60 d of ensiling.

2.2. *Chemical Composition and Fermentation Characteristic and Microbial Enumeration*

Fresh alfalfa was dried at 65 °C for 48 h in a forced-air oven to estimate the dry matter (DM) content and then ground through 0.20 mm mesh for chemical composition analysis. The WSC and buffer capacity (BC) were measured using anthrone colorimetry [12] and hydrochloric acid–sodium hydroxide methods [13], respectively. The crude protein (CP) was analyzed using the AOAC method [14]. An ANKOM 2000 fiber analyzer (Ankom Technology, Fairport, NY, USA) was used for the determination of neutral detergent fiber (NDF) and acid detergent fiber (ADF) throughout according to the method of Van Soest et al. [15].

For fermentation characteristic and microbial enumeration analysis, 10 g of fresh or ensiled alfalfa was homogenized with 90 mL of sterilized distilled water. Extracts were filtered through four layers of medical gauze and qualitative filter paper (pore size 15–20 μm; Hangzhou Special Paper Co., Ltd., Hangzhou, China) for pH and fermentation products determination. The pH was measured with a glass electrode pH meter (S20K; Mettler Toledo, Greifensee, Switzerland), and the ammoniacal nitrogen (NH_3-N) was determined based on the phenol–hypochlorite colorimetry method [16]. The filtrate was further processed with a dialyzer of 0.22 μm to determine the organic acid contents, including lactic acid (LA), acetic acid (AA), and butyric acid (BA), using high-performance liquid chromatography (LC-10A; Shimadzu, Tokyo, Japan) as we described previously [4].

Extracts of fresh or ensiled alfalfa were serially diluted (10^{-1} to 10^{-5}) in sterilized water for microbial enumeration by plate culture [17]. The LAB counts were measured on de Man, Rogosa, and Sharpe agar (Difco Laboratories, Detroit, MI, USA) incubated at 37 °C for 48 h under anaerobic conditions (Anaerobic box, TE-HER Hard Anaerobox, ANX-1; Hirosawa Ltd., Tokyo, Japan). Blue light broth agar (Nissui Ltd., Tokyo, Japan) and nutrient agar (Nissui Ltd.) were used for enterobacteria and aerobic bacteria enumeration, respectively, and incubated at 37 °C for 48 h. Yeast was enumerated on potato dextrose agar (Nissui Ltd.), which was acidified to pH 3.5 with sterilized tartaric acid, and incubated at 30 °C for 48 h. Before Clostridia enumeration, extracts were heated at 80 °C for 10 min to inactivate the vegetative cells and to trigger the germination of spores. The Clostridia were counted on Clostridia count agar (Nissui Ltd.) using the Hungate technique [18] and incubated at 37 °C for 7 d.

2.3. *Identification and Correlation Analysis of Key Clostridia and LAB Species*

2.3.1. DNA Extraction and SMRT Sequencing

In order to obtain microbial cells, 10 g of fresh or ensiled alfalfa was homogenized with 40 mL of sterilized distilled water and filtered through two layers of medical gauze; the filtrate was then centrifuged at 10,000 rpm and 4 °C for 15 min. The sample was used for total genomic DNA extraction using the FastDNA™ SPIN Kit for Soil (MP Biomedicals, Santa Ana, CA, USA). The full-length bacterial 16S rRNA gene was amplified with primer pairs 27F (5′-AGRGTTYGATYMTGGCTCAG-3′) and 1492R (5′-RGYTACCTTGTTACGACTT-3′). The PCR system was as follows: 4 μL of 5 $\times$ FastPfu buffer, 2 μL of dNTPs (2.5 mmol/L), 0.8 μL of each primer (5 μmol/L), 0.4 μL of FastPfu DNA polymerase, 0.2 μL BSA (2 ng/μL), 10 ng of template DNA, and double-distilled water was added to obtain a final volume of 20 μL. The PCR program was as follows: 95 °C for 3 min to denature the DNA, with amplification proceeding for 27 cycles at 95 °C for 30 s, 60 °C for 30 s, and 72 °C for 30 s; a final extension of 10 min at 72 °C was added to ensure complete amplification and was held at 4 °C. DNA libraries were constructed based on purified amplification products using SMRTbell® Express Template Prep Kit 2.0 (Pacifc Biosciences, Menlo Park, CA, USA). Thereafter, single-molecule real-time (SMRT) sequencing was performed on the Pacbio Sequel II System (Pacifc Biosciences). The sequencing data were submitted to the NCBI Sequence Read Archive database (accession: PRJNA1146988).

2.3.2. Sequencing Analysis

Circular consensus sequencing (CCS) reads were collected using SMRT Link (V8.0), and raw barcode-CCS reads were obtained using Lima (V1.7.0). Raw barcode-CCS reads were barcode-identified and length-filtered to obtain Clean-CCS reads using cutadapt (V1.9.1). UCHIME (V4.2) was then used to identify and remove chimeric sequences to obtain effective CCS reads. High-quality reads were clustered into operational taxonomic units (OTUs) at 97% similarity using USEARCH (V10.0), and the representative OTU sequences were used for the annotation of taxonomic information based on the Silva database (V13.2) [19]. The richness index of Chao1 and the diversity index of Shannon were calculated using QIIME (V1.9.1) and displayed with R (V2.15.3) to evaluate the alpha diversity. Principal coordinate analysis (PCoA) was performed based on OTUs to assess community dissimilarity among samples using QIIME (V1.9.1). Microbial networks were carried out using Spearman correlation to statistically investigate the correlations between dominant LAB and Clostridia species [20]. R (V3.6.3) was used to calculate Spearman correlation coefficients and generate a heatmap to explore relationships between the relative abundances of dominant LAB and Clostridia species and fermentation parameters [21].

2.4. Statistical Analysis

All microbial counts were $\log_{10}$ transformed to obtain log-normal distributed data. Data for chemical composition, fermentation characteristics, microbial count, and relative abundance of dominant Clostridia and LAB species were analyzed using a two-way ANOVA for a 3 × 3 (cultivar × additive) factorial arrangement of treatments using the GLM procedures of SAS 9.1 (SAS Institute, Cary, NC, USA). Significant differences between treatments were analyzed using Tukey's test and were declared at $p < 0.05$.

3. Results

3.1. Characteristics of Fresh Alfalfa

As described in Table 1, the chemical compositions and microbial populations of alfalfa before ensiling were markedly affected by the cultivar. The highest ($p < 0.05$) CP content and the lowest ($p < 0.05$) WSC content and BC were found in SD, while the lowest ($p < 0.05$) DM content and the highest ($p < 0.05$) WSC content and BC were found in SW. In addition, the population of epiphytic LAB was significantly higher ($p < 0.05$) in SD compared to the other two cultivars. The Clostridia count attached to fresh alfalfa was below the detection limit. It is worth noting that the total number of top ten epiphytic LAB was numerically higher ($p = 0.06$) in SD than in the other two cultivars.

Table 1. Chemical compositions, epiphytic microbial counts, and dominant epiphytic LAB species of fresh alfalfa.

Item	SD	CE	SW	SEM	p-Value
Chemical compositions (g/kg DM)					
DM (g/kg FM)	186.12 b	210.80 a	179.49 c	4.813	<0.001
CP	204.03 a	179.67 b	184.17 b	3.856	<0.001
WSC	41.23 c	44.43 b	52.13 a	1.656	<0.001
NDF	402.13	401.70	401.00	1.800	0.975
ADF	304.10 b	333.07 a	332.33 a	4.827	<0.001
BC (mEq/kg DM)	478 c	526 b	551 a	10.7	<0.001
Microbial counts ($\log_{10}$ cfu/g FM)					
LAB	4.26 a	4.14 b	4.12 b	0.023	0.001
Clostridia	ND	ND	ND	-	-
Enterobacteria	5.44 b	5.64 a	5.42 b	0.041	0.011
Aerobic bacteria	5.77	5.93	5.79	0.033	0.088
Yeast	3.64 a	3.50 b	3.50 b	0.026	0.004
Top ten epiphytic LAB species (%)					
Lactobacillus gasseri	0.045	0.003	0.011	0.0103	0.212
Enterococcus faecalis	0.011	0.001	0.008	0.0033	0.495
Enterococcus pernyi	0.010	0.021	0.000	0.0046	0.162
Enterococcus casseliflavus	0.007	0.002	0.001	0.0013	0.107
Enterococcus silesiacus	0.006	0.006	0.016	0.0031	0.408
Lactococcus garvieae	0.006	0.001	0.002	0.0011	0.107
Lactococcus raffinolactis	0.003	0.001	0.000	0.0009	0.546
Lactiplantibacillus pentosus	0.001	0.002	0.001	0.0006	0.819
Lactiplantibacillus plantarum	0.001	0.000	0.001	0.0005	0.619
Leuconostoc pseudomesenteroides	0.001	0.001	0.001	0.0006	0.971
Total	0.093	0.040	0.038	0.0115	0.060

SD, CE, and SW indicate alfalfa cultivars Sanditi, Celsius, and SW5909, respectively; DM, dry matter; FM, fresh matter; CP, crude protein; WSC, water-soluble carbohydrates; NDF, neutral detergent fiber; ADF, acid detergent fiber; BC, buffering capacity; LAB, lactic acid bacteria; SEM, standard error of mean; ND, not detected. Means with different letters in the same rows are significantly different at the level of $p < 0.05$.

3.2. Fermentation Characteristics and Microbial Counts of Alfalfa Silage

As shown in Figure 1, the results indicate that the alfalfa cultivar, additive treatment, and their interaction significantly ($p < 0.05$) affected the fermentation characteristics. Irrespective of alfalfa cultivar, undesirable clostridial fermentation occurred in silage CK and LB compared to SC treatment, as indicated by higher ($p < 0.05$) pH, AA, BA, and NH_3-N concentrations. Furthermore, for silage CK, higher ($p < 0.05$) BA and NH_3-N concentrations and lower ($p < 0.05$) LA production were apparent in CE and SW compared to SD. As expected, regardless of alfalfa cultivar, SC treatment effectively accelerated LA fermentation, resulting in a higher ($p < 0.05$) LA/AA ratio and lower ($p < 0.05$) pH and BA and NH_3-N concentrations compared to silage CK. Therefore, clostridial fermentation was effectively inhibited due to SC treatment, and the lowest ($p < 0.05$) BA and NH_3-N concentrations occurred in SD.

Compared to CK, LB and SC treatments increased ($p < 0.05$) LAB count, and the highest number was observed in SC treatment regardless of alfalfa cultivar (Figure 2). In terms of alfalfa cultivar, a higher ($p < 0.05$) LAB count was found in alfalfa SD compared to CE and SW when the same additive treatment was applied. The proliferation of Clostridia in alfalfa silage was effectively inhibited ($p < 0.05$) because of LB and SC treatments, especially for the SDSC and CESC groups in which Clostridia was undetectable.

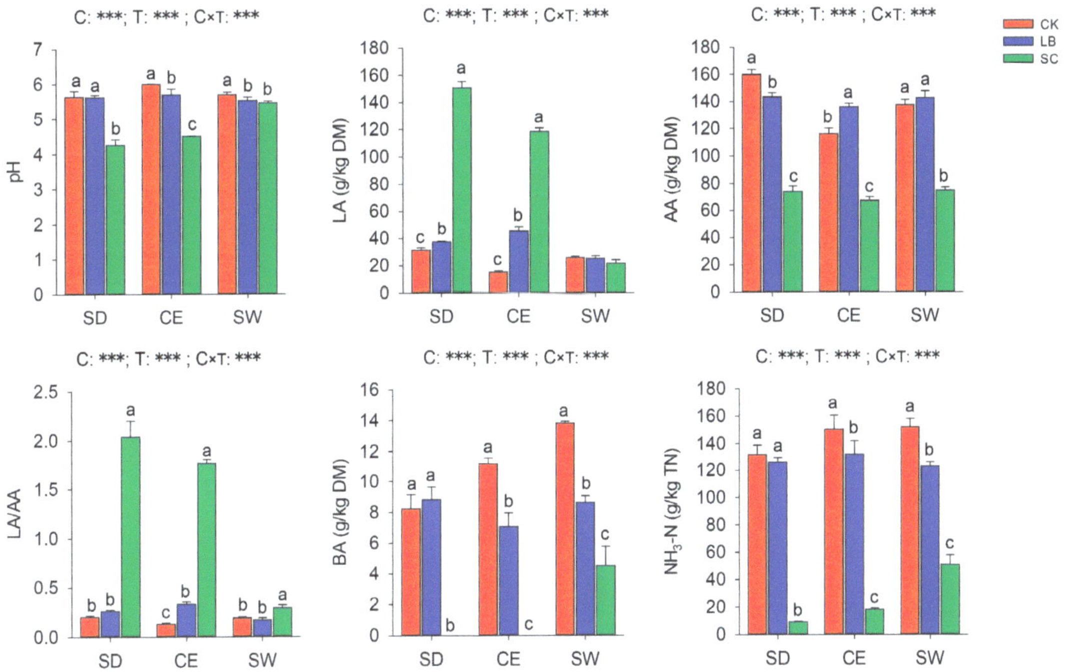

Figure 1. Fermentation characteristics of alfalfa silages prepared with different cultivars and additives. SD, CE, and SW indicate alfalfa cultivars Sanditi, Celsius, and SW5909, respectively. CK, control; LB, silage treated with LAB inoculant; SC, silage treated with sucrose addition. LA, lactic acid; AA, acetic acid; LA/AA, lactic to acetic acid ratio; BA, butyric acid; NH_3-N, ammonia nitrogen; TN, total nitrogen. Means with different letters indicate significant differences among additive treatments of the same alfalfa cultivar ($p < 0.05$). C, alfalfa cultivar; T, additive treatment; C $\times$ T, interaction between alfalfa cultivar and additive treatment. ***, $p < 0.001$.

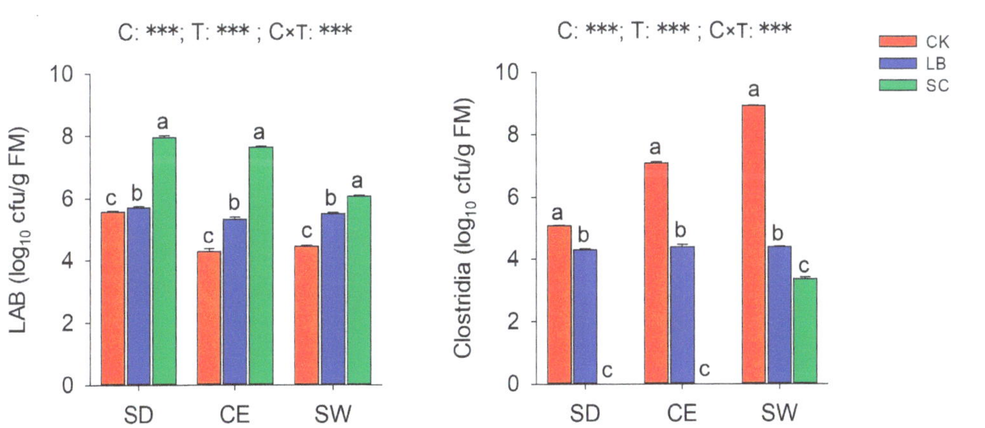

Figure 2. LAB and Clostridia counts in alfalfa silages prepared with different cultivars and additives. SD, CE, and SW indicate alfalfa cultivars Sanditi, Celsius, and SW5909, respectively. CK, control; LB, silage treated with LAB inoculant; SC, silage treated with sucrose addition. LAB, lactic acid bacteria; FM, fresh matter. Means with different letters indicate significant differences among additive treatments of the same alfalfa cultivar ($p < 0.05$). C, alfalfa cultivar; T, additive treatment; C $\times$ T, interaction between alfalfa cultivar and additive treatment. ***, $p < 0.001$.

3.3. Bacterial Community of Alfalfa Silage

The alpha diversity of the bacterial community is displayed in Figure 3. Compared to silage CK, SC treatment decreased ($p < 0.05$) the indexes of Shannon and Chao1, while these indexes both increased due to LB treatment, although not all differences were significant. The variance in the bacterial communities among the different silage groups was clearly reflected by the PCoA. Component 1 and component 2 could explain 20.89–25.80% and 14.11–19.18% of the total variance, respectively. In terms of alfalfa SD and CE, the bacterial community in silage SC was clearly separated from silages CK and LB, while there was an intersection between the CK and LB samples. The bacterial community in alfalfa SW silage could be clearly divided into three groups according to various additive treatments.

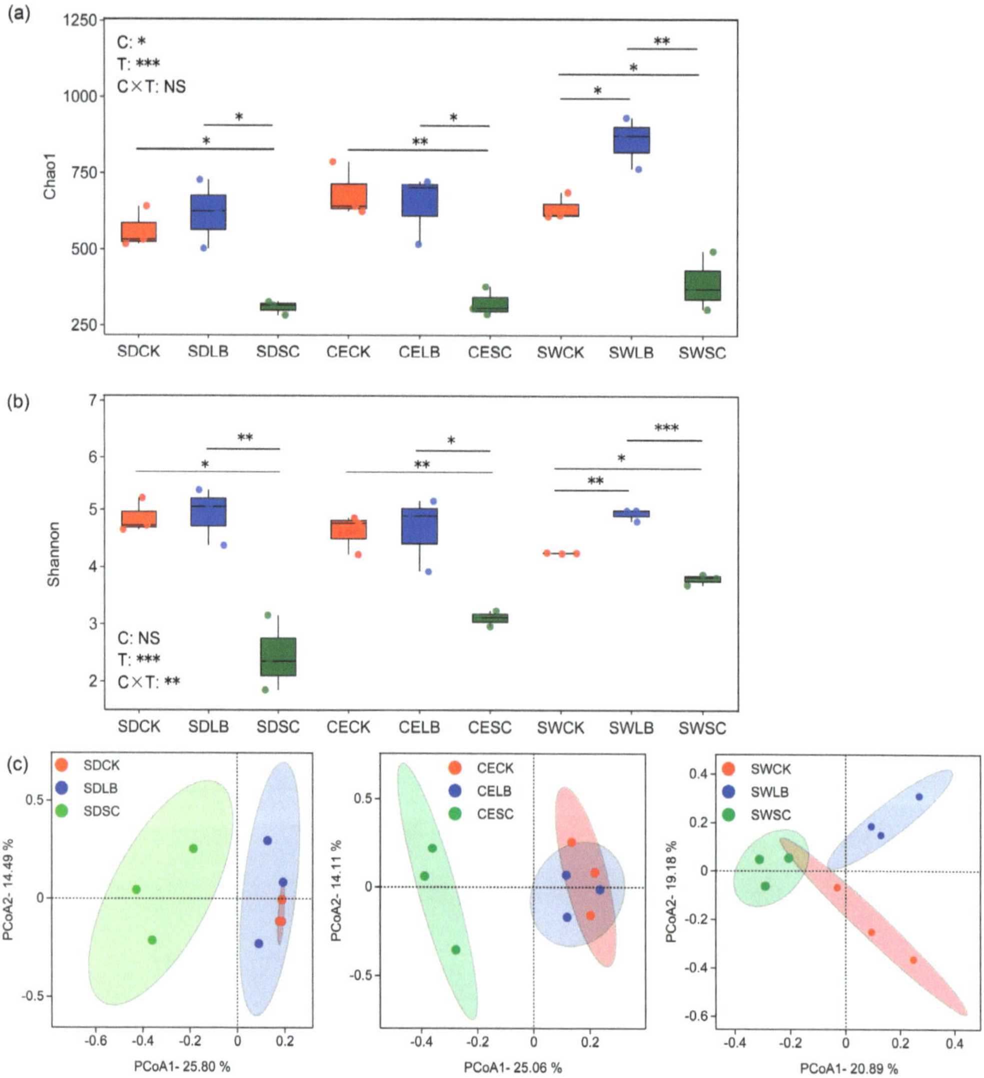

Figure 3. Chao1 index (**a**), Shannon index (**b**), and principal coordinates analysis (PCoA) (**c**) of the bacterial community in alfalfa silages prepared with different cultivars and additives. NS, *, ** and ***, represent $p > 0.05$, $0.01 < p < 0.05$, $0.001 < p < 0.01$ and $p < 0.001$, respectively.

The bacterial community at the genus and species levels is illustrated in Figure 4. Undesirable *Enterobacter*, unclassified Lachnospiraceae, *Bifidobacterium*, *Rhabdanaerobium*, *Clostridium* sensu_stricto_12, and *Caproiciproducens* dominanted the microbiota in silage CK, with the total relative abundance ranging from 31.72% to 59.15%. These undesirable bacteria were also detected in silage LB, with the total relative abundance ranging from 40.37% to 45.17%, although the relative abundance of favorable *Lentilactobacillus*, *Lactobacillus*, and *Lactiplantibacillus* increased ($p < 0.05$) due to LB treatment compared to silage CK. It is worth noting that *Clostridium* sensu_stricto_12 was promoted ($p < 0.05$) because of LB treatment compared to silage CK. After SC treatment, anaerobic fermentation underwent a shift from undesirable bacteria to *Lactobacillus* and *Lentilactobacillus*. In the SDSC and CESC groups in particular, the relative abundance of *Lentilactobacillus* was 86.61% and 74.97%, respectively.

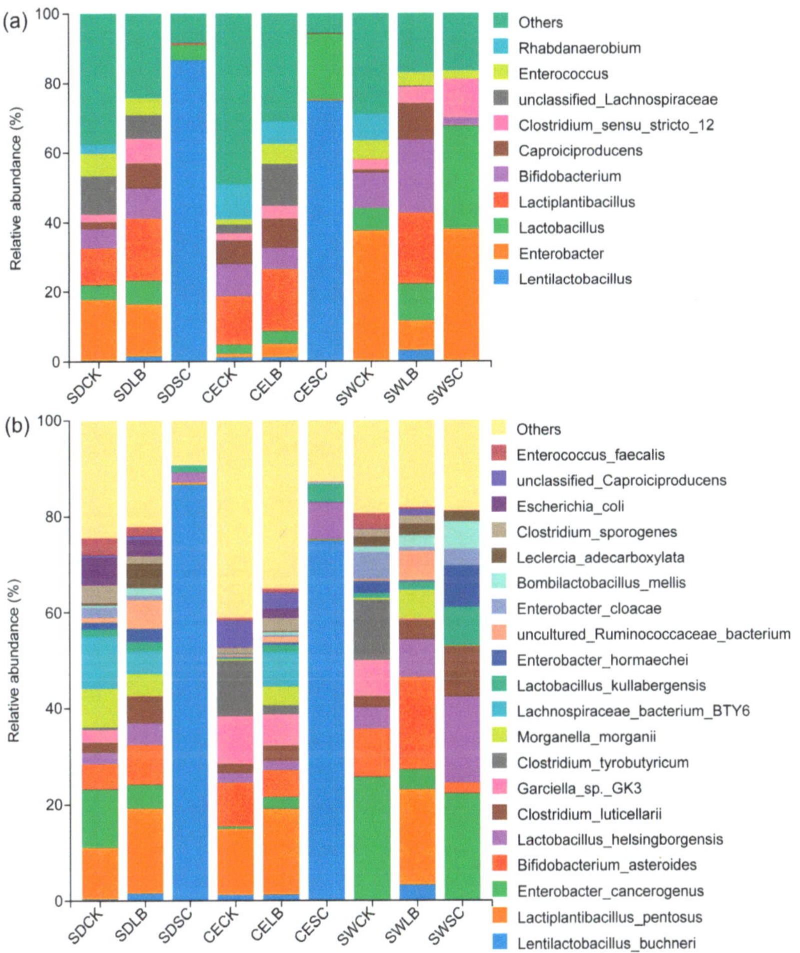

Figure 4. Relative abundance of the top ten genera (**a**) and top twenty species (**b**) of the bacterial community in alfalfa silages prepared with different cultivars and additives.

Species that are unfavorable for silage fermentation accounted for a large proportion in silage CK, although the bacterial community at the species level varied among the different alfalfa cultivars. The total relative abundance of *Enterobacter cancerogenus*, *Bifidobacterium asteroides*, *Clostridium luteellii*, *Garciella* sp._GK3, and *C. tyrobutyricum* in the SDCK, CECK, and SWCK groups was as high as 22.68%, 33.02%, and 58.01%, respectively. The relative abundance of desirable *Lactiplantibacillus pentosus* in silage LB was as low as 17.76% to 19.97%, although *L. pentosus* was stimulated ($p < 0.05$) due to LB treatment and became the first dominant species. A large relative abundance of the above-mentioned undesirable bacterial species was also observed in silage LB, which ranged from 18.81% to 27.52%. *L. buchneri*, as the representative bacterial species, became the first dominant specie in the SDSC and CESC groups with the relative abundances of 86.58% and 74.94%, respectively, followed by *Lactobacillus helsingborgensis* and *Lactobacillus kullabergensis*. However, *E. cancerogenus* dominated the bacterial community in the SWSC group.

3.4. Key Clostridia and LAB Species of Alfalfa Silage

Astatistical comparison of the top six Clostridia and LAB species of alfalfa silage is illustrated in Figure 5. *L. buchneri*, *L. pentosus*, *L. helsingborgensis*, *L. kullabergensis*, *E. faecalis*, and *Lactiplantibacillus plantarum* were the most dominant LAB species, and their relative abundance varied across the different alfalfa cultivars and additive treatments. *L. buchneri* became more abundant ($p < 0.05$) in the silage SC of alfalfa SD and CE, which resulted in the high fermentation quality of these silages. LB treatment promoted ($p < 0.05$) *L. pentosus*, rather than *L. buchneri* and *L. plantarum* in the inoculant, compared to silage CK and SC. The relative abundance of *L. helsingborgensis* and *L. kullabergensis* also increased ($p < 0.05$) due to LB treatment compared to silage CK. It is worth noting that *L. helsingborgensis* and *L. kullabergensis* were the top three LAB species in silage SC, which might be associated with LA fermentation. The relative abundance of *E. faecalis* in alfalfa SD and SW silages was in the following order: CK > LB > SC. *L. plantarum* was rarely detected in all alfalfa silages, with a relative abundance of lower than 1%.

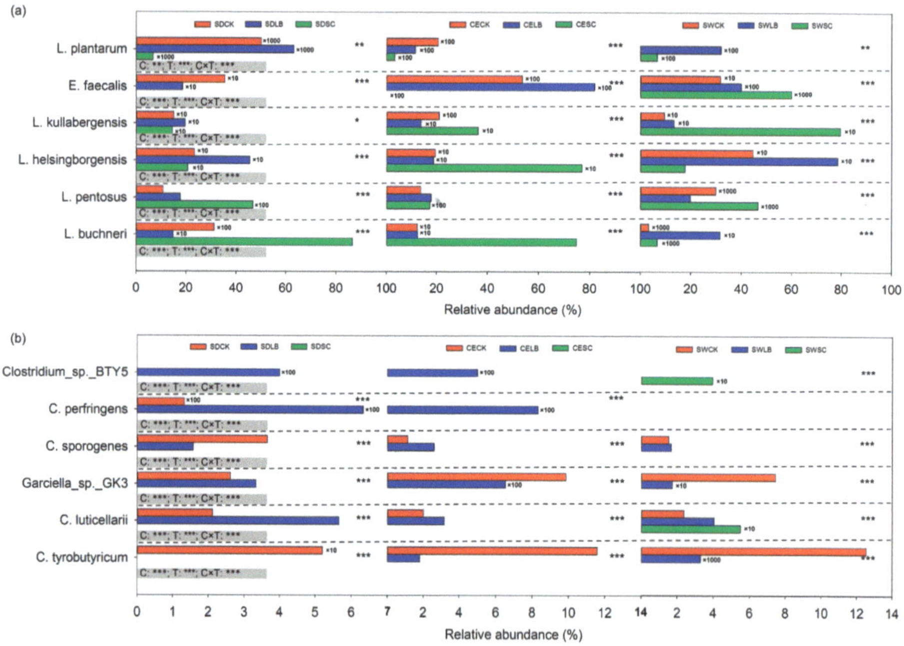

Figure 5. Significant differences in the top six LAB (**a**) and Clostridia (**b**) species in alfalfa silages prepared with different cultivars and additives. *, ** and ***, represent $0.01 < p < 0.05$, $0.001 < p < 0.01$ and $p < 0.001$, respectively.

C. tyrobutyricum, *Clostridium luticellarii*, *Garciella* sp._GK3, *Clostridium sporogenes*, *C. perfringens*, and *Clostridium* sp._BTY5 were the most dominant Clostridia species in alfalfa silage. *C. tyrobutyricum*, with a relative abundance as high as 11.57% and 12.53%, dominated the clostridial community of the CECK and SWCK groups, respectively, with them encountering the most intense clostridial fermentation. The first dominant Clostridia species associated with the poorly fermented SDCK group was detected as *C. sporogenes*, with the proportion of 3.65%. The next most abundant Clostridia species in silage CK regardless of alfalfa cultivar was *Garciella* sp._GK3, which might also play a crucial role in clostridial fermentation. *C. tyrobutyricum* in alfalfa CE and SW and *C. sporogenes* in alfalfa SD were all inhibited ($p < 0.05$) due to LB treatment compared to silage CK. Instead, *C. luticellarii* and *Garciella* sp._GK3 became the dominant Clostridia species in silage LB. As expected, these six dominant Clostridia species were less abundant or not detected in silage SC.

The correlation networks among the top six most abundant Clostridia and LAB species are shown in Figure 6a. LAB species that are negatively ($p < 0.05$) correlated with *C. tyrobutyricum*, *Garciella* sp._GK3, and *C. lutecellarii* were *E. faecalis*, *L. pentosus*, and *L. buchneri*. However, LAB species that are negatively ($p < 0.05$) correlated with *C. sporogenes* were not identified. An antagonistic relationship between LAB and Clostridia species was also discovered, such as the fact that *C. sporogenes* showed significant negative correlations ($p < 0.05$) with *C. lutecellarii*.

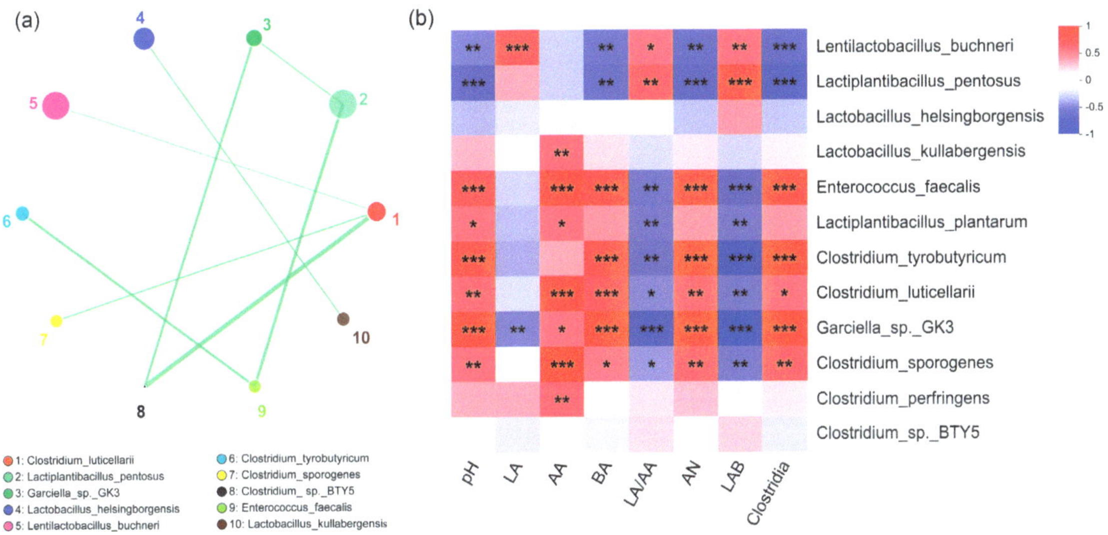

Figure 6. Interaction networks of the top six Clostridia and LAB species (**a**) and Spearman correlation heatmap of the top six Clostridia and LAB species and fermentation parameters (**b**). Nodes along the red line are positively correlated, while nodes along the green line are negatively correlated. The size of the nodes is directly proportional to the abundance of the species in all samples. *, ** and ***, represent $0.01 < p < 0.05$, $0.001 < p < 0.01$ and $p < 0.001$, respectively.

Spearman correlation further illustrated the relationships between dominant Clostridia species, LAB species, and silage fermentation parameters (Figure 6b). pH, BA concentration, NH_3-N concentration, and Clostridia count were positively ($p < 0.05$) correlated with *E. faecalis*, *C. tyrobutyricum*, *C. luticellarii*, *Garciella* sp._GK3, and *C. sporogenes* but negatively ($p < 0.05$) correlated with *L. buchneri* and *L. pentosus*. LA concentration was positively ($p < 0.05$) correlated with *L. buchneri* but negatively ($p < 0.05$) correlated with *Garciella* sp._GK3. LA/AA and LAB count were positively ($p < 0.05$) correlated with *L. buchneri* and *L. pentosus* but negatively ($p < 0.05$) correlated with *E. faecalis*, *L. plantarum*, *C. tyrobutyricum*, *C. luticellarii*, *Garciella* sp._GK3, and *C. sporogenes*.

4. Discussion

It is well established that the ensiling process is strongly affected by the chemical compositions and epiphytic microorganisms of crops prior to ensiling [3,22]. Suitable DM contents (>300 g/kg FM), sufficient WSC contents (60–70 g/kg DM) and LAB populations (>10^5 cfu/g FM), low BC, and undesirable microbial populations are critical for successful ensiling [23]. In this study, higher LAB count and lower BC were found in alfalfa SD compared to the other two cultivars. It is possible that the depression in BC is associated with a reduction in the organic acid fraction [24], although higher CP content was found in alfalfa SD. It has been reported that herbage BC mainly results from potassium, calcium, and magnesium salts of organic acids, while only about 10% to 20% is attributed to plant proteins [25]. LAB are not usually part of the normal microflora of the growing plant, and harvesting and chopping operations can provide a low level of LAB inoculation (10^2 to 10^4 cfu/g) on alfalfa prior to ensiling, which is probably caused by LAB growth on plant juices [26,27]. Therefore, it has been suggested that alfalfa SD is more suitable for recruiting considerable specific LAB species and promoting their proliferation. This hypothesis was further confirmed by our SMRT sequencing results, as it indicated that there were differences in LAB species attachment on various pre-ensiled alfalfa cultivars, *Lactobacillus gasseri*, and *E. faecalis* that preferred alfalfa SD.

The preservation of silage nutrients is dependent on their subsequent fermentations. Silage encountered in clostridial fermentation typically has a high pH (>5), contains little or no LA, has a high BA concentration (>5 g/kg DM) with a typical BA smell, is high in NH_3-N (>120 g/kg TN), and has high concentrations of amines [25,28]. Consistent with previous results [29], silage CK underwent severe clostridial fermentation as indicated by high pH, BA, and NH_3-N concentrations. In addition, similar to the results of Zhao et al. [8], clostridial fermentation was not effectively inhibited due to LB treatment. We hypothesize that two underlying mechanisms may contribute to this result: (1) excessive moisture causes a decrease in the concentration of soluble sugars in silage, which might propagate harmful bacteria and inhibit LAB reproduction [2], and (2) the LAB strains used in this study showed poor adaptability to alfalfa ensiling conditions and were less competitive than the native LAB population [8,30]. In this study, the dominant LAB in silage LB might possess a heterofermentative pathway according to the LA/AA of less than 2. As it was reported previously, the occurrence of clostridial fermentation can metabolize sugars and LA to BA and considerably hydrolyze proteins, leading to a significant loss of nutrients [2,31]. Significant hygiene risks also occurred in silage that underwent clostridial fermentation as indicated by the large amount of Clostridia in silages CK and LB. Therefore, this study once again emphasizes the importance of inhibiting clostridial fermentation in alfalfa silage.

Fortunately, SC treatment stimulated LA fermentation for alfalfa SD and CE, thereby inhibiting clostridial fermentation. These results confirm the hypothesis that a more restrictive effect on silage fermentation occurred due to the absence of a fermentation substrate compared to the lack of LAB [2]. Our results are consistent with the results reported by Lin et al. [3], who found that differences in the chemical compositions and epiphytic bacterial communities of various alfalfa cultivars significantly affected silage fermentation. In this study, therefore, the higher BC and lower LA-producing cocci (including *Enterococcus faecalis*, *Enterococcus pernyi*, *Enterococcus casseliflavus*, *Enterococcus silesiacus*, *Lactococcus garvieae*, *Lactococcus raffinolactis*, and *Leuconostoc pseudomesenteroides*) count might be the main reasons for the poor fermentation of alfalfa SW compared to the other two cultivars regardless of additive treatments. However, how the microbial population responds to different cultivars and additives during ensiling is still unclear.

A decline in the alpha diversity of a bacterial community is usually associated with successful silage fermentation, and the possible reason for this is that antibacterial compounds and organic acids produced by sufficient LA fermentation inhibit undesirable bacteria [6,32,33]. Similarly, in this study, the bacterial abundance and diversity decreased because of SC treatment compared to CK and LB treatments. The PCoA results confirmed that there were significant differences in the bacterial community structure among different fermentation types [7,34]. The bacterial communities of silages CK and LB were highly overlapped due to the occurrence of clostridial fermentation, while the distribution of the bacterial community was mainly shifted by SC treatment.

High-moisture alfalfa silage characterized by poor (but natural) fermentation was frequently dominated by *Enterobacter, Clostridium,* or heterofermentative LAB [2]. Consistent with previous results, compared to silage SC, a higher relative abundance of *Enterobacter* and *Clostridium* was detected in poorly fermented silages CK and LB, as well as undesirable *Bifidobacterium*. Silage BA and NH_3-N production with a resulting high pH is taken as a reliable index of clostridial fermentation. It should be noted that Clostridia is the main producer of BA and NH_3-N, although *Enterobacter* has also been shown to produce NH_3-N [28]. In addition, a higher abundance of *Lentilactobacillus* might be the main reason for the high fermentation quality in silage SC of alfalfa SD and CE. Therefore, a more detailed analysis of the dominant Clostridia and LAB species in silage was conducted in this study.

Similar to our previous findings [4,10,35], *C. tyrobutyricum, Garciella* sp._GK3, *C. sporogenes, C. perfringens,* and *Clostridium* sp._BTY5 were identified as the key Clostridia species accounting for clostridial fermentation of alfalfa silage. It is worth noting that *C. luticellarii,* which was isolated from a mud cellar used for producing strong aromatic liquors, was first reported in silage in this study. *C. tyrobutyricum* became the first dominant Clostridia species in the CECK and SWCK groups, showing the most intense clostridial fermentation and was positively correlated with pH, BA concentration, NH_3-N concentration, and Clostridia count. Therefore, *C. typobutyricum* was considered as the principal responsible for clostridial fermentation of alfalfa silage. The characteristic of acid tolerance and capacity to ferment lactate accounted for its role in stimulating clostridial fermentation [28]. In addition, *C. tyrobutyricum* adapted to unfavorable environmental conditions such as those encountered during silage fermentation due to the ability of vegetative cells to sporulate and adopt metabolic pathways, leading to the production of non-protein-free amino acids [36].

C. luticellarii were the first or second dominant Clostridia species in silage LB, leading to clostridial fermentation. This was supported by the results of Spearman correlation analysis, in which *C. luticellarii* were positively correlated with pH, BA concentration, NH_3-N concentration, and Clostridia count. Similarly, Wang et al. [37] found that BA was the major organic acid product when *C. luticellarii* were grown on reinforced clostridium medium culture. Changes in LAB community and microenvironment during the fermentation of silage LB might have stimulated *C. luticellarii,* since Clostridia contain different phylogenetic and basic characteristics. During the brewing process of Chinese liquors in mud cellars, diverse anaerobic and facultatively aerobic microbes including LAB and Clostridia accumulated [38]. Therefore, it is reasonable to speculate that *C. lutecellarii* may have formed a synergistic mechanism with certain LAB species during the long-term evolutionary process. In addition, the elevated AA concentration in silage LB might have elevated BA production by *C. luticellarii* as suggested by Mariën et al. [39].

Spearman correlation results also confirmed that clostridial fermentation was also stimulated by *Garciella* sp._GK3, which was labeled as the next most abundant Clostridia species in silage CK regardless of alfalfa cultivar. These findings were consistent with the results reported by Fan et al. [40] and Zhang et al. [41], who found that *Garciella* became one of the dominant genera in poor-fermented alfalfa silage. *Garciella,* belonging to cluster XII of the order Clostridiales, were first discovered and named in 2003 [42]. However, unfortunately, *Garciella* detected in alfalfa silage have not been identified at the species level so far, which seriously limits their mechanistic evaluation in clostridial fermentation. According to the search results based on the NCBI database, the only identified species of this genus is *Garciella nitrarireducens*. The capacity of *G. nitrarireducens* with LA-utilizing and nitrate-reducing abilities might help to explain the role of *Garciella* in stimulating Clostridia activity, as we suggested previously [4].

L. buchneri, L. pentosus, E. faecalis, L. plantarum, L. helsingborgensis, and *L. kullabergensis* are commonly found in alfalfa silage [4,34,43,44], while only *E. faecalis, L. pentosus,* and *L. buchneri* showed significant inhibition of the dominant Clostridia species in this study. As we know, LA-producing cocci (*Weissella, Leuconostoc, Pediococcus, Lactococcus,* and *Enterococcus*) initiate LA fermentation at the early stages of the ensiling process, while LA-producing rod *Lactobacillus,* with low pH resistance ability, play an important role in pH reduction at the later stages of ensiling [45,46]. Owing to their desirable functions, *E. faecalis* is therefore often applied alone or in combination with *Lactobacillus* to promote silage fermentation [44,47,48]. However, in this study, *E. faecalis* was negatively correlated with LAB count, and positively correlated with pH, BA concentration, and Clostridia count. The rapid acidification process during the early ensiling stage contributed by *E. faecalis* helps to inhibit Clostridia, while the high presence of *E. faecalis* in undesirable silages CK and LB indicated unfinished or incomplete fermentation [21,43].

L. pentosus became the dominant LAB in silage CK of alfalfa SD and CE, indicating its strong environmental adaptability to naturally fermented alfalfa silage. This conclusion was further confirmed because the dominant LAB in silage LB was also *L. pentosus,* rather than *L. plantarum* and *L. buchneri* present in the additive. Considering the unfavorable ensiling conditions of insufficient sugar content, it was likely that *L. pentosus* had a competitive edge over other LAB species for sugar substances during alfalfa silage fermentation. The main monosaccharides in alfalfa are glucose and xylose, and it has been reported that *L. pentosus* is able to ferment xylose to produce LA, but other common LAB species in silage such as *L. coryniformis, L. casei,* and *L. plantarum* could not sufficiently utilize xylose [34]. Previous research has shown that the fermentation quality and nutrient preservation of alfalfa silage were promoted due to *L. pentosus* inoculation [49]. This study further indicated that the adaptability of exogenous LAB was not as good as that of local LAB, especially under adverse ensiling conditions [11]. In recent years, an increasing number of studies have also confirmed that epiphytic LAB are more effective than exogenous LAB when applied to silage, considering the interaction of epiphytic LAB with plants during silage fermentation [50]. In this study, therefore, *L. pentosus* might be the major species leading to LA fermentation, as indicated by the result that *L. pentosus* was positively correlated with LA concentration while being negatively correlated with pH.

It is worth noting that *L. buchneri* was greatly promoted because of SC treatment and became the absolute dominant LAB associated with promoting LA fermentation in well-fermented silage SC of alfalfa SD and CE. Consistently, Lin et al. [3] found that *Lentilactobacillus* was the dominant genera in well-fermented alfalfa silage, with a relative abundance as high as 67.10%. Therefore, it has been suggested that *L. buchneri* is one of the ideal functional bacteria during ensiling and could be used as an inoculant to improve silage quality, although it is commonly introduced to address the problem of aerobic stability [51]. There are two possible explanations for this phenomenon. One is related to the hypothesis that the dominant homofermentative LAB stimulates LA fermentation at the early stage, while the slow-growing and acid-tolerance *L. buchneri* gradually becomes dominant when the fermentation process stabilizes at the late stage. This hypothesis was partly confirmed by Guo et al. [52], who found that *L. buchneri* showed less competitive colonizing behavior at the earlier fermentation stage than *L. plantarum* and played a role in the late stage of alfalfa silage fermentation. *Lentilactobacillus* is labeled as a novel slow-growing *Lactobacillus,* with the genus *Lactobacillus* being extremely reclassified at phenotypic, ecological, and genotypic levels, and the type species of this genus is *L. buchneri* [53]. The other reason is probably that *L. buchneri* is a facultative heterofermentative LAB and tends to undergo homofermentation under sufficient WSC conditions. Chen et al. [50] reported that heterofermentative LAB are commonly used as silage inoculants, and today, most of the bacteria in this group are considered facultative heterofermentative LAB.

41. Zhang, X.Y.; Zhao, S.S.; Wang, Y.P.; Yang, F.Y.; Wang, Y.; Fan, X.M.; Feng, C.S. The Effect of *Lactiplantibacillus plantarum* ZZU203, cellulase-producing *Bacillus methylotrophicus*, and their combinations on alfalfa silage quality and bacterial community. *Fermentation* **2023**, *9*, 287. [CrossRef]
42. Miranda-Tello, E.; Fardeau, M.L.; Sepulveda, J.; Fernandez, L.; Cayol, J.L.; Thomas, P.; Ollivier, B. *Garciella nitratireducens* gen. nov., sp nov., an anaerobic, thermophilic, nitrate- and thiosulfate-reducing bacterium isolated from an oilfield separator in the Gulf of Mexico. *Int. J. Syst. Evol. Microbiol.* **2003**, *53*, 1509–1514. [CrossRef]
43. Li, P.; Zhao, W.J.; Yan, L.J.; Chen, L.Y.; Chen, Y.L.; Gou, W.L.; You, M.H.; Cheng, Q.M.; Chen, C. Inclusion of abandoned rhubarb stalk enhanced anaerobic fermentation of alfalfa on the Qinghai Tibetan Plateau. *Bioresour. Technol.* **2022**, *347*, 126347. [CrossRef]
44. Na, N.; Qili, M.; Wu, N.R.; Sun, L.; Xu, H.W.; Zhao, Y.; Wei, X.B.; Xue, Y.L.; Tao, Y. Bacterial community and fermentation quality of ensiling alfalfa with commercial lactic acid bacterial additives. *Front. Microbiol.* **2022**, *13*, 836899. [CrossRef] [PubMed]
45. Jiang, D.; Li, B.N.; Zheng, M.L.; Niu, D.Z.; Zuo, S.S.; Xu, C.C. Effects of *Pediococcus pentosaceus* on fermentation, aerobic stability and microbial communities during ensiling and aerobic spoilage of total mixed ration silage containing alfalfa (*Medicago sativa* L.). *Grassl. Sci.* **2020**, *66*, 215–224. [CrossRef]
46. Li, X.M.; Chen, F.; Wang, X.K.; Xiong, Y.; Liu, Z.Y.; Lin, Y.L.; Ni, K.K.; Yang, F.Y. Innovative utilization of herbal residues: Exploring the diversity of mechanisms beneficial to regulate anaerobic fermentation of alfalfa. *Bioresour. Technol.* **2022**, *360*, 127429. [CrossRef] [PubMed]
47. Li, J.F.; Yuan, X.J.; Desta, S.T.; Dong, Z.H.; Mugabe, W.; Shao, T. Characterization of *Enterococcus faecalis* JF85 and *Enterococcus faecium* Y83 isolated from Tibetan yak (*Bos grunniens*) for ensiling *Pennisetum sinese*. *Bioresour. Technol.* **2018**, *257*, 76–83. [CrossRef]
48. Tian, J.; Huang, L.Y.; Tian, R.; Wu, J.Y.; Tang, R.X.; Zhang, J.G. Fermentation quality and bacterial community of delayed filling stylo silage in response to inoculating lactic acid bacteria strains and inoculating time. *Chem. Biol. Technol. Agric.* **2023**, *10*, 44. [CrossRef]
49. Guo, L.N.; Wang, X.K.; Chen, H.L.; Li, X.M.; Xiong, Y.; Zhou, H.Z.; Xu, G.; Yang, F.Y.; Ni, K.K. Exploring the fermentation quality, bacterial community and metabolites of alfalfa ensiled with mugwort residues and *Lactiplantibacillus pentosus*. *Chem. Biol. Technol. Agric.* **2023**, *10*, 107. [CrossRef]
50. Chen, L.J.; Wang, Y.L.; Li, X.; MacAdam, J.W.; Zhang, Y.H. Interaction between plants and epiphytic lactic acid bacteria that affect plant silage fermentation. *Front. Microbiol.* **2023**, *14*, 1164904. [CrossRef]
51. Muck, R.E. Silage microbiology and its control through additives. *Rev. Bras. Zootec.* **2010**, *39*, 183–191. [CrossRef]
52. Guo, X.S.; Ke, W.C.; Ding, W.R.; Ding, L.M.; Xu, D.M.; Wang, W.W.; Zhang, P.; Yang, F.Y. Profiling of metabolome and bacterial community dynamics in ensiled *Medicago sativa* inoculated without or with *Lactobacillus plantarum* or *Lactobacillus buchneri*. *Sci. Rep.* **2018**, *8*, 357. [CrossRef]
53. Zheng, J.S.; Wittouck, S.; Salvetti, E.; Franz, C.; Harris, H.M.B.; Mattarelli, P.; O'Toole, P.W.; Pot, B.; Vandamme, P.; Walter, J.; et al. A taxonomic note on the genus *Lactobacillus*: Description of 23 novel genera, emended description of the genus *Lactobacillus* Beijerinck 1901, and union of *Lactobacillaceae* and *Leuconostocaceae*. *Int. J. Syst. Evol. Microbiol.* **2020**, *70*, 2782–2858. [CrossRef]

5. Conclusions

Compared to alfalfa CE and SW, alfalfa SD was more likely to result in well-preserved silage. Our results also suggested that, compared to silage CK, LB treatment did not change but SC treatment considerably inhibited clostridial fermentation. Bacterial abundance and diversity decreased due to SC treatment compared to silages CK and LB. *C. tyrobutyricum*, *C. luticellarii*, *Garciella* sp._GK3, *C. sporogenes*, *C. perfringens*, and *Clostridium* sp._BTY5 were identified as the dominant Clostridia species accounting for clostridial fermentation. In addition, through correlation network analysis, *E. faecalis*, *L. buchneri*, and *L. pentosus* were first screened as the key LAB species with significant inhibitory effects on *C. tyrobutyricum*, *C. luticellarii*, and *Garciella* sp._GK3, respectively. The results of Spearman correlation among dominant Clostridia, dominant LAB, and fermentation parameters confirmed the crucial role of the three abovementioned Clostridia or LAB species in stimulating or inhibiting clostridial fermentation, respectively. Therefore, this study improves our understanding of the clostridial fermentation mechanism of silage and might be helpful for screening target-based LAB strains to produce high-quality alfalfa silage.

Author Contributions: Conceptualization, M.Z.; methodology, M.Z.; investigation, M.Z., Q.L., P.M., and X.T.; resources, M.Z. and P.M.; data curation, Q.L., P.M., and X.T.; writing—original draft preparation, M.Z.; writing—review and editing, M.Z., Y.G. and L.M.; supervision, Y.G. and L.M.; project administration, M.Z.; funding acquisition, M.Z. All authors have read and agreed to the published version of the manuscript.

Funding: This research was funded by the National Natural Science Foundation of China (grant number 32201461) and the Youth Research Fund of BAAFS (grant number QNJJ202310).

Institutional Review Board Statement: Not applicable.

Data Availability Statement: All data generated or analyzed during this study are included in this published article.

Conflicts of Interest: The authors declare no conflicts of interest.

References

1. Li, R.R.; Hu, M.; Jiang, X.; Xu, C.C. Metagenomic insights into the microbiota involved in lactate and butyrate production and manipulating their synthesis in alfalfa silage. *J. Appl. Microbiol.* **2023**, *134*, lxad197. [CrossRef] [PubMed]
2. Jiang, H.; Wang, H.R.; Bao, B.H.; Qu, H.; Wang, J.; Sun, L.; Liu, B.; Gao, F.Q. Effect of compound additives on nutritional composition, fermentation quality, and bacterial community of high-moisture alfalfa silage. *Fermentation* **2023**, *9*, 453. [CrossRef]
3. Lin, J.Y.; Li, G.H.; Sun, L.; Wang, S.; Meng, X.; Sun, L.C.; Yuan, L.; Xu, L.B. Varieties and ensiling: Impact on chemical composition, fermentation quality and bacterial community of alfalfa. *Front. Microbiol.* **2023**, *13*, 1091491. [CrossRef]
4. Zheng, M.L.; Niu, D.Z.; Jiang, D.; Zuo, S.S.; Xu, C.C. Dynamics of microbial community during ensiling direct-cut alfalfa with and without LAB inoculant and sugar. *J. Appl. Microbiol.* **2017**, *122*, 1456–1470. [CrossRef] [PubMed]
5. Zheng, M.L.; Niu, D.Z.; Zuo, S.S.; Mao, P.C.; Meng, L.; Xu, C.C. The effect of cultivar, wilting and storage period on fermentation and the clostridial community of alfalfa silage. *Ital. J. Anim. Sci.* **2018**, *17*, 336–346. [CrossRef]
6. Bai, J.; Xu, D.M.; Xie, D.M.; Wang, M.S.; Li, Z.Q.; Guo, X.S. Effects of antibacterial peptide-producing *Bacillus subtilis* and *Lactobacillus buchneri* on fermentation, aerobic stability, and microbial community of alfalfa silage. *Bioresour. Technol.* **2020**, *315*, 123881. [CrossRef]
7. Yang, F.Y.; Wang, Y.P.; Zhao, S.S.; Wang, Y. *Lactobacillus plantarum* inoculants delay spoilage of high moisture alfalfa silages by regulating bacterial community composition. *Front. Microbiol.* **2020**, *11*, 1989. [CrossRef]
8. Zhao, S.S.; Wang, Y.P.; Yang, F.Y.; Wang, Y.; Zhang, H. Screening a *Lactobacillus plantarum* strain for good adaption in alfalfa ensiling and demonstrating its improvement of alfalfa silage quality. *J. Appl. Microbiol.* **2020**, *129*, 233–242. [CrossRef]
9. Zhang, Z.; Nair, S.; Tang, L.; Zhao, H.; Hu, Z.; Chen, M.; Zhang, Y.; Kao, S.J.; Jiao, N.; Zhang, Y. Long-term survival of *Synechococcus* and heterotrophic bacteria without external nutrient supply after changes in their relationship from antagonism to mutualism. *mBio* **2021**, *12*, e01614-21. [CrossRef]
10. Zheng, M.L.; Niu, D.Z.; Jiang, D.; Li, R.R.; Meng, L.; Xu, C.C. Metagenome analyses reveal the role of *Clostridium perfringens* in alfalfa silage anaerobic deterioration. *FEMS Microbiol. Lett.* **2020**, *367*, fnaa052. [CrossRef]
11. Huang, F.Q.; Wang, T.W.; Zhang, J.Q.; Tahir, M.; Sun, J.H.; Liu, Y.Y.; Yun, F.F.; Xia, T.Q.; Teng, K.L.; Wang, J.W.; et al. Exploring the bacterial community succession and metabolic profiles of *Lonicera japonica* Thunb. residues during anaerobic fermentation. *Bioresour. Technol.* **2023**, *367*, 128264. [CrossRef] [PubMed]

12. Owens, V.N.; Albrecht, K.A.; Muck, R.E.; Duke, S.H. Protein degradation and fermentation characteristics of red clover and alfalfa silage harvested with varying levels of total nonstructural carbohydrates. *Crop Sci.* **1999**, *39*, 1873–1880. [CrossRef]
13. Playne, M.J.; McDonald, P. The buffering constituents of herbage and of silage. *J. Sci. Food Agric.* **1966**, *17*, 264–268. [CrossRef]
14. AOAC. *Official Methods of Analysis*, 15th ed.; Association of Official Analytical Chemists: Arlington, VA, USA, 1990.
15. Van Soest, P.J.; Robertson, J.B.; Lewis, B.A. Methods for dietary fiber, neutral detergent fiber, and nonstarch polysaccharides in relation to animal nutrition. *J. Dairy Sci.* **1991**, *74*, 3583–3597. [CrossRef]
16. Broderick, G.A.; Kang, J.H. Automated simultaneous determination of ammonia and total amino-acids in ruminal fluid and in vitro media. *J. Dairy Sci.* **1980**, *63*, 64–75. [CrossRef]
17. Wang, H.L.; Ning, T.T.; Hao, W.; Zheng, M.L.; Xu, C.C. Dynamics associated with prolonged ensiling and aerobic deterioration of total mixed ration silage containing whole crop corn. *Asian Australas. J. Anim. Sci.* **2016**, *29*, 62–72. [CrossRef]
18. Hungate, R.E. Chapter IV a roll tube method for cultivation of strict anaerobes. *Method. Microbiol.* **1969**, *3*, 117–132. [CrossRef]
19. Edgar, R.C. UPARSE: Highly accurate OTU sequences from microbial amplicon reads. *Nat. Methods.* **2013**, *10*, 996–998. [CrossRef]
20. Bai, J.; Ding, Z.T.; Ke, W.C.; Xu, D.M.; Wang, M.S.; Huang, W.K.; Zhang, Y.X.; Liu, F.; Guo, X.S. Different lactic acid bacteria and their combinations regulated the fermentation process of ensiled alfalfa: Ensiling characteristics, dynamics of bacterial community and their functional shifts. *Microb. Biotechnol.* **2021**, *14*, 1171–1182. [CrossRef]
21. Zheng, M.L.; Mao, P.C.; Tian, X.X.; Meng, L. Effects of exogenous lactic acid bacteria and maize meal on fermentation quality and microbial community of *Orychophragmus violaceus* silage. *Front. Microbiol.* **2023**, *14*, 1276493. [CrossRef]
22. Wang, S.R.; Wang, Y.X.; Liu, H.P.; Li, X.B.; Zhao, J.; Dong, Z.H.; Li, J.F.; Kaka, N.A.; Nazar, M.; Shao, T. Using PICRUSt2 to explore the functional potential of bacterial community in alfalfa silage harvested at different growth stages. *Chem. Biol. Technol. Agric.* **2022**, *9*, 98. [CrossRef]
23. Guo, G.; Shen, C.; Liu, Q.; Zhang, S.L.; Wang, C.; Chen, L.; Xu, Q.F.; Wang, Y.X.; Huo, W.J. Fermentation quality and in vitro digestibility of first and second cut alfalfa (*Medicago sativa* L.) silages harvested at three stages of maturity. *Anim. Feed Sci. Technol.* **2019**, *257*, 114274. [CrossRef]
24. McDonald, P.; Henderson, A.R. Buffering capacity of herbage samples as a factor in ensilage. *J. Sci. Food Agric.* **1962**, *13*, 395–400. [CrossRef]
25. Buxton, D.R.; O'Kiely, P. Preharvest plant factors affecting ensiling. In *Silage Science and Technology*; Buxton, D.R., Muck, R.E., Harrison, J.H., Eds.; American Society of Agronomy: Madison, WI, USA, 2003; pp. 199–250.
26. Stirling, A.C.; Whittenbury, R. Sources of the lactic acid bacteria occurring in silage. *J. Appl. Bacteriol.* **1963**, *26*, 86–90. [CrossRef]
27. Muck, R.E. Initial bacterial numbers on lucerne prior to ensiling. *Grass Forage Sci.* **1989**, *44*, 19–25. [CrossRef]
28. Pahlow, G.; Muck, R.E.; Driehuis, F.; Elferink, S.J.W.H.O.; Spoelstra, S.F. Microbiology of ensiling. In *Silage Science and Technology*; Buxton, D.R., Muck, R.E., Harrison, J.H., Eds.; American Society of Agronomy: Madison, WI, USA, 2003; pp. 31–93.
29. Sun, J.J.; Wang, J.; Bai, C.S.; Zhao, J.M.; Yun, Y.; Yu, Z.; Xue, Y.L.; Zhang, T.W.; Bao, W.L. Natural fermentation quality, bacteria, and functional profiles of three cuttings of alfalfa silage in a year in Inner Mongolia, China. *Front. Microbiol.* **2023**, *14*, 1083620. [CrossRef]
30. Contreras-Govea, F.E.; Muck, R.E.; Broderick, G.A.; Weimer, P.J. *Lactobacillus plantarum* effects on silage fermentation and in vitro microbial yield. *Anim. Feed Sci. Technol.* **2013**, *179*, 61–68. [CrossRef]
31. Muck, R.E. Recent advances in silage microbiology. *Agr. Food Sci.* **2013**, *22*, 3–15. [CrossRef]
32. Zhang, Y.; Liu, Y.; Meng, Q.; Zhou, Z.; Wu, H. A mixture of potassium sorbate and sodium benzoate improved fermentation quality of whole-plant corn silage by shifting bacterial communities. *J. Appl. Microbiol.* **2020**, *128*, 1312–1323. [CrossRef]
33. Li, P.; Liao, C.S.; Yan, L.J.; Zhang, C.B.; Chen, L.Y.; You, M.H.; Cheng, Q.M.; Chen, C. Effects of small-scale silo types and additives on silage fermentation and bacterial community of high moisture alfalfa on the Qinghai-Tibetan Plateau. *Anim. Feed Sci. Technol.* **2023**, *299*, 115594. [CrossRef]
34. Huo, W.J.; Zhang, Y.J.; Zhang, L.Y.; Shen, C.; Chen, L.; Liu, Q.; Zhang, S.L.; Wang, C.; Guo, G. Effect of lactobacilli inoculation on protein and carbohydrate fractions, ensiling characteristics and bacterial community of alfalfa silage. *Front. Microbiol.* **2022**, *13*, 1070175. [CrossRef] [PubMed]
35. Li, R.R.; Jiang, D.; Zheng, M.L.; Tian, P.J.; Xu, C.C. Microbial community dynamics during alfalfa silage with or without clostridial fermentation. *Sci. Rep.* **2020**, *10*, 17782. [CrossRef] [PubMed]
36. D'Incecco, P.; Pellegrino, L.; Hogenboom, J.A.; Cocconcelli, P.S.; Bassi, D. The late blowing defect of hard cheeses: Behaviour of cells and spores of *Clostridium tyrobutyricum* throughout the cheese manufacturing and ripening. *LWT-Food Sci. Technol.* **2018**, *87*, 134–141. [CrossRef]
37. Wang, Q.; Wang, C.D.; Li, C.H.; Li, J.G.; Chen, Q.; Li, Y.Z. *Clostridium luticellarii* sp nov., isolated from a mud cellar used for producing strong aromatic liquors. *Int. J. Syst. Evol. Microbiol.* **2015**, *65*, 4730–4733. [CrossRef]
38. Wang, C.D.; Chen, Q.; Wang, Q.; Li, C.H.; Leng, Y.Y.; Li, S.G.; Zhou, X.W.; Han, W.J.; Li, J.G.; Zhang, X.H.; et al. Long-term batch brewing accumulates adaptive microbes, which comprehensively produce more flavorful Chinese liquors. *Food Res. Int.* **2014**, *62*, 894–901. [CrossRef]
39. Mariën, Q.; Regueira, A.; Ganigué, R. Steerable isobutyric and butyric acid production from CO_2 and H_2 by *Clostridium luticellarii*. *Microb. Biotechnol.* **2024**, *17*, e14321. [CrossRef]
40. Fan, X.M.; Zhao, S.S.; Yang, F.Y.; Wang, Y.; Wang, Y.P. Effects of lactic acid bacterial inoculants on fermentation quality, bacterial community, and mycotoxins of alfalfa silage under vacuum or nonvacuum treatment. *Microorganisms* **2021**, *9*, 2614. [CrossRef]

MDPI AG
Grosspeteranlage 5
4052 Basel
Switzerland
Tel.: +41 61 683 77 34

Agriculture Editorial Office
E-mail: agriculture@mdpi.com
www.mdpi.com/journal/agriculture

www.ingramcontent.com/pod-product-compliance
Lightning Source LLC
Chambersburg PA
CBHW041353210926
51850CB00013B/3066

* 9 7 8 3 7 2 5 8 3 4 1 1 2 *